Advanced Techniques for Ground Penetrating Radar Imaging

Advanced Techniques for Ground Penetrating Radar Imaging

Editors

Yuri Álvarez-López
María García-Fernández

MDPI • Basel • Beijing • Wuhan • Barcelona • Belgrade • Manchester • Tokyo • Cluj • Tianjin

Editors
Yuri Álvarez-López
Universidad de Oviedo
Spain

María García-Fernández
Universidad de Oviedo
Spain

Editorial Office
MDPI
St. Alban-Anlage 66
4052 Basel, Switzerland

This is a reprint of articles from the Special Issue published online in the open access journal *Remote Sensing* (ISSN 2072-4292) (available at: https://www.mdpi.com/journal/remotesensing/special_issues/GPRI).

For citation purposes, cite each article independently as indicated on the article page online and as indicated below:

LastName, A.A.; LastName, B.B.; LastName, C.C. Article Title. *Journal Name* **Year**, *Volume Number*, Page Range.

ISBN 978-3-0365-2149-7 (Hbk)
ISBN 978-3-0365-2150-3 (PDF)

Cover image courtesy of Yuri Álvarez López and María García Fernández

Contents

About the Editors

Yuri Álvarez-López (Dr.) received his M.S. and Ph.D. degrees in telecommunication engineering from the University of Oviedo (Spain), in 2006 and 2009, respectively. He was a Visiting Scholar at the Department of Electrical Engineering, Syracuse University, Syracuse, USA, in 2008; a Visiting Postdoc at the ALERT Center of Excellence, Northeastern University, Boston, USA, from 2011 to 2014; and a Visiting Postdoc at ELEDIA Research Center, Trento, Italy, in 2015. He has been with the Signal Theory and Communications research group of the University of Oviedo since 2006, where he is currently Professor. His research interests include antenna diagnostics, antenna measurement techniques, inverse scattering and imaging techniques, and phaseless methods for antenna diagnostics and imaging. Dr. Alvarez received the 2011 Regional and National Awards to the Best Ph.D. Thesis on Telecommunication Engineering (category: security and defense).

María García-Fernández (Dr.) received her M.Sc. and Ph.D. degrees in Telecommunication Engineering from the University of Oviedo (Spain) in 2016 and 2019, respectively. Since 2013, she has been involved in several research projects with the Signal Theory and Communications Research Group, TSC-UNIOVI, University of Oviedo. She was a Visiting Student with Stanford University, Palo Alto, CA, USA, in 2013 and 2014, a Visiting Scholar with the Gordon Center for Subsurface Sensing and Imaging Systems, Northeastern University, Boston, MA, USA, in 2018, and a Visiting Researcher at the Radar Department of TNO, The Hague, The Netherlands, in 2019. Her current research interests include inverse scattering, remote sensing, radar systems, imaging techniques, antenna measurement and diagnostics, and non-invasive measurement systems on board unmanned aerial vehicles. Dr. García was the recipient of one National Award for the Best Ph.D. Thesis on Telecommunication Engineering in 2020.

Preface to "Advanced Techniques for Ground Penetrating Radar Imaging"

Ground penetrating radar (GPR) has become one of the key technologies in subsurface sensing and, in general, in non-destructive testing (NDT), since it is able to detect both metallic and nonmetallic targets. GPR for NDT has been successfully introduced in a wide range of sectors, such as mining and geology, glaciology, civil engineering and civil works, archaeology, and security and defense.

In recent decades, improvements in georeferencing and positioning systems have enabled the introduction of synthetic aperture radar (SAR) techniques in GPR systems, yielding GPR–SAR systems capable of providing high-resolution microwave images. In parallel, the radiofrequency front-end of GPR systems has been optimized in terms of compactness (e.g., smaller Tx/Rx antennas) and cost. These advances, combined with improvements in autonomous platforms, such as unmanned terrestrial and aerial vehicles, have fostered new fields of application for GPR, where fast and reliable detection capabilities are demanded. In addition, processing techniques have been improved, taking advantage of the research conducted in related fields like inverse scattering and imaging. As a result, novel and robust algorithms have been developed for clutter reduction, automatic target recognition, and efficient processing of large sets of measurements to enable real-time imaging, among others.

This Special Issue provides an overview of the state of the art in GPR imaging, focusing on the latest advances from both hardware and software perspectives.

Yuri Álvarez-López, María García-Fernández
Editors

Editorial

Editorial for the Special Issue "Advanced Techniques for Ground Penetrating Radar Imaging"

Yuri Álvarez López * and María García-Fernández

Area of Signal Theory and Communications, University of Oviedo, Edificio Polivalente, 1st Floor, Campus Universitario de Gijón, 33203 Gijon, Spain; garciafmaria@uniovi.es
* Correspondence: alvarezyuri@uniovi.es; Tel.: +34-985-182281

Citation: López, Y.Á.; García-Fernández, M. Editorial for the Special Issue "Advanced Techniques for Ground Penetrating Radar Imaging". *Remote Sens.* **2021**, *13*, 3696. https://doi.org/10.3390/rs13183696

Received: 1 September 2021
Accepted: 13 September 2021
Published: 16 September 2021

Publisher's Note: MDPI stays neutral with regard to jurisdictional claims in published maps and institutional affiliations.

1. Introduction

Ground Penetrating Radar (GPR) has become one of the key technologies in subsurface sensing and, in general, in Non-Destructive Testing (NDT), since it is able to detect both metallic and nonmetallic targets. Furthermore, it can also provide images from the underground, thus improving detection capabilities. GPR for NDT has been successfully introduced in a wide range of sectors, such as mining and geology (detection of cavities, mineral deposits), glaciology (measurement of ice thickness), civil engineering and civil works (detection of cracks and defects in infrastructure), archaeology, and security and defense (detection of buried landmines and Improvised Explosive Devices, IEDs).

Improvements in georeferring and positioning systems have enabled the introduction of Synthetic Aperture Radar (SAR) techniques in GPR, yielding GPR–SAR systems capable of providing high-resolution microwave images. In parallel, the radiofrequency front-end of GPR systems has been optimized in terms of compactness (e.g., smaller Tx/Rx antennas) and cost. These advances, combined with improvements in autonomous platforms, such as unmanned terrestrial and aerial vehicles, have fostered new fields of application for GPR, where fast and reliable detection capabilities are demanded. In addition, processing techniques have been improved, putting together advances in the field of inverse scattering and imaging and in the area of machine learning and artificial intelligence. As a result, novel and robust algorithms have been developed for noise and clutter reduction, automatic target recognition, and efficient processing of large sets of measurements to enable real-time imaging, among others.

This Special Issue comprises a set of contributions covering both hardware and software improvements for enhanced GPR imaging. The techniques described in these contributions have been successfully applied to a wide variety of areas, such as archaeology, landmine detection, snow thickness monitoring, and buried infrastructure location.

The scope of the Special Issue contributions can be classified into three main groups: High-resolution GPR systems [1–4], noise mitigation in GPR measurements [5,6], and GPR data processing enhancement [7–10].

2. High-Resolution GPR Systems

Contributions [1–4] introduce high-resolution GPR architectures for landmine and IEDs detection [1,2] snow thickness monitoring [3], and location of drainage pipes [4]. In general, compact Ultra-Wide-Band (UWB) radar modules working within the 150 MHz to 6 GHz are used, as this frequency band provides a good trade-off between range resolution and penetration depth.

Airborne-based GPR systems have become a research area of great interest due to their capability of contactless location and imaging of buried objects. This is especially desirable in the field of landmine and IED detection in order to minimize the risk of accidental detonation. An airborne Down-Looking GPR (DLGPR) architecture is presented in [1],

1

proving the capability of conducting autonomous scanning with centimeter-level accuracy thanks to the use of Real Time Kinematics (RTK) and LIDAR sensors.

DLGPR systems provide good dynamic range thanks to the short distance between the radar and the ground, but at the expense of strong clutter coming from the air-ground interface. This limits the capacity of detecting shallow targets. Forward-Looking GPR (FLGPR) systems are able to minimize air-ground clutter. However, the dynamic range is greatly reduced with respect to DLGPR. Combination of FLGPR and DLGPR architectures yields a GPR system with reduced air-ground clutter, but without compromising the dynamic range. In [2] an experimental validation of a hybrid FLGPR-DLGPR architecture is presented. Practical implementation of the hybrid FLGPR-DLGPR has been made possible thanks to the use of transmitting and receiving radar modules working in the 3–5 GHz frequency band that use a wireless link for synchronization.

A GPR application devoted to measure the available water in snow cover is described in [3]. The radar module is implemented using a Software Defined Radio (SDR) and two UWB Vivaldi antennas. Measurements acquired in real conditions are processed by means of an approximated method derived from an electromagnetic model used to calculate the reflectance of snowpacks, proving the accuracy of the presented technique to retrieve the Snow Water Equivalent (SWE) parameter. Besides, a 120-GHz Frequency Modulated Continuous Wave (FMCW) radar is introduced to accurately measure the snow thickness.

An application of a GPR system to detect buried infrastructure is presented in [4]. In particular, the research focuses on tailoring the issue of multiple reflections that occur when two large targets are buried close to each other (two drainage pipes in this case). GPR-SAR processing is also applied together with ground removal resulting in improved along-track resolution of the images given by the GPR systems with respect to raw radargrams. In the post-processed images, the upper and lower bounds of the buried drainage pipes can be detected.

3. Noise Mitigation in GPR Measurements

Besides air-ground clutter, GPR measurements can be corrupted by different sources of noise. Different strategies have been proposed to mitigate the impact of noise in GPR measurements, thus improving detection capabilities. These strategies range from machine learning techniques [5] to spectral domain filtering approaches [6].

Aiming at improving the GPR image quality, [5] introduces a methodology to reduce random GPR noise. The technique presented by the authors is based on a neural network-based structure for denoising autoencoders (Convolutional Denoising AutoEncoders, CDAEs), introducing several improvements such as a dropout regularization layer, an atrous convolution layer, and a residual-connection structure. Validation of the method presented in [5] has been conducted using both simulation-based datasets and field measurements, proving that this technique not only reduces GPR noise, but also minimizes the degradation of the original waveform data.

Clutter can be partially mitigated by filtering the GPR measurements in the frequency domain. Sometimes, optimal choice of filter parameters must be selected based on a trial-and-error procedure. A new methodology is proposed in [6] to make filter parameterization easier, based on a Singular Value Decomposition (SVD) method applied in the two-dimensional spectral domain. The proposed filtering method has been validated using a three-dimensional GPR dataset, resulting in an increased geometric sharpness of GPR images.

4. GPR Data Processing Enhancement

Feature extraction from GPR data can be improved by means of advanced GPR processing techniques. On the one hand, processing methods are improved by accurate characterization of the measurement scenario. For example, if the composition of the soil is available, an accurate estimation of propagation velocity can be performed, thus improving GPR image focusing [1,7]. On the other hand, image processing techniques combined with machine learning approaches result in a more efficient detection of the features and targets of interest. The latter is of special interest when the surveyed volume is much greater than the targets and features, so visual inspection of the volume is extremely time-consuming and challenging. Besides, machine learning techniques have been also introduced to improve GPR image quality, e.g., by means of sidelobe suppression [9].

Two coherency functionals, the Complex Matched Coherency Measure, and the Complex Matched Analysis, are proposed in [7] to improve the Signal-to-Noise Ratio (SNR) of GPR data and to accurately retrieve the propagation velocity. Range migration algorithms are proved to perform better when considering a spatial-dependent propagation velocity rather than using a constant or only along-track estimation of propagation velocity.

Conventional GPR imaging is sensitive to changes in the measured amplitude of the signal, so resulting GPR images correspond to changes in the reflectivity. In attribute analysis, phase and frequency information is considered, resulting in GPR coherence maps. In [8] multi-trace attribute analysis is conducted to enhance GPR imaging, proving that, under certain conditions, improved data visualizations are achievable. The proposed GPR trace coherence imaging is applied to GPR data sets taken at archaeological sites, showing the capability to detect targets and features that conventional GPR imaging cannot resolve.

The presence of sidelobes in radar images may result in false detections or missed targets or features. A wide variety of techniques for sidelobe suppression has been proposed, ranging from windowing methods (e.g., Hamming window) to Convolutional Neural Network (CNN)-based techniques. The latter can be affected by the fact that the Point Spread Function (PSF) in the radar images can be sometimes spatially variant. A Spatial-Variant CNN (SV-CNN) with spatial-variant convolutional kernels is proposed in [9] to overcome this issue, proving its better performance compared to the conventional CNN in realistic scenarios.

Improvements in autonomous terrestrial and aerial systems have enabled fast scanning of large areas using GPR systems, where the resulting 3D data sets have to be processed automatically. Machine learning methods are quite efficient in properly identifying features and targets of interest, but at the expense of requiring a large number of data sets for training purposes. Ref. [10] introduces a machine learning framework based on wavelet scattering networks, which are functionally equivalent to CNN. The main goal is to detect the features corresponding to buried pipelines in GPR datasets. Results presented in [10] yield a classification accuracy greater than 95% in the presented examples.

5. Conclusions

Contributions of this Special Issue illustrate part of the advances in GPR technology for non-destructive inspection and imaging. The development of compact low-cost sensors has enabled the implementation of novel GPR architectures. Besides, recent advances in signal processing, machine learning, and data science, are being introduced in GPR data post-processing to improve feature extraction and automatic target recognition, even when dealing with larger GPR datasets. These machine learning techniques have also been applied for noise mitigation and artifact suppression in GPR measurements.

Author Contributions: Conceptualization, Y.Á.L. and M.G.-F.; writing—original draft preparation, Y.Á.L. and M.G.-F.; writing—review and editing, Y.Á.L. and M.G.-F. Both authors have read and agreed to the published version of the manuscript.

Funding: The preparation of this Special Issue has been partially supported by the Ministerio de Defensa—Gobierno de España and the University of Oviedo under Contract 2019/SP03390102/00000204/CN-19-002 ("SAFEDRONE").

Acknowledgments: The Academic Editors would like to thank the authors of the submitted contributions for the time and effort required to prepare these articles; the reviewers, who have provided valuable comments, corrections, and suggestions to improve que quality of the submitted manuscript.

Conflicts of Interest: The authors declare no conflict of interest.

References

1. Garcia-Fernandez, M.; Alvarez-Lopez, Y.; Las Heras, F. Autonomous Airborne 3D SAR Imaging System for Subsurface Sensing: UWB-GPR on Board a UAV for Landmine and IED Detection. *Remote Sens.* **2019**, *11*, 2357. [CrossRef]
2. García-Fernández, M.; Álvarez-Narciandi, G.; Álvarez-López, Y.; Fernando Las-Heras Andrés, F. Analysis and validation of a hybrid Forward-Looking Down-Looking Ground Penetrating Radar architecture. *Remote Sens.* **2021**, *13*, 1206. [CrossRef]
3. Alonso, R.; García Del Pozo, J.-M.; Buisán, S.; Álvarez, J.-A. Analysis of Snow Water Equivalent in AEMet-Formigal Field Laboratory (Spanish Pyrenees) During the 2019/2020 Winter Season Using a Stepped Frequency Continuous Wave Radar (SFCW). *Remote Sens.* **2021**, *13*, 616. [CrossRef]
4. Iftimie, N.; Savin, A.; Steigmann, R.; Dobrescu, G.S. Underground Pipeline Identification into a Non-Destructive Case Study Based on Ground-Penetrating Radar Imaging. *Remote Sens.* **2021**, *13*, 3494. [CrossRef]
5. Feng, D.; Wang, X.; Wang, X.; Ding, S.; Zhang, H. Deep Convolutional Denoising Autoencoders with Network Structure Optimization for the High-Fidelity Attenuation of Random GPR Noise. *Remote Sens.* **2021**, *13*, 1761. [CrossRef]
6. Oliveira, R.J.; Caldeira, B.; Teixidó, T.; Borges, J.F. GPR Clutter Reflection Noise-Filtering through Singular Value Decomposition in the Bidimensional Spectral Domain. *Remote Sens.* **2021**, *13*, 2005. [CrossRef]
7. Stucchi, E.; Ribolini, A.; Tognarelli, A. High-Resolution Coherency Functionals for Improving the Velocity Analysis of Ground-Penetrating Radar Data. *Remote Sens.* **2020**, *12*, 2146. [CrossRef]
8. Trinks, I.; Hinterleitner, A. Beyond Amplitudes: Multi-Trace Coherence Analysis for Ground-Penetrating Radar Data Imaging. *Remote Sens.* **2020**, *12*, 1583. [CrossRef]
9. Dai, Y.; Jin, T.; Song, Y.; Sun, S.; Wu, C. Convolutional Neural Network with Spatial-Variant Convolution Kernel. *Remote Sens.* **2020**, *12*, 2811. [CrossRef]
10. Jin, Y.; Duan, Y. Wavelet Scattering Network-Based Machine Learning for Ground Penetrating Radar Imaging: Application in Pipeline Identification. *Remote Sens.* **2020**, *12*, 3655. [CrossRef]

Article

Autonomous Airborne 3D SAR Imaging System for Subsurface Sensing: UWB-GPR on Board a UAV for Landmine and IED Detection

Maria Garcia-Fernandez *, Yuri Alvarez-Lopez and Fernando Las Heras

Area of Signal Theory and Communications, University of Oviedo, 33203 Gijón, Spain;
alvarezyuri@uniovi.es (Y.A.-L.); flasheras@uniovi.es (F.L.H.)
* Correspondence: garciafmaria@uniovi.es

Received: 3 September 2019; Accepted: 7 October 2019; Published: 11 October 2019

Abstract: This work presents an enhanced autonomous airborne Synthetic Aperture Radar (SAR) imaging system able to provide full 3D radar images from the subsurface. The proposed prototype and methodology allow the safe detection of both metallic and non-metallic buried targets even in difficult-to-access scenarios without interacting with the ground. Thus, they are particularly suitable for detecting dangerous targets, such as landmines and Improvised Explosive Devices (IEDs). The prototype is mainly composed by an Ultra-Wide-Band (UWB) radar module working from Ultra-High-Frequency (UHF) band and a high accuracy dual-band Real Time Kinematic (RTK) positioning system mounted on board an Unmanned Aerial Vehicle (UAV). The UAV autonomously flies over the region of interest, gathering radar measurements. These measurements are accurately geo-referred so as to enable their coherent combination to obtain a well-focused SAR image. Improvements in the processing chain are also presented in order to deal with some issues associated to UAV-based measurements (such as non-uniform acquisition grids) as well as to enhance the resolution and the signal to clutter ratio of the image. Both the prototype and the methodology were validated with measurements, showing their capability to provide high-resolution 3D SAR images.

Keywords: Ground Penetrating Radar (GPR); Unmanned Aerial Vehicles (UAVs); Synthetic Aperture Radar (SAR); Real Time Kinematic (RTK); Ultra-Wide-Band (UWB); landmine and IED detection; non-destructive testing

---

1. Introduction

In the last years, there has been a massive development of new applications using Unmanned Aerial Vehicles (UAVs) [1–4]. A considerable amount of these applications is based on integrating electromagnetic sensors on board the UAVs (e.g., power detectors for antenna measurement [5], or radars for earth observation [6] and subsurface sensing [7]). Safety, measurement speed and the ability to fly over difficult-to-access areas are some of the advantages that have contributed to widely increase the usage of UAVs for security and defense applications, such as landmine and Improvised Explosive Device (IED) detection.

Several Non-Destructive Testing (NDT) techniques have been employed for subsurface sensing applications, since they allow extracting information of the subsurface and detecting possible buried targets without interacting with them. Focusing on the field of landmine and IED detection, the most common NDT techniques are, according to their physical principle [8]: electromagnetic induction, Nuclear Quadrupole Resonance (NQR), thermal imaging and Ground Penetrating Radar (GPR). GPR has been found to be a useful strategy for this application since it is able to provide high resolution images from the subsurface and, as a result, it makes possible the detection of both metallic and dielectric buried targets [9,10]. However, GPR capabilities are considerably affected by the soil heterogeneity, the roughness of the air–soil interface and the possible low signal to clutter ratio (especially for non-metallic targets) [11].

There are several criteria to classify GPR systems. The first criterion takes into account whether the GPR antennas are placed directly in contact with the ground (ground-coupled) or above the ground (air-launched). The former architecture usually provides a better signal to clutter ratio, due to a higher penetration into the ground and weaker reflections from the air–soil interface (assuming a well-matching between the antennas and the soil impedance). The main drawback of this configuration is that the antennas are directly over the soil, which prevents the use of ground-coupled systems in applications which require a safety stand-off distance such as the aforementioned landmine and IED detection. The latter architecture avoids the contact with the ground, but suffers from stronger clutter, which makes the target detection more difficult. Another criterion classifies GPR systems according to the orientation of the antennas with respect to the soil surface, distinguishing between Forward (or Side)-Looking GPR (FLGPR, SLGPR) and Down-Looking GPR (DLGPR). In FLGPR [12] and SLGPR [13], the antennas are usually oriented looking ahead, which contributes to minimize the reflection coming from the air–soil interface. Nevertheless, they suffer from lower sensitivity and resolution (compromising the distinction between targets under or above the ground). On the other hand, in DLGPR [14], the antennas are placed looking downwards, providing higher resolution but receiving stronger reflections from the air–soil interface.

As aforementioned, in landmine and IED detection, the scanning system must be placed at a safe stand-off distance to prevent accidental detonations, thus FLGPR architectures have been usually proposed due to the difficulties to keep a safe distance with a DLGPR system. However, in the last years, a new approach based on a UAV-mounted radar has been proposed [7,15], allowing the use of a DLGPR configuration in safe conditions.

Although several UAV-mounted GPR systems have been presented [7,16,17], the system and processing techniques presented in this contribution have several novelties with respect to existing state-of-the-art systems. The most important feature is its ability to provide high resolution well-focused 3D images of the underground from radar measurements gathered during autonomous flights. This requires integrating a high accuracy real-time positioning system on board the UAV because: (i) the UAV needs accurate positioning to follow the predefined flight path; and (ii) the radar measurements are coherently combined using a Synthetic Aperture Radar (SAR) algorithm, which imposes a geo-referring accuracy several times lower than the smallest working wavelength. Furthermore, the positions of the radar measurements are carefully processed to discard data that could cause unfocusing or resolution degradation in the SAR image (for instance, due to oversampling in some areas [18], since it is very difficult to perform a uniformly sampled acquisition with the UAV). Regarding the processing of the radar data, some improvement steps are also proposed to enhance the resolution and the target discrimination. In addition, the lowest working frequency of the selected radar has been notably decreased (down to Ultra-High-Frequency, UHF, band), allowing the detection of targets buried in higher loss soils. Both the prototype and the processing techniques were tested with experimental flights, comparing the resulting 3D SAR images with the ground-truth. A video summarizing the features of the system and a brief application example is provided as supplementary material.

2. System Architecture

The main goal of the proposed system is that the UAV autonomously flies over a region of interest collecting radar measurements, which are geo-referred and sent in real-time to a ground control station, to produce high-resolution 3D SAR images. The UAV-mounted GPR prototype, which is shown in Figure 1, comprises the following subsystems:

- The flight control subsystem is composed of the UAV flight controller and common positioning sensors on board UAVs. These sensors are: an Inertial Measurement Unit (IMU), a barometer, and a Global Navigation Satellite System (GNSS) receiver.

- The enhanced positioning system consists of a laser rangefinder and a dual-band Real Time Kinematic (RTK) system.
- The radar subsystem includes the radar module, and the transmitter and receiver antennas.
- The communication subsystem has a receiver at 433 MHz for the link with the pilot remote controller and a wireless local area network (WLAN) transceiver at 5.8 GHz to exchange data with the ground station. Both frequencies were selected to minimize possible interferences with the radar subsystem during the experimental campaigns.

Figure 1. UAV-mounted GPR prototype.

The architecture of the system is similar to the one adopted by the prototype proposed in [7]. The main improved features are: (i) the radar subsystem, which works at considerably lower frequencies; and (ii) the enhanced positioning system, whose accuracy has been greatly increased. As a result, with the prototype presented in this contribution, targets buried in soils with higher losses can be detected and, due to the enhanced positioning accuracy, well-focused 3D SAR images of the underground can be obtained.

Regarding the radar subsystem, an M-sequence Ultra-Wide-Band (UWB) radar covering a frequency range from 100 MHz to 6 GHz was selected [19]. This radar transmits a type of pseudo-random binary signal called Maximum Length Binary Sequence (MLBS) periodically, thus the received backscattered signal must be correlated with the ideal MLBS to obtain the impulse response. Two UWB Vivaldi antennas, working from 600 MHz to 6 GHz, were selected for transmitting and receiving. For the experimental validation shown in this contribution, only the frequency band from $f_{\min} = 600$ MHz to $f_{\max} = 3$ GHz was selected for processing since the soil losses in the measured scenario produce too much attenuation at higher frequencies. As shown in Figure 1, 3D printed structures were designed and fabricated to properly mount the radar and the antennas on board the UAV.

Concerning the enhanced positioning system, to coherently combine all the measurements gathered with the prototype, they must be accurately geo-referred. To avoid artifacts and obtain a well-focused SAR image, this accuracy should be better than $\lambda_{\min}/4 = 2.5$ cm in the horizontal plane (i.e., in cross-range) and $\lambda_{\min}/8 = 1.25$ cm in the vertical direction (i.e., in range), where $\lambda_{\min} = 10$ cm is the smaller wavelength. A dual-band multiconstellation RTK system was selected to fulfill this requirement. This system is composed of two RTK beacons [20], one mounted on the UAV and another one working as base station static on the ground. The latter sends real-time correction data to the former to achieve cm-level positioning accuracy. The reason a dual-band RTK was chosen is to obtain better accuracy and availability (that is, percentage of time that corrected coordinates are provided), more robustness (especially when working in challenging environments, e.g., with limited sky view) and faster deployment time (since less time is required to resolve carrier phase ambiguity), compared to single-band RTKs previously used. The selected dual-band RTK was integrated into the UAV to provide these accurate coordinates in real time for both the UAV navigation and the measurements geo-referring. The expected accuracy is 0.5 cm in the horizontal plane and 1 cm in the vertical direction, which corresponds to $\lambda_{min}/20$ and $\lambda_{min}/10$, respectively. Nevertheless, even better relative positioning accuracies were observed during the measurement campaigns.

3. Methodology

The data acquired with the prototype (composed by the georeferred radar measurements) are processed according to the flowchart shown in Figure 2. The boxes in gray correspond to the processing of the geo-referring data gathered with the positioning subsystem, blue boxes represent the basic GPR-SAR processing, and green ones highlight the steps corresponding to improvements in the GPR-SAR processing. The final result is a 3D high-resolution SAR reflectivity image (yellow box).

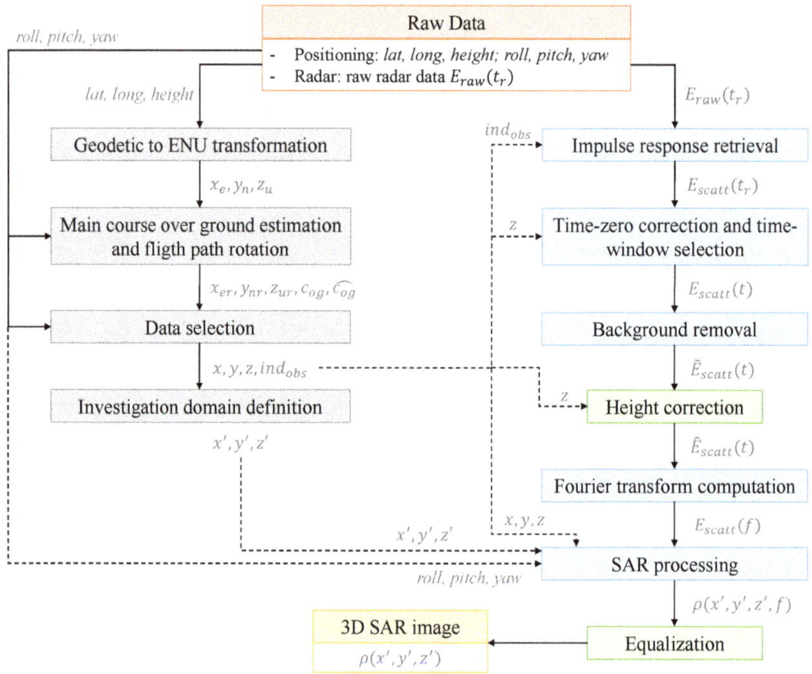

Figure 2. Flowchart of the data processing.

3.1. Positioning Data Processing

The flowchart of the positioning data processing is shown in Figure 2 (left). The positioning data to be processed is composed by: latitude (*lat*), longitude (*lon*) and height from the RTK system, and roll, pitch and yaw from the IMU. Due to the high accuracy of the RTK system in both horizontal and vertical directions, the laser rangefinder is not used in the processing. It is used, however, in the UAV navigation to help the UAV to keep an approximately constant distance to the soil surface.

This geo-referring data are processed mainly: (i) to select which measurements will be processed (discarding those that do not provide valuable information); (ii) to compute the position of the radar antennas in a cartesian coordinate system; and (iii) to define the investigation domain (i.e., where the SAR image is obtained) according to the flight path.

As explained above, the UAV autonomously flies over the region of interest. The flight path (i.e., the measurement grid) is a planar rectangular surface at a constant height and heading, where geo-referred radar measurements are continuously collected. However, it must be noticed that the proposed positioning data processing does not require prior knowledge of the predefined measurement grid, thus allowing the processing of measurements gathered in planar grids with both autonomous and manual flights.

The first step consists of transforming the coordinates from the geodetic system (latitude, longitude and height) to a local ENU (East–North–Up) system (x_e, y_n, z_u). This requires selecting a reference point as origin of the ENU system. In particular, the position of the UAV when it is first turned on is used as reference: the latitude and the longitude are given by the RTK, and the height is given by the physical distance from the RTK antenna to the ground (so that at the reference position, $z_u = 0$ m).

Then, the main course over the ground or main ground track (denoted as $\widehat{c_{og}}$) is estimated and the flight path is rotated according to this value. After the rotation, the flight path will be almost aligned with the x- and y-axis, which helps to facilitate the visualization of the measurements and the definition of the investigation domain. $\widehat{c_{og}}$ will also be used in the next step (data selection) to discard measurements acquired when the course over the ground is too far from its main value.

The main ground track estimation is obtained from the set of UAV positions as follows: (i) the course over the ground c_{og} (discarding the sense) is computed according to Equation (1), where v_n and v_e are the velocities in the north and east directions and mod denotes the modulo operation; (ii) c_{og} is discretized in bins of 1°-width and the mode is computed to obtain a rough estimation of the main course over the ground $\widetilde{c_{og}}$; and (iii) the estimation is improved by computing the mean of all measurements in the range $\widetilde{c_{og}} \pm 10°$, yielding $\widehat{c_{og}}$. If the flight path is performed autonomously (as in this contribution), this value is almost the same as the desired course defined with the waypoints. It must be noticed that the mean of c_{og} is a worse estimation of the main course over the ground than $\widehat{c_{og}}$, since it is greatly influenced by the course of the UAV when it changes sense or when it is almost still.

$$c_{og}[rad] = \mathrm{mod}\left(\mathrm{atan}\left(\frac{v_n}{v_e}\right), \pi\right) \approx \mathrm{mod}\left(\mathrm{atan}\left(\frac{\partial y_n}{\partial x_e}\right), \pi\right) \tag{1}$$

Figure 3 shows the histogram of c_{og} of an autonomous flight defined with waypoints (with a desired course of 65°). The mode gives the rough estimation of $\widetilde{c_{og}} = 65.5°$. After improving this estimation as explained before, $\widehat{c_{og}} = 65.09°$, which is almost the same as the desired course. The original and rotated flight paths are shown in Figure 4. As expected, the rotated path (x_{er}, y_{nr}, z_{ur}) is now aligned with the x- and y-axis.

The next step aims to discard the data that do not provide valuable information (e.g., when there is a sudden change in attitude) or that could degrade the SAR image. Regarding the latter, it must be noticed that the spatial sampling should be as uniform as possible to avoid artifacts in the SAR image. If there is an oversampled region (where much more measurements were taken), the SAR image pixels close to this region will have a high amplitude, thus masking the detection of targets in other parts of the image. In measurements taken with UAVs, it is very difficult to perform a uniform acquisition, since the UAV deviates from the ideal flight path mainly due to wind conditions and positioning systems uncertainties. Furthermore, in the experimental flights, it was observed that, when the UAV changes sense of movement (i.e., when it changes from moving in one direction to moving in the opposite one), the speed usually decreases, thus resulting in oversampling. It must be remarked that the UAV heading is fix to a constant value (making it coincident with the desired main course over the ground).

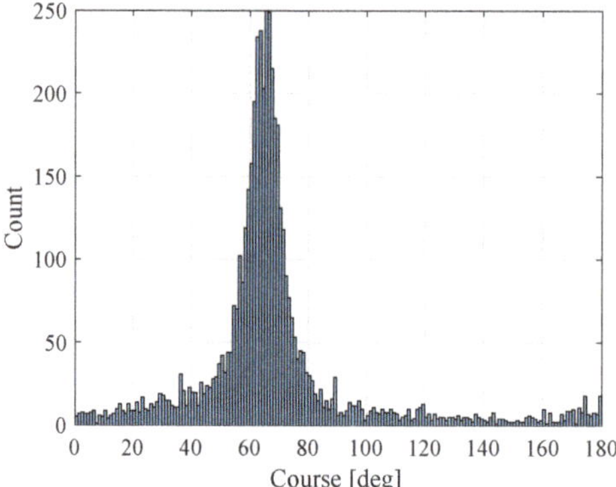

Figure 3. Histogram of the course over the ground.

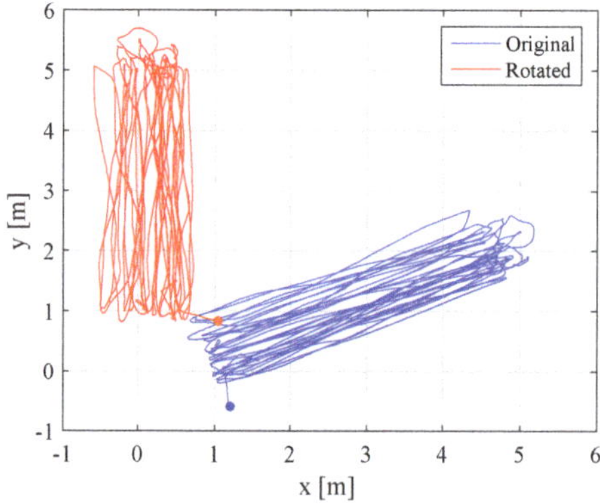

Figure 4. Original and rotated flight path (in blue and red,, respectively).

The data that are kept for further processing must satisfy the following constraints, where $th_{()}$ denotes a fix threshold:

- $\Delta_{xy} = \sqrt{\Delta x_{er}^2 + \Delta y_{nr}^2} > th_\Delta$, where Δx_{er} and Δy_{er} are the differences between adjacent values of x_{er} and y_{er}, respectively. This condition ensures that the UAV is actually moving.
- $|roll - \overline{roll}| < th_{roll}$, $|pitch - \overline{pitch}| < th_{pitch}$, and $|yaw - \overline{yaw}| < th_{yaw}$, where $\overline{()}$ denotes the mean value. These conditions are used to filter out the positions where there was a noticeable change in attitude.
- $|\text{rem}(c_{og} - \widehat{c_{og}}, \pi)| < th_c$, where rem denotes the remainder operator. Only measurements with the course over ground close to the main one are kept. This means that, if the UAV path deviates noticeably from the main path, these measurements are discarded.
- $|z_{ur} - \overline{z_{ur}}| < th_z$. This condition helps to get rid of considerable changes in height.

The following thresholds were employed: $th_\Delta = \lambda_{\min}/6 \approx 0.017$ m (where $\lambda_{\min}$ is the smaller wavelength), $th_{att} = \overline{att} + 3\text{std}(att)$ where std is the standard deviation and att is each of the attitude angles (roll, pitch and yaw), $th_c = 20°$ and $th_z = 0.25$ m. They were found to provide a good compromise to avoid oversampling and discard unnecessary data, while keeping enough measurements to obtain a well-focused SAR image.

In the flight used for illustrating the positioning data processing, the UAV is mainly moving back and forth between $y = 1$ m and $y = 5$ m. If only one of each four consecutive UAV positions are plotted (Figure 5a), it can be clearly seen the oversampling when the UAV changes sense of movement (at $y = 1$ m and $y = 5$ m). Figure 5b shows the full flight path (after the rotation performed in the previous step) and the selected flight path. Most of the discarded data correspond to these oversampled regions (as can be inferred from Figure 5c), whereas the rest of discarded data mainly correspond to sudden changes in movement due to wind.

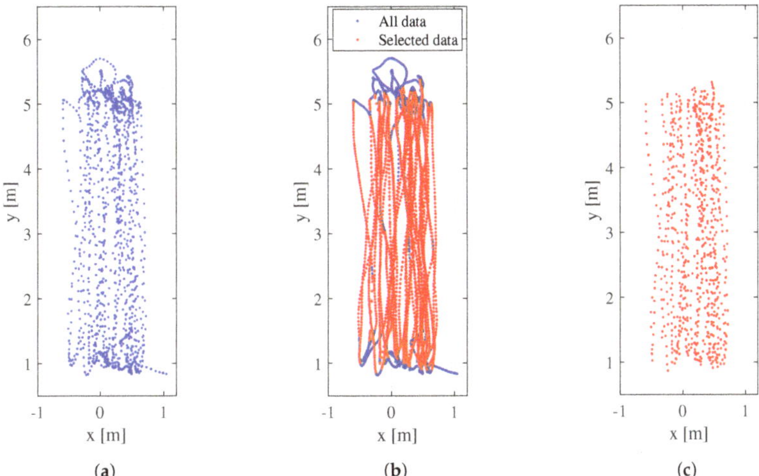

Figure 5. One of each four consecutive UAV positions (**a**); full and selected UAV positions (in blue and red, respectively) (**b**); and one of each four consecutive selective UAV positions (**c**).

This data selection step also helps to reduce the number of measurements that are processed, thus helping to decrease the computational time. In the previously shown example, the full flight path contains 4993 measurements points, whereas the selected path has 3118 points. This means that only 62.5% of the measurements will be processed.

As a result of this step, the indexes of the measurements that will be processed (ind_{obs}) are stored in a vector. Thus, the observation domain coordinates (i.e., the positions where these measurements were taken) are: $r = (x, y, z)$, where $x = x_{er}(ind_{obs})$, $y = y_{nr}(ind_{obs})$ and $z = z_{ur}(ind_{obs})$.

Finally, the investigation domain, $r' = (x', y', z')$, where the 3D SAR image will be calculated must be defined. The investigation plane (x' and y' coordinates) is defined according to the observation plane (x and y coordinates) as follows:

- First, the minimum area bounding rectangle that encloses the observation plane (called bounding box) is retrieved. To find it, the convex hull of the observation plane coordinates (i.e., the smallest convex polygon containing the observation plane) is computed. Then, the bounding box can be easily obtained since it always contains an edge of the convex hull.
- Then, the maximum axis-aligned rectangle inside the bounding box is computed, since it is easy to define and work with an axis-aligned investigation domain and the observation domain is almost aligned (due to the rotation according to the main course over the ground previously performed).
- Finally, the investigation plane is defined by shrinking this rectangle by a scale factor of sf_t and sf_{ct} in the track and across-track directions (to avoid edge effects in the SAR image) and sampling it every δ_t and δ_{ct}, respectively.

An example of the computation of the investigation plane following this procedure is shown in Figure 6, where the bounding box is depicted with a solid black line, the maximum axis-aligned rectangle inside it is drawn with a dash-dot red line and the edges of the investigation plane are shown in solid red. For the results shown in this contribution, the investigation plane is obtained by shrinking the rectangle with the scale factors $sf_t = 0.95$ and $sf_{ct} = 0.85$ (which prevents edge effects in the SAR image) and sampling it every $\delta_t = \delta_{ct} = \lambda_{\min}/4 = 0.025$ m. Since the soil surface is around $z = 0$ m, z' is defined between -0.6 m and 0.4 m, and is sampled every $\delta_z = 0.02$ m. These values can be adjusted by the user as desired (e.g., if coarse sampling of the investigation domain is considered enough for the specific application considered).

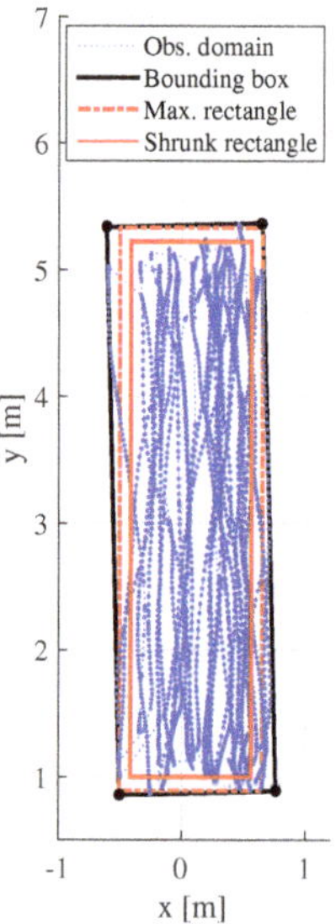

Figure 6. Definition of the investigation plane edges (in solid red line) from the observation plane coordinates (in dotted blue line).

It is worth noting that the investigation domain coordinates, $r' = (x', y', z')$, where the 3D SAR image is calculated, can be transformed back to the geodetic system of coordinates by applying the inverse of the previous operations (i.e., rotating them an angle $-\widehat{c_{og}}$ and then applying the transformation from the local east–north–up system to the geodetic system).

3.2. Radar Data Processing

The flowchart of the radar data processing is shown in Figure 2 (right). As aforementioned, the boxes in blue correspond to the basic processing and the boxes in green are optional steps to improve the resulting SAR image quality.

Since the radar used in this contribution transmits a pseudorandom binary sequence, the first step consists of cross-correlating the raw radar data $E_{raw}(t_r)$ with the ideal transmitted binary sequence to obtain an approximation of the impulse response function $E_{scatt}(t_r)$ (where t_r is the time axis). As explained above, it must be noticed that only the selected measurements (given by ind_{obs}) will be processed.

The next step deals with adjusting to a common time-zero and selecting the time-window of interest. Estimating the time-zero is important in order to remove the effect of the wires and the radar internal delays as well as to obtain a well-focused image. The estimation is performed with a calibration measurement at the beginning of the experimental campaign. The prototype is placed over the ground at a known distance (d_{rg}) and this distance is also estimated with the radar ($d_{eg} = ct_g/2$, where c is the light speed and t_g is the time instant where the ground is detected). Therefore, the time-zero is given by $t_0 = 2(d_{eg} - d_{rg})/c$, the corrected time-axis is $t_c = t_r - t_0$ and the measurements at $t_c < 0$ will be discarded. Besides, the time window of interest is selected so as to reduce the data size (since measurements at larger range do not provide valuable information). In particular, the used criterion selects the time-window t that corresponds to a range of $\overline{r_0} \pm 1$ m, where $\overline{r_0}$ is the mean distance between the radar antennas and the soil. It must be remarked that the time-zero is estimated only once, and then the same time-zero correction is applied to all measurements (discarding the data at $t_c < 0$).

The radar measurements of the flight shown in Section 3.1 after applying this step are depicted in Figure 7, where the range axis is given by $r_{ng} = ct/2$.

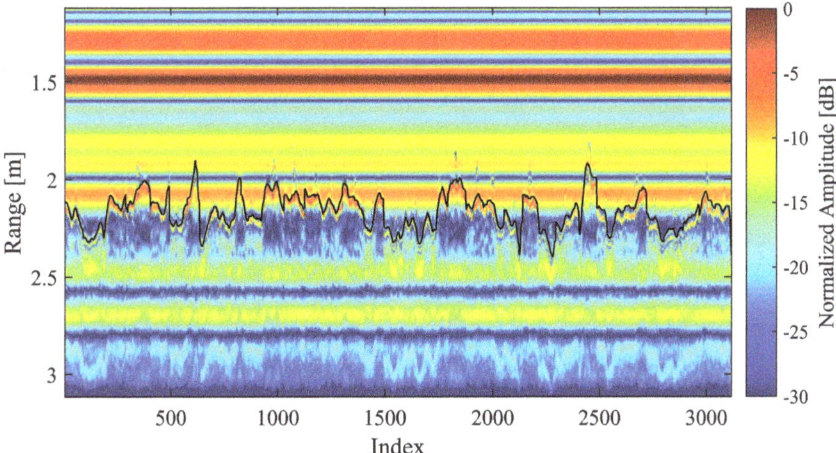

Figure 7. Normalized radar measurements ($E_{scatt}(r_{ng})$) after time-zero correction and time-window selection. The distance between the radar antennas and the soil is depicted in black on top of the measurement.

To improve the signal to clutter ratio [21], the background should be estimated and removed from the radar measurements. In this contribution, the background is estimated as the average of all measurements and thus, the average is subtracted from each measurement. This helps to reduce the clutter, such as the coupling effects between the antennas.

The resulting improved radar measurements $\widetilde{E}_{scatt}(t)$ are shown in Figure 8, where a clear improvement can be noticed comparing this image with the original radar data. This enhancement can be better observed taking a closer look to both original and improved radar measurements (Figure 9). Furthermore, there is a good agreement between the location of the air–soil interface in the radar measurements and the distance between the radar antennas and the soil provided by the positioning system. The small discrepancies are mainly due to the soil being not perfectly flat (besides the scattering effects due to the presence of grass on the ground), and the errors in the positioning system.

After this step, a height correction can be applied to the data to enhance the SAR image quality (mainly focusing and resolution). This correction consists of first shifting the radar measurements by $z - \bar{z}$, so that the resulting data $\widehat{E}_{scatt}(t)$ look as if the measurements were taken at constant height.

The radar measurements after the height shifting are shown in Figure 10 and a closer look is depicted in Figure 11. As expected, the strong reflection at the air–soil interface is now almost at the same range for all measurements. In addition, it can also be concluded that there is a target at around 15 cm over the ground.

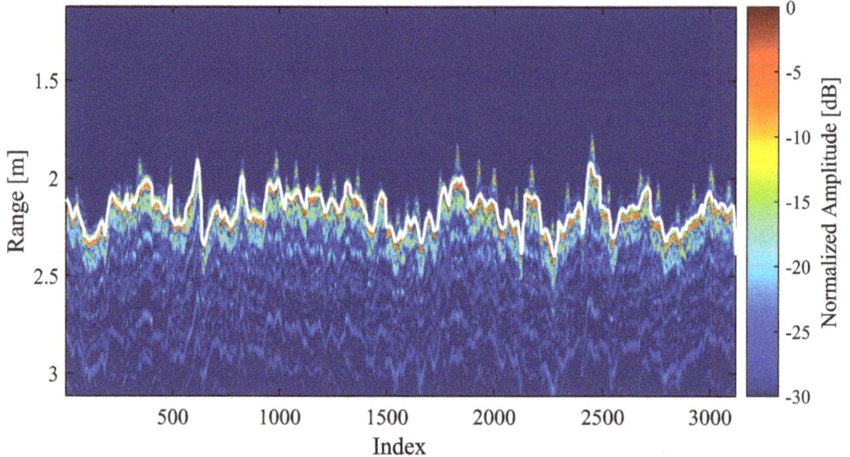

Figure 8. Normalized radar measurements ($\widetilde{E}_{scatt}(r_{ng})$) after background subtraction. The distance between the radar antennas and the soil is depicted in white on top of the measurement.

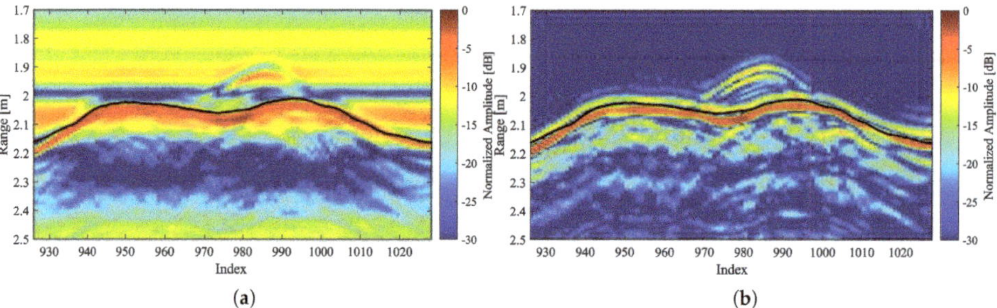

Figure 9. Closer look to the radar measurements before (**a**) and after (**b**) background subtraction.

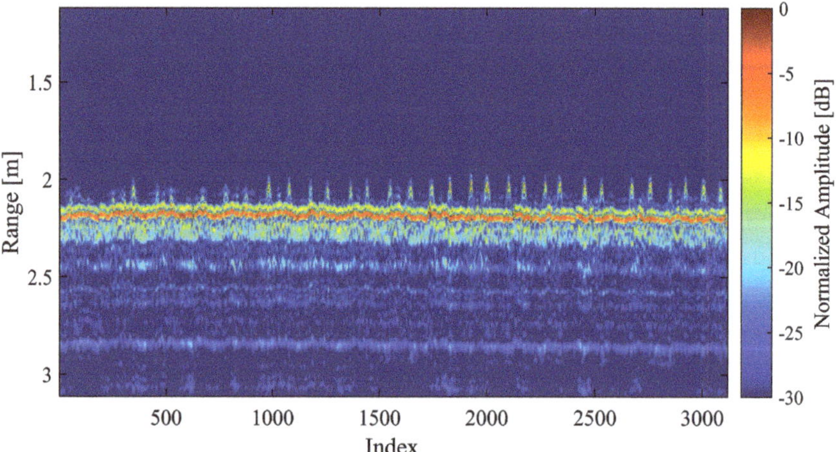

Figure 10. Normalized radar measurement ($\widehat{E}_{scatt}(r_{ng})$) after height shifting.

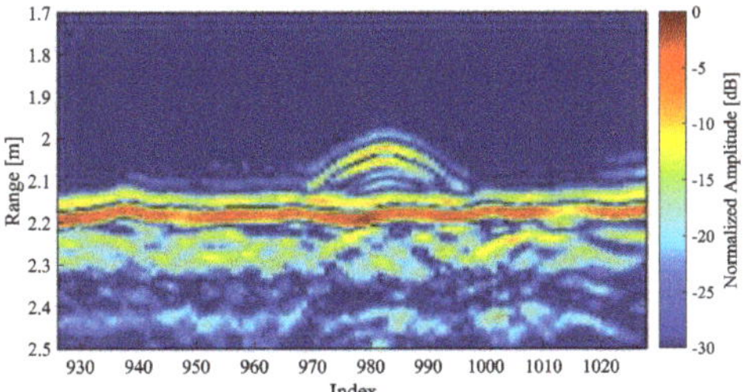

Figure 11. Closer look to the radar measurement ($\widehat{E}_{scatt}(r_{ng})$) after height shifting.

After the shifting, a second background subtraction is applied to mitigate the reflection from the air–soil interface and further enhance the signal to clutter ratio.

The next step consists of applying the Fourier transform to obtain $E_{scatt}(f)$ since the SAR processing will be performed in the frequency domain. Before the SAR processing, the positions of the transmitter and receiver antennas, denoted as $r_t = (x_t, y_t, z_t)$ and $r_r = (x_r, y_r, z_r)$, must be calculated since the observation domain coordinates are given with respect to the RTK antenna. These positions are computed through translation and rotation operations taking into account the attitude angles, the observation domain coordinates and the distances between the RTK antenna, the radar antennas and the IMU. If height correction has been applied, this must also be taken into account since in this case $z = \bar{z}$.

Then, a SAR algorithm [22] is applied to coherently combine the measurements and obtain a well-focused image. SAR reflectivity at pixel r'_p is given by Equation (2), where f_n are the selected frequencies of interest; r_t^m and r_r^m denote the position of the transmitter and receiver antennas, respectively, at the mth measurement point; $k_{0,n}$ is the wavenumber in free-space at the nth frequency; and $R_{p,m}$ is the total path length between the transmitter antenna, the pixel where the SAR reflectivity is computed and the receiver antenna.

$$\rho(r'_p) = \sum_{n=1}^{N} \rho(r'_p, n) = \sum_{n=1}^{N} \sum_{m=1}^{M} E_{scatt}(f^n, r_t^m, r_r^m) \exp(+jk_{0,n} R_{p,m}) \tag{2}$$

If free-space propagation is assumed (i.e., the soil composition is not taken into account), then $R_{p,m} = \|r_t^m - r'_p\| + \|r_r^m - r'_p\|$. Thus, if an object is buried at d_{obj} depth and the soil permittivity is ε_r, it will be detected at $\sqrt{\varepsilon_r} d_{obj}$ in the SAR image. To consider the soil composition when calculating the path length for $z'_p < 0$ m (i.e., under the soil surface), $R_{p,m}$ is computed according to (3), where $n_s = \sqrt{\varepsilon_r - 1} - \sqrt{\varepsilon_r}$ and the other parameters are shown in Figure 12 [23].

$$R_{p,m} = 2d\sqrt{\varepsilon_r - 1} + \frac{d_t(d_t - dn_s \cos(2\phi_t))}{d_t + dn_s \sin(2\phi_t)^2} + \frac{d_r(d_r - dn_s \cos(2\phi_r))}{d_r + dn_s \sin(2\phi_r)^2} \tag{3}$$

To improve the resolution of the SAR image, an equalization of the frequency response can be performed. When working with UWB radars and antennas, the amplitude levels of the SAR image show a great variation across the whole frequency band (mainly due to the fact that the antenna behaviour notably changes with the frequency) [19]. Usually, the data at lower frequencies mask the data at higher frequencies, yielding a SAR image with worse resolution than expected. To overcome this issue, instead of computing the SAR image for all the frequencies directly, the SAR image is computed for each nth frequency and is normalized by the maximum of its absolute value. Then, all nth SAR images are added to obtain the final SAR image, according to Equation (4).

$$\rho(r'_p) = \sum_{n=1}^{N} \frac{\rho(r'_p, n)}{max\{|\rho(r'_p, n)|\}} \tag{4}$$

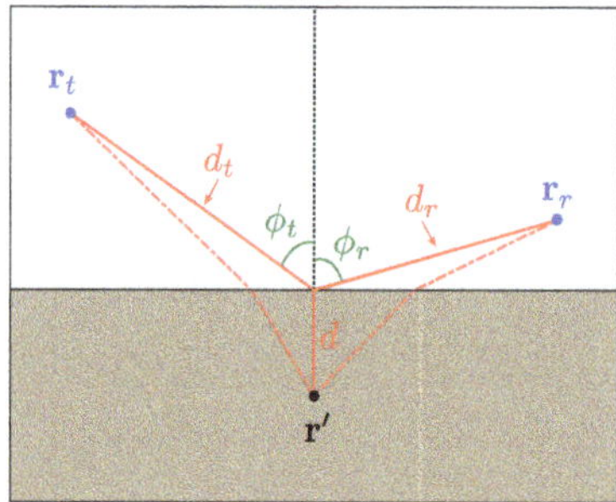

Figure 12. Main parameters involved in the estimation of the path length when the soil permittivity is taken into account (dashed line represents the true ray path).

4. Results and Discussion

The proposed system was validated at the airfield of the University of Oviedo (Figure 13). Two measurements were performed with the setup shown in Figure 14. In both cases, a metallic disk (of 9 cm radius) was placed on top of a plastic briefcase (with 14 cm height). In the first measurement, an open cylindrical metallic box (of 9.5 cm radius) was placed inside of a small hole of 8 cm depth (without soil covering it). For the second measurement, the box was covered with soil (i.e., the box was buried 8 cm under the soil surface).

Figure 13. Measurement scenario.

Figure 14. Setup for the measurement (**a**); first with the 9.5 cm radius metallic box uncovered (**b**); and second with this box buried and covered with soil (**c**).

In both measurements, the flight was performed autonomously. The predefined flight path was a rectangle of dimensions $\Delta x_p = 1$ m and $\Delta y_p = 4$ m sampled at $\delta x_p = 0.03$ m and $\delta y_p = 0.25$ m to define the waypoints positions, where x_p denotes the cross-track direction and y_p the along-track directions. The height was fixed at 2.3 m distance to ground (from the laser rangefinder). It must be remarked that radar measurements are continuously acquired during the flight (i.e., they are acquired not at each waypoint, but all over the flight path). The resulting flight path for the second measurement was used to illustrate the processing of the position data described in Section 3. To facilitate the comparison of the results shown in this section, both flights were rotated according to the same $\widehat{c_{og}}$ (in particular, $\widehat{c_{og}} = 65.09°$ as estimated for the second measurement) and the same investigation domain was used ($x' \in [-0.4, 0.6]$ m, $y' \in [1, 5]$ m and $z' \in [-0.6, 0.4]$ m).

In the following subsections, different slices of the resulting SAR images are depicted. Please note that a YZ plane is an along-track view (i.e., an along-track vs. depth slice), a YX plane is a top view (i.e., an along-track vs. across-track slice) and an XZ plane is an across-track view (i.e., across-track vs. depth slice).

4.1. Basic Processing

Both measurements were processed according to the procedure explained in Section 3, but first only applying the basic radar processing steps (without the improvement steps and without taking into account the soil composition). The resulting 3D SAR images were compared and the most relevant slices are shown in Figures 15 and 16. These slices were normalized by the maximum value of the full 3D SAR image to facilitate the comparison between different slices of the same measurement.

First, the slices at the position where the cylindrical box is detected are depicted in Figure 15, on the left for the first measurement (where the box is not covered by soil, Figure 14b) and on the right for the second measurement (where the box is covered by soil, i.e., buried under the ground, Figure 14c). In the former, the box is detected at approximately its true position, at $(x, y, z) = (-0.05, 4.55, -0.08)$ m. In the later, as expected, the box is detected deeper (at $z = -0.14$ m) since the soil composition had not been taken into account yet. In addition, in the along-track view (i.e., YZ plane), it can also be noticed that the target is not clearly distinguished from the soil surface interface, due to the strong reflection at the soil interface and the resolution in the z-axis not being high enough. This issue is overcome with the improvement steps in the radar processing. Furthermore, the amplitude of the SAR image at the box position is around 10 dB smaller when the box is buried, due to the high losses of the soil.

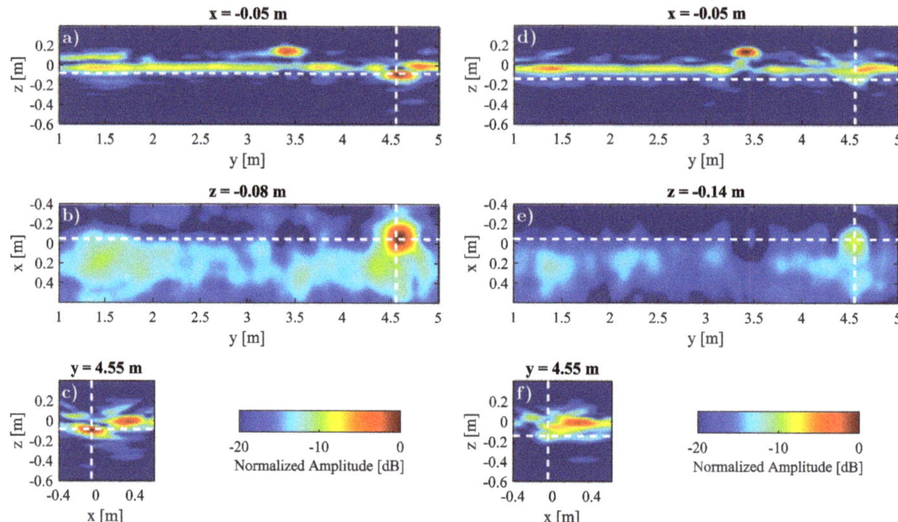

Figure 15. Slices of the 3D SAR image at the position where the buried metallic cylindrical box is detected. Results for the the first measurement (uncovered buried metallic box) are depicted on the left: along-track view at $x = -0.05$ m (**a**); top view at $z = -0.08$ m (**b**); and across-track view at $y = 4.55$ m (**c**). Results for the the second measurement (buried metallic box covered with soil) are depicted on the right: along-track view at $x = -0.05$ m (**d**); top view at $z = -0.14$ m (**e**); and across-track view at $y = 4.55$ m (**f**).

Then, the slices at the position where the disk (placed on top of a briefcase) is detected are depicted on the left part of Figure 16. Only the slices of the second measurement are shown since the 3D SAR images of both measurements are almost the same in all the investigation domain except in the area where the cylindrical box is buried. The disk is detected at its true position, at $(x, y, z) = (-0.075, 3.4, 0.14)$ m.

Finally, the slices of a region without targets at one position at the air–soil interface are represented on the right part of Figure 16. The air–soil interface is clearly detected and, as the actual physical air–soil interface, it is not perfectly flat.

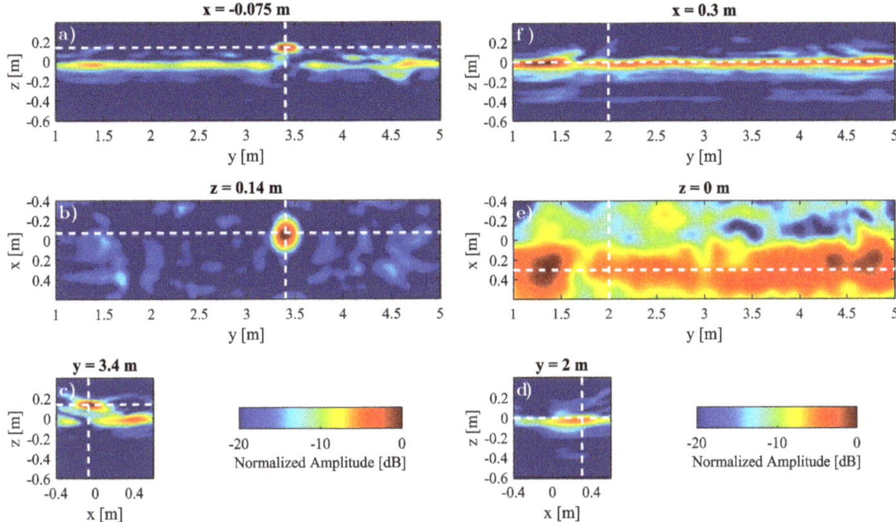

Figure 16. Slices of the 3D SAR image for the second measurement at the position where the cylindrical disk on top of a briefcase is detected (along-track view at $x = -0.075$ m (**a**); top view at $z = 0.14$ m (**b**); and across-track view at $y = 3.4$ m (**c**)) and at a position at the soil surface interface without targets (along-track view at $x = 0.3$ m (**d**); top view at $z = 0$ m (**e**); and across-track view at $y = 2$ m (**f**)).

4.2. Enhanced Processing

As explained above, the SAR images obtained with the basic processing should be improved mainly to obtain a better resolution and a higher signal to clutter ratio. In particular, this is especially important in order to detect shallow buried targets and distinguish them from the air–soil interface. The slices of the SAR image obtained for the second measurement at the position where the cylindrical box is detected are shown to analyze the effect of the proposed improvements.

4.2.1. Height Correction

The results obtained applying the height correction described in Section 3.2 are shown on the left part of Figure 17. Comparing them with the results without the height correction (right part of Figure 15), the cylindrical metallic box has become clearly distinguishable from the air–soil interface. Furthermore, the reflection of the air–soil interface has been mitigated thanks to the second background subtraction applied in the height correction step. This mitigation results in less amplitude in the SAR image around $z = 0$ m, which is especially evident in the YZ slice (where some areas corresponding to the interface have a normalized amplitude smaller than -30 dB). As a result, there is an improvement in the signal to clutter ratio. As a side effect, there is also a considerable amplitude in the SAR image approximately under the metallic disk, which might be due to some secondary reflections.

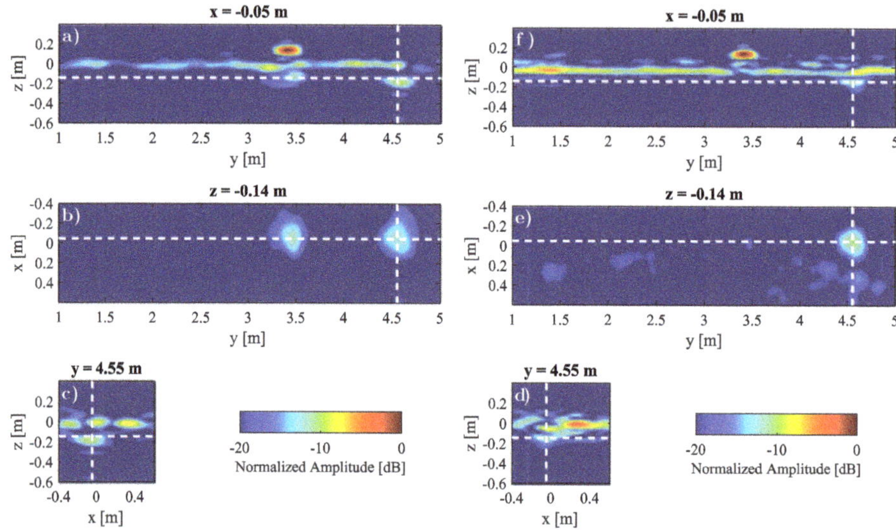

Figure 17. Slices of the 3D SAR image at the position where the cylindrical box is detected when height correction is applied (on the left) and when the SAR image is equalized (on the right): (**a**,**d**) along-track views at $x = -0.05$ m; (**b**,**e**) top views at $z = -0.14$ m; and (**c**,**f**) across-track views at $y = 4.55$ m.

4.2.2. Equalization

The effect of applying the equalization can be observed comparing the SAR image slices when equalization is applied (right part of Figure 17) with the results of the basic processing (right part of Figure 15). As explained above, the goal of the equalization is to effectively use the whole bandwidth, preventing the low frequency data masking the high frequency one. Thus, the range resolution, which corresponds in this case to the resolution in the z-axis, should improve with the equalization. This improvement is clearly visible since the width of the high reflectivity areas corresponding to the disk, the box and the air–soil interface is narrower. As a result, the buried box can be now distinguished from the air–soil interface.

4.2.3. Height Correction and Equalization

If both height correction and equalization are applied, the resulting SAR image (shown in left part of Figure 18) has better resolution and higher signal to clutter ratio than the one obtained with the basic processing. Furthermore, the equalization also helps to remove the artifact that appears under the metallic disk after applying the height correction. Thus, the combination of these improvements helps to enhance the range resolution (especially the equalization) and the signal to clutter ratio (mainly by reducing the reflection at the air–soil interface with the height correction).

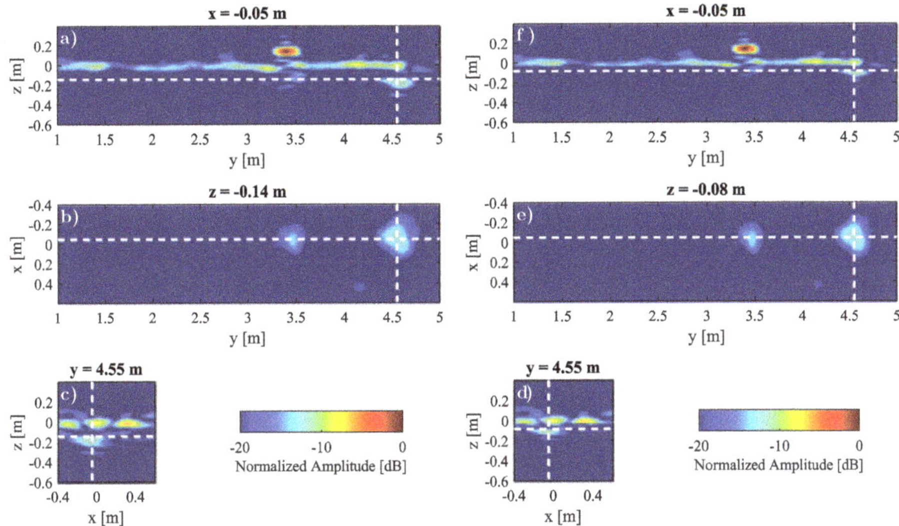

Figure 18. Slices of the 3D SAR image at the position where the cylindrical box is detected when height correction and equalization are applied. Results on the left are obtained without taking into account the soil composition, whereas results on the right are obtained assuming $\varepsilon_r = 3$: (**a,d**) along-track views at $x = -0.05$ m; (**b,e**) top views at $z = -0.14$ m and $z = -0.08$ m,, respectively; and (**c,f**) across-track views.

4.2.4. Soil Composition Consideration

Finally, the soil composition must be taken into account in order to obtain a well-focused SAR image with the buried targets detected at their true depth. Soil permittivity can be estimated from the radar measurements or from the characteristics of the scenario (temperature, soil material components and volumetric water content). A relative permittivity of $\varepsilon_r = 3$ has been assumed, based on previous soil characterization results from [24]. This agrees with the fact that, when free-space propagation is considered, the box is detected at a depth of approximately $d_{box}\sqrt{\varepsilon_r} = 0.14$ m (being $d_{box} = 0.08$ m the true depth).

The resulting SAR images are shown in Figure 18b. The box is now detected around its true depth (that is, the depth where the box is detected when it is not covered by soil).

4.3. Comparison

To further compare the results, the histograms of the 3D SAR images amplitudes when using basic and enhanced processing (height correction, equalization and soil composition consideration) are shown in Figure 19 (in blue and orange respectively) for the second measurement (i.e., when there is a buried target). In this figure, the vertical lines indicate the maximum amplitude of the buried target in the corresponding 3D SAR images. This representation allows quantitatively assessing the influence of the proposed enhanced processing. When applying this enhanced processing, the distance between the target amplitude (depicted in the vertical line) and the clutter is increased, which helps to improve the detection capabilities of the system and to reduce the false alarm rate.

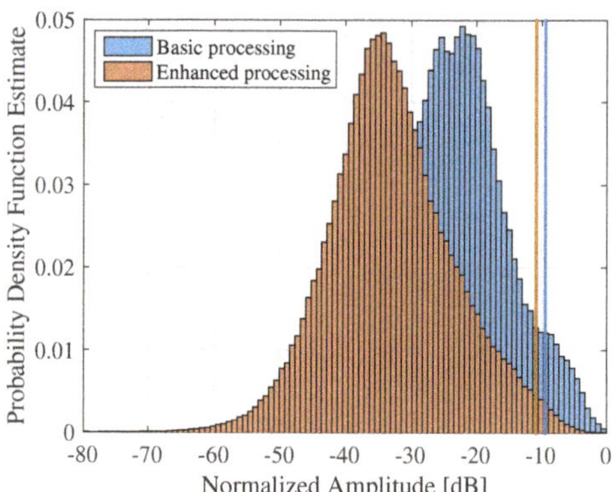

Figure 19. Histograms of the 3D SAR images amplitudes of the second measurement when using basic and enhanced processing (height correction, equalization and soil composition consideration). Vertical lines indicate the maximum amplitude of the buried target in the corresponding 3D SAR images.

5. Conclusions

An enhanced UAV-mounted SAR imaging system to obtain 3D high-resolution radar images (where both underground and overground targets can be detected) is presented. The ultimate goal is the detection of buried hazards, mainly landmines and IEDs, although it can be used for a wide range of application involving both subsurface sensing and terrain observation.

First, several improvements have been implemented in the proposed prototype, mainly to increase the accuracy of the positioning system and the penetration of the electromagnetic waves into the soil. As shown in the results, this allows the coherent combination of measurements gathered at arbitrary positions (3D measurement grid), providing high-resolution radar images, and it also enables the detection of targets buried in higher losses soils.

Concerning the methodology, the processing chain for both the positioning and radar data is thoroughly explained. This methodology can be used to process data collected in both manual and autonomous flight mode, since it selects which measurements will be processed and the location of the investigation domain taking into account the set of UAV positions where measurements were acquired during the flight (i.e., the observation domain). Furthermore, several enhancements in the radar processing are presented to improve the resulting 3D images (mainly in terms of resolution and target discrimination).

Both the prototype and the methodology were experimentally validated with autonomous measurement flights, proving the capability of the system to provide high-resolution 3D SAR images even when using a basic processing strategy. In addition, the effectiveness of the enhanced radar processing was also proved, showing that it yields better resolution and signal to clutter ratio. As expected, results also confirm that, when an estimation of the soil composition is taken into account the buried targets are detected at their true depth.

6. Patents

The work presented in this contribution is related to the patent "Airborne Systems and Detection Methods Localisation and Production of Images of Buried Objects and Characterisation of the Composition of the Subsurface". Publication No. WO/2017/125627. International Application No. PCT/ES2017/000006. Priority date: 21 January 2016. International filing date: 18 January 2017. University of Oviedo, University of Vigo. Available online: https://patentscope.wipo.int/search/en/detail.jsf?docId=WO2017125627 (accessed on 1 September 2019).

Supplementary Materials: A video of the system is available online at http://www.mdpi.com/2072-4292/11/20/2357/s1.

Author Contributions: Conceptualization, M.G.-F., Y.A.-L. and F.L.H.; methodology, M.G.-F., Y.A.-L. and F.L.H.; software, M.G.-F. and Y.A.-L.; hardware, M.G.-F. and Y.A.-L.; validation, M.G.-F., Y.A.-L. and F.L.H.; data curation, M.G.-F.; writing—original draft preparation, M.G.-F.; writing—review and editing, Y.A.-L. and F.L.H.; and funding acquisition, F.L.H. and Y.A.-L.

Funding: This research was funded by the Ministerio de Educación—Gobierno de España under Grant FPU15/06341; by the Ministerio de Ciencia, Innovación y Universidades—Gobierno de España under Project RTI2018-095825-B-I00 ("MilliHand"); by the Government of the Principality of Asturias (PCTI) and European Union (FEDER) under Grant IDI/2018/000191; and by the Instituto Universitario de Investigación Industrial de Asturias (IUTA) under Project SV-19-GIJON-1-17 ("RadioUAV").

Conflicts of Interest: The authors declare no conflict of interest.

References

1. Yao, H.; Qin, R.; Chen, X. Unmanned Aerial Vehicle for Remote Sensing Applications—A Review. *Remote Sens.* **2019**, *11*, 1443. [CrossRef]
2. Hildmann, H.; Kovacs, E. Review: Using Unmanned Aerial Vehicles (UAVs) as Mobile Sensing Platforms (MSPs) for Disaster Response, Civil Security and Public Safety. *Drones* **2019**, *3*, 59. [CrossRef]
3. Manfreda, S.; McCabe, M.F.; Miller, P.E.; Lucas, R.; Pajuelo Madrigal, V.; Mallinis, G.; Ben Dor, E.; Helman, D.; Estes, L.; Ciraolo, G.; et al. On the Use of Unmanned Aerial Systems for Environmental Monitoring. *Remote Sens.* **2018**, *10*, 641. [CrossRef]
4. Jeziorska, J. UAs for Wetland Mapping and Hydrological Modeling. *Remote Sens.* **2019**, *11*, 1997. [CrossRef]
5. Garcia-Fernandez, M.; Alvarez-Lopez, Y.; Las Heras, F. On the use of Unmanned Aerial Vehicles for antenna and coverage diagnostics in mobile networks. *IEEE Commun. Mag.* **2018**, *56*, 72–78. [CrossRef]
6. Llort, M.; Aguasca, A.; Lopez-Martinez, C.; Martinez-Marin, T. Initial evaluation of SAR capabilities in UAV multicopter platforms. *IEEE J. Sel. Top. Appl. Earth Obs. Remote Sens.* **2018**, *11*, 127–140. [CrossRef]
7. Garcia-Fernandez, M.; Alvarez-Lopez, Y.; Arboleya-Arboleya, A.; Gonzalez-Valdes, B.; Rodriguez-Vaqueiro, Y.; Las Heras, F.; Pino, A. Synthetic Aperture Radar imaging system for landmine detection using a Ground Penetrating Radar on board an Unmanned Aerial Vehicle. *IEEE Access* **2018**, *6*, 45100–45112. [CrossRef]
8. Robledo, L.; Carrasco, M.; Mery, D. A survey of land mine detection technology. *Int. J. Remote Sens.* **2009**, *30*, 2399–2410. [CrossRef]
9. Jol, H.M. *Ground Penetrating Radar Theory and Applications*; Elsevier Science: Amsterdam, The Netherlands, 2009.
10. Daniels, D.J. A review of GPR for landmine detection. *Sens. Imaging Int. J.* **2006**, *7*, 90–123. [CrossRef]
11. Rappaport, C.; El-Shenawee, M.; Zhan, H. Suppressing GPR clutter from randomly rough ground surfaces to enhance nonmetallic mine detection. *Subsurf. Sens. Technol. Appl.* **2003**, *4*, 311–326. [CrossRef]
12. Wang, T.; Keller, J.M.; Gader, P.D.; Sjahputera, O. Frequency subband processing and feature analysis of forward-looking ground-penetrating radar signals for land-mine detection. *IEEE Trans. Geosci. Remote Sens.* **2007**, *45*, 718–729. [CrossRef]
13. Peichl, M.; Schreiber, E.; Heinzel A.; Kempf, T. TIRAMI-SAR—A Synthetic Aperture Radar approach for efficient detection of landmines and UXO. In Proceedings of the 10th European Conference on Synthetic Aperture Radar—EuSAR, Berlin, Germany, 3–5 June 2014.
14. Comite, D.; Galli, A.; Catapano, I.; Soldovieri, F. Advanced imaging for down-looking contactless GPR systems. In Proceedings of the International Applied Computational Electromagnetics Society Symposium—Italy (ACES), Florence, Italy, 26–30 March 2017.
15. Gonzalez-Valdes, B.; Alvarez-Lopez, Y.; Arboleya, A.; Rodriguez-Vaqueiro, Y.; Garcia-Fernandez, M.; Las Heras, F.; Pino, A. Airborne Systems and Methods for the Detection, Localisation and Imaging of Buried Objects, and Characterization of the Subsurface Composition. ES Patent 2,577,403 (WO2017/125627), 21 January 2016.
16. Fasano, G.; Renga, A.; Vetrella, A.R.; Ludeno, G.; Catapano, I.; Soldovieri, F. Proof of concept of micro-UAV-based radar imaging. In Proceedings of the International Conference on Unmanned Aircraft Systems (ICUAS), Miami, FL, USA, 13–16 June 2017.

17. Schartel, M.; Bähnemann, R.; Burr, R.; Mayer, W.; Waldschmidt, C. Position acquisition for a multicopter-based synthetic aperture radar. In Proceedings of the 20th International Radar Symposium (IRS), Ulm, Germany, 26–28 June 2019.

18. Alvarez-Narciandi, G.; Lopez-Portuges, M.; Las-Heras, F.; Laviada, J. Freehand, Agile and High-Resolution Imaging With Compact mm-Wave Radar. *IEEE Access* **2019**, *7*, 95516–95526. [CrossRef]

19. Gonzalez-Diaz, M.; Garcia-Fernandez, M.; Alvarez-Lopez, Y.; Las Heras, F. Improvement of GPR SAR-based techniques for accurate detection and imaging of buried objects. *IEEE Trans. Instrum. Meas.* **2019**. [CrossRef]

20. Topcon B111 Receiver. Available online: https://www.topconpositioning.com/oem-components-technology/gnss-components/b111 (accessed on 31 August 2019).

21. Solimene, R.; Cuccaro, A.; Dell'Aversano, A.; Catapano, I; Soldovieri, F. Ground clutter removal in gpr surveys. *IEEE J. Sel. Top. Appl. Earth Obs. Remote Sens.* **2014**, *7*, 792–798. [CrossRef]

22. Johansson, E.M.; Mast, J.E. Three-dimensional ground-penetrating radar imaging using synthetic aperture time-domain focusing. In Proceedings of the International Symposium on Optics, Imaging, and Instrumentation, International Society for Optics and Photonics, San Diego, CA, USA, 14 September 1994.

23. Fuse, Y.; Gonzalez-Valdes, B.; Martinez-Lorenzo, J.A.; Rappaport, C.M. Advanced SAR imaging methods for forward-looking ground penetrating radar. In Proceedings of the European Conference on Antennas and Propagation, Davos, Switzerland, 10–15 April 2016.

24. Alvarez, Y.; Garcia-Fernandez, M.; Arboleya-Arboleya, A.; Gonzalez-Valdes, B.; Rodriguez-Vaqueiro, Y.; Las-Heras, F.; Pino, A. SAR-based technique for soil permittivity estimation. *Int. J. Remote Sens.* **2017**, *38*, 5168–5186. [CrossRef]

Article

Beyond Amplitudes: Multi-Trace Coherence Analysis for Ground-Penetrating Radar Data Imaging

Immo Trinks [1,2,*] and Alois Hinterleitner [1,3]

1   Ludwig Boltzmann Institute for Archaeological Prospection and Virtual Archaeology, Hohe Warte 38, 1190 Vienna, Austria; alois.hinterleitner@archpro.lbg.ac.at
2   Institute for Prehistory and Historical Archaeology, University of Vienna, Franz-Klein-Gasse 1, 1190 Vienna, Austria
3   Central Institute for Meteorology and Geodynamics, Hohe Warte 38, 1190 Vienna, Austria
*   Correspondence: immo.trinks@archpro.lbg.ac.at; Tel.: +43-699-1520-6508

Received: 29 April 2020; Accepted: 14 May 2020; Published: 16 May 2020

Abstract: Under suitable conditions, ground-penetrating radar (GPR) measurements harbour great potential for the non-invasive mapping and three-dimensional investigation of buried archaeological remains. Current GPR data visualisations almost exclusively focus on the imaging of GPR reflection amplitudes. Ideally, the resulting amplitude maps show subsurface structures of archaeological interest in plan view. However, there exist situations in which, despite the presence of buried archaeological remains, hardly any corresponding anomalies can be observed in the GPR time- or depth-slice amplitude images. Following the promising examples set by seismic attribute analysis in the field of exploration seismology, it should be possible to exploit other attributes than merely amplitude values for the enhanced imaging of subsurface structures expressed in GPR data. Coherence is the seismic attribute that is a measure for the discontinuity between adjacent traces in post-stack seismic data volumes. Seismic coherence analysis is directly transferable to common high-resolution 3D GPR data sets. We demonstrate, how under the right circumstances, trace discontinuity analysis can substantially enhance the imaging of structural information contained in GPR data. In certain cases, considerably improved data visualisations are achievable, facilitating subsequent data interpretation. We present GPR trace coherence imaging examples taken from extensive, high-resolution archaeological prospection GPR data sets.

Keywords: GPR; coherence; semblance; attribute analysis; imaging; GPR trace; high-resolution data; large-scale survey; archaeological prospection

---

1. Introduction

The ground-penetrating radar (GPR) method enjoys increasing popularity in the field of archaeological prospection [1,2]. Among all near-surface geophysical survey methods, GPR has the potential to offer under suitable ground conditions 3D information on buried structures in unrivalled spatial imaging resolution. For archaeological prospection purposes, it is common today to record numerous closely spaced, parallel, vertical GPR sections using single-channel GPR antenna systems towed by hand in a sledge or pushed in a cart, or even motorised multi-channel GPR antenna arrays for efficient extensive high-resolution surveys [3]. While ordinary, manually conducted archaeological prospection GPR surveys usually extend across areas covering some hundreds to a few thousand square metres, modern high-resolution motorised GPR surveys can cover areas measuring several square kilometres. The mean frequency of the GPR antennae systems deployed for archaeological prospection is commonly between 200 and 600 MHz. Under favourable ground conditions, the penetration depth of such GPR measurements reach mostly 1.5 to 2 m. In certain cases, it can exceed 3 m depth in dry, sandy soils [4]. Cross-line GPR trace spacing for high-resolution surveys is normally not wider than

25 cm, and can be as little as 8 cm in the case of very dense multi-channel GPR array measurements, with inline GPR trace spacing being in the range of 2–5 cm. The recorded GPR data is commonly processed section by section and then merged and interpolated to form a 3D data volume.

Current state-of-the-art in GPR data visualisation for archaeological prospection surveys is the generation of GPR amplitude maps in the form of GPR time- or depth-slices [5] that sometimes are referred to as C-scans ([6], p. 248). These horizontal data slices are often averaged over a vertical time- or depth-range in order to enhance the visual appearance of structures and anomalies of interest. The generation of such thick slices, in contrast to single-sample slices [7], requires the envelope computation through Hilbert transformation of the GPR traces in order to prevent possible amplitude cancellations when averaging positive and negative amplitude values within a time- or depth-window. This process inevitably reduces the vertical resolution of the GPR data. In order to exploit the full vertical resolution of the data, full-amplitude single-sample GPR depth-slices can be generated, resulting in slices of as little as 6 mm thickness in the case of 400 MHz impulse GPR data. However, the frequent polarity reversal associated with single-sample, full-amplitude slice visualisations, when moving up and down through the data volume and stack of slices, respectively, can obscure the structural information and complicate the understanding of the imaged anomalies.

Three-dimensional GPR data volumes and the contained individual GPR traces comprise beyond simple amplitude values much more information, information that currently is rarely utilised by common GPR data analysis and visualisation techniques. Since GPR data resemble in many respects reflection seismic data [8], it suggests itself to adapt effective algorithms developed for the processing of exploration seismic data to GPR data imaging. In particular, post-stack seismic trace attribute analysis [9] appears to offer promising avenues for the advancement of GPR data visualisation and high-resolution subsurface imaging.

2. Material and Methods

2.1. Seismic Trace Attribute Analysis

In the field of reflection- or exploration-seismology, seismic traces are recorded time series of the acoustic energy that has returned to the surface after reflection from interfaces and discontinuities in the subsurface. Seismic traces comprise many wavelets with different amplitude, frequency and phase shifts. Seismic attributes are quantities that can be extracted or derived from the seismic data to enhance information that may not be obvious in standard amplitude based data visualisations, thus permitting improved possibilities for data analysis and interpretation [9–14]. Seismic attributes can be differentiated into several groups. Taner suggested a classification of seismic attributes into physical and geometric attributes [15], with physical relating to wave propagation, lithology, and other physical parameters, and geometrical relating to dip and azimuth of maximum coherency directions. In reflection seismology it is common to differentiate between pre-stack and post-stack seismic attributes. According to Taner, both pre-stack and post-stack attributes classify into the two sub-classes of instantaneous attributes and wavelet attributes. An alternative classification divides between amplitude attributes, time/horizon attributes, and frequency attributes [16]. Amplitude attributes are mean amplitude, average energy, root mean square amplitude, maximum magnitude, amplitude versus offset (AVO) attributes, and the anelastic attenuation factor. Time/horizon attributes are coherence, dip, azimuth, and curvature. Multi-trace seismic attributes are computed from more than one input trace. They are used to determine coherence, dip and azimuth, or curvature. In general, for most GPR data analogies, only seismic post-stack attributes are of interest. Geophysicists Kurt Marfurt from the University of Oklahoma and Satinder Chopra are the authorities when it comes to seismic attribute and coherence analysis and imaging [17]. The study by Kurt Marfurt headed Research Consortium on *Attribute-Assisted Seismic Processing & Interpretation* (AASPI) has made a rich set of presentations and course notes on seismic attribute analysis available online [18].

The goal of exploration seismology, as much as GPR measurements for archaeological prospection, is the detection and identification of specific anomalous regions and structures in the data, which in general are caused by discontinuities in the subsurface. In the case of reflection seismic surveys, the geological stratification of sedimentary deposits may have been disturbed for instance by erosion channels, intrusions such as salt domes, or fracturing of the rock. The corresponding anomalies can be of great interest for hydrocarbon exploration and reservoir studies. Therefore, the most common attribute used for seismic data analysis, aside from amplitude mapping, is coherence, semblance or discontinuity. According to Chopra [19], coherence is a seismic attribute that measures the similarity between neighbouring seismic traces in 2D or 3D seismic data. It is mainly used to image faults and lateral structural and stratigraphic discontinuities in the data. Coherence mapping is used to enhance stratigraphic information that otherwise may be difficult to extract from the data [20].

On archaeological sites, the subsurface has been disturbed by past human activities, such as the construction of public, private and defensive architecture, including infrastructure installations (e.g., track ways, road pavings, channels, pipes), monuments, fortifications, the excavation of graves, pits, trenches, or the deposition of waste, dirt, bodies, and objects. Any such disturbance could result in discontinuities in the corresponding geophysical prospection data, which thus can be observed as so-called anomalies [21]. This is why the discontinuity, expressed through the coherence or semblance attribute, is of particular interest when it comes to the analysis of GPR data acquired for archaeological prospection. Earlier applications of seismic attribute analysis to the processing of GPR data have been reported by Grasmueck [22], Henryk and Golebiowski [23], Tronicke and Böninger [15,24–27], Zhao et al. [28–30], and Morris et al. [31]. Zhao et al. [29] made use of instantaneous phase, instantaneous amplitude, frequency slope fall and median frequency to image a prehistoric canoe surveyed by GPR. Böninger and Tronicke first convincingly demonstrated the feasibility of attribute-based GPR data processing using amplitude, energy, coherency and similarity to image tombs in relatively small high-resolution GPR data sets, covering areas of 6 × 12 m and 8 × 14 m size [26]. In another publication [25], they describe the use of similarity, energy, analytical signal, topographic slope, and similarity for the analysis of GPR data collected at the palace garden of Paretz in Germany, covering an area of some 30 × 40 m. In a further study [27], symmetry and principal component analysis polarisation attributes are used for the imaging of GPR data acquired across an area of 24 × 32 m. Tronicke and Böninger [15] discuss how attribute analysis can be a valuable technique for the analysis and interpretation of GPR data, illustrating it with a data example covering an area of 20 × 60 m. In this paper, we will focus on the computation and visualisation of the coherence attribute for extensive high-resolution GPR data volumes, and the subsequent data analysis.

2.2. GPR Coherence and Discontinuity Analysis

GPR reflection amplitudes are suited to image and map structures in the subsurface that are strongly reflecting, such as stones, stone layers or remains of stone or brick walls, or that alternatively show increased signal absorption, such as the filling of pits, trenches or postholes. Furthermore, interfaces between layers with different dielectric properties can give rise to increased or decreased reflection amplitudes of the transmitted GPR source pulse. Depending on the electromagnetic properties of the traversed material, frequency dependent absorption and damping of the GPR signal can occur.

Continuity, coherence or semblance (cf. *resemblance, similarity*) exists in GPR data when adjacent GPR traces are highly equal in regard to amplitude, phase shift and frequency content. As stated above, inhomogeneities in the subsurface can cause discontinuities in the recorded GPR data. Thus, it is possible and sensible to compare neighbouring GPR traces in the 3D data volume in order to image areas and zones of discontinuity, dissimilarity or incoherence. Kington [32,33] presented different methods used to compute semblance, coherence and discontinuity attributes for seismic data analysis. He described first the work of Bahorich and Farmer [34] who used simple cross-correlation between three traces. Kington [33] notes that this method is sensitive to noise, and that Marfurt et al. [35] generalised this approach to an arbitrary number of input traces, referring to it as *semblance-based coherence* computation. He explains that this new method is computationally faster and less susceptible to noise, but it is sensitive to lateral amplitude changes and phase differences. Therefore, according to Kington, Gersztenkorn and Marfurt [36] proposed an Eigenstructure-based algorithm to remove amplitude sensitivity for the enhanced imaging of subtle structures. In order to differentiate between dipping reflectors and local discontinuities, as such caused by fault zones, Marfurt [37] proposed a structural dip correction that can be applied to the Kington mentioned methods. Kington describes finally the Gradient Structure Tensor (GST) method proposed by Randen et al. [38] as an alternative approach to discontinuity analysis. The GST approach can be used for the automated interpretation of seismic data based on the analysis of texture attributes by analysing "the variability in the local gradient within a moving window" [33]. The method estimates a gradient vector, a local gradient covariance matrix, and uses then a principal component analysis where the principal eigenvector represents the normal to the dip and azimuth of the local reflection.

As a first attempt, we have implemented in the in-house developed GPR processing software *ApRadar* a simple cross-correlation approach involving adjacent GPR traces for the computation of the coherence [C]. For each sample of each GPR trace, first a correlation [COR] with another neighbouring trace over a certain time window is computed, normally for one (±0.5) or two (±1.0) wavelengths, with the wavelength defined via the nominal frequency of the GPR antenna used. Then, the coherence is obtained by subtracting the correlation from 1.0:

$$C = 1.0 - COR \tag{1}$$

The correlation is in the range of $[-1, \ldots, +1]$: it is $+1$ for two identical time series and -1 in the case of a phase shift of $180°$. Therefore, the coherence computes to the range of $[0, \ldots, +2]$, with a value of zero meaning no deviation between traces, and $+2$ in case of a very large deviation.

We have implemented two geometrically different ways to compute the coherence with regard to neighbouring GPR traces, termed *inline* and *in+crossline* coherence, respectively (Figure 1). In the case of the *inline* coherence [C_{inline}], three traces are used: the central trace ($\bullet$), the one before and the one after ($\bullet$). Thus, two coherences are computed, C_{before} and C_{after}, to determine C_{inline}:

$$C_{inline} = (C_{before} + C_{after})/2.0. \tag{2}$$

The *in+crossline* coherence [$C_{in+crossline}$] is computed by using the three traces of the *inline* coherence as well as the two traces next to the central trace in the neighbouring channels. Thus, in total, five traces are used:

$$C_{in+crossline} = (C_{before} + C_{after} + C_{left} + C_{right})/4.0. \tag{3}$$

The particular challenges arising for coherence data processing are the necessity for the exact determination of time-zero, and, in the case of inline and in+crossline coherence calculations of data acquired with multi-channel systems, as in the examples shown and discussed below, the consideration of the slightly different signal shapes transmitted/recorded by different adjacent antennae. Time-zero determination is much more critical for coherence calculations than in the case of conventional amplitude slice visualisations because of its effect on phase shifts of the individual traces. If time-zero is not determined precisely for each individual trace, then deviations in coherence will result for all time samples of the trace in question. For each trace, time-zero is calculated individually with sub-pixel accuracy by adjusting a trace averaged from the measurement data by coherence calculation. This elaborate computation is necessary because of the constantly ever so slightly changing coupling of the antennas in the multi-antenna array boxes in the case of real measurements, which cause slightly different waveforms of the first wavelet. The zero crossing of the near-field signal is used, which is the crossing between the maxima and minima, respectively, of the primary and secondary pulse. In the case of the16-channel 400 MHz MALÅ Imaging Radar Array (MIRA) commonly employed by us, the recorded signature is highly stable, permitting accurate determinations of the zero crossing. A precise time-zero determination is particularly important for the calculation of the crossline coherence, which takes adjacent channels recorded with different antennae into account.

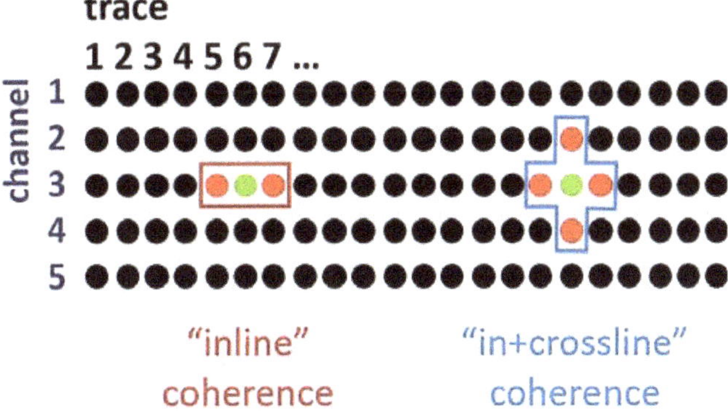

Figure 1. Top-view of the ground-penetrating radar (GPR) traces used for inline and in+crossline coherence computation.

The data processing prior to the computation of the coherence is basically the same as for the generation of Hilbert-transformed GPR depth-slices, though of course omits the envelope trace calculation itself. First, a dewow filter is applied to remove the DC component from the GPR traces. Then, the data are bandpass filtered to eliminate frequency content below half the nominal antenna frequency, as well as above twice its value. Although the GPR antennae should in general be shielded to prevent unwanted reflections from structures located above the ground surface, the recorded GPR signal is actually affected by disturbances that do not originate from the subsurface. Spurious reflections can be caused by the antenna construction itself, the mounting frame of the antenna, and in the case of motorised measurement systems, from the towing or pushing vehicle. These disturbances are predominantly constant and thus can be removed through computation and subtraction of an average trace. To this purpose, a moving average trace is calculated for a range of ±10–20 m before and after the central trace. Finally, if desired, a Kirchhoff migration can be performed.

For visualisation, the GPR amplitude values are replaced with the coherence values prior to computation of the depth-slices. Individual channels are adjusted by multiplication in order to adjust for slight differences between the channels, levelling them on the same mean value, before they are projected into the 3D data volume. The data are then exported as GeoTiff images and visualised in such a way that breaklines and changes in the waveform are displayed as black or dark grey values, while coherent areas show as white or light grey values. Thick depth-slices are generated by summing coherence values over a certain depth-range. Migration of the data can focus the images and can reduce linear noise and distortions that mostly occur in the inline direction of the measurements. As in the case of amplitude maps, migration can result in geometrically more realistic images and thereby facilitate data interpretation.

3. Results—Coherence Imaging Examples

Below, three examples for coherence imaging taken from extensive high-resolution GPR data sets illustrate the potential of this approach. Conventional GPR depth-slice images in the form of amplitude maps are presented alongside coherence maps for the same depth-ranges. All examples shown here are from Scandinavian sites where our coherence imaging tests resulted in notable benefits. These single-phase sites and corresponding archaeological remains appear to be more suitable for coherence imaging than common, multi-phase continental sites with buried Medieval or Roman architectural remains that often come along with complex overburden and strongly inhomogeneous strata. The subtle changes associated with the stratification of filled pits, postholes, and trenches encountered on Scandinavian Iron Age sites seem to be appropriate for the imaging of discontinuities expressed in phase and frequency shifts rather than strong amplitude contrasts. The coherence imaging approach is currently tested on a variety of high-resolution GPR data sets from different regions, covering diverse types of archaeological remains. The outcome of this study will be published separately.

3.1. Rysensten—Denmark

In 2014, the LBI ArchPro in collaboration with Holstebro Museum conducted a geophysical archaeological prospection pilot study at archaeological sites in West Jutland, Denmark. Among others, the site of an abandoned medieval village near Rysensten was surveyed [4,39]. The medieval village of Rysensten is thought to have existed between AD 1050 and AD 1536, and its remains are situated in sandy soil. The site, located close to the coast of the North Sea and the manor of Rysensten on a plateau between the River Ramme and Nissum Fjord, had been discovered by aerial photography [40]. The crop-marks showed rows of postholes, pits and boundary trenches belonging to more than 15 long houses. Here, in total 11.6 ha of the area was surveyed with a motorised 16-channel 400 MHz MIRA system with 8 cm crossline channel spacing and 4 cm inline trace spacing. The data have been processed to form a 3D data volume with 8 × 8 cm horizontal voxel size.

In general, the GPR amplitude maps from Rysensten are dominated by broad, long-wavelength, reflective and absorbing bands associated with geological interfaces between different soil layers (Figures 2A and 3A). The geometry of these sandy soil layers appears to be caused by dune like structures. Aside from these extensive anomalies originating from geological inhomogeneities in the subsurface, broad linear absorbing anomalies associated with human-made drainage trenches, some of which running parallel or perpendicular to each other, dominate the amplitude maps. The archaeological remains, which predominantly appear to be filled pits, postholes or trenches, have been dug into these soil layers. In the conventional GPR amplitude depth-slices it is possible to see these features, as expected, as absorbing anomalies in areas of increased reflectivity, down to 1–1.2 m depth.

The corresponding coherence maps show the large banded anomalies caused by the natural soil stratification hardly at all (Figures 2B and 3B). Instead, these alternative data visualisations highlight anomalies associated with short-wavelength structures and abrupt discontinuities in the subsurface, irrespective of the reflection amplitudes caused by these structures. The level of detail contained in the coherence maps can be best appreciated when zooming into the data images (Figure 3). Small-scale anomalies that are barely visible in the amplitude maps and that are caused by distinct discontinuities across short distances, associated with pits, postholes and small trenches that have been dug or cut into the natural soil layering, stand out in the coherence slices, in which lowest coherence is indicated with black colour. The geometrical arrangement of the coherence anomalies visible in the data from Rysensten, forming rows of anomalies presumably associated with remains of postholes or pits, some of which are aligned in parallel, indicates the presence of several building remains. The coherence maps show the archaeological remains in greater quantity and clarity compared to the amplitude maps.

A possibility to jointly utilise and analyse both amplitude and coherence maps is the combination of corresponding data visualisations through image fusion [41,42]. For this purpose, the different data are visualised using different colour representations: one data set is mapped to 254 grey-scale values while the other is visualised using a symmetrical seismic colour-map, blending linearly from dark red via red to white to blue and then dark blue. Figure 4 illustrates this approach, in which the grey-scale amplitude data has been mapped to a seismic colour scale (Figure 4A), while the coherence data has been visualised using a grey-scale representation (Figure 4B). The fused image (Figure 4C) combines the information contained in both data visualisations, facilitating the understanding of the subsurface situation. In this case, a simple normal blending method with an alpha value of 0.425 for the top image (A) with subsequent contrast normalisation has been used.

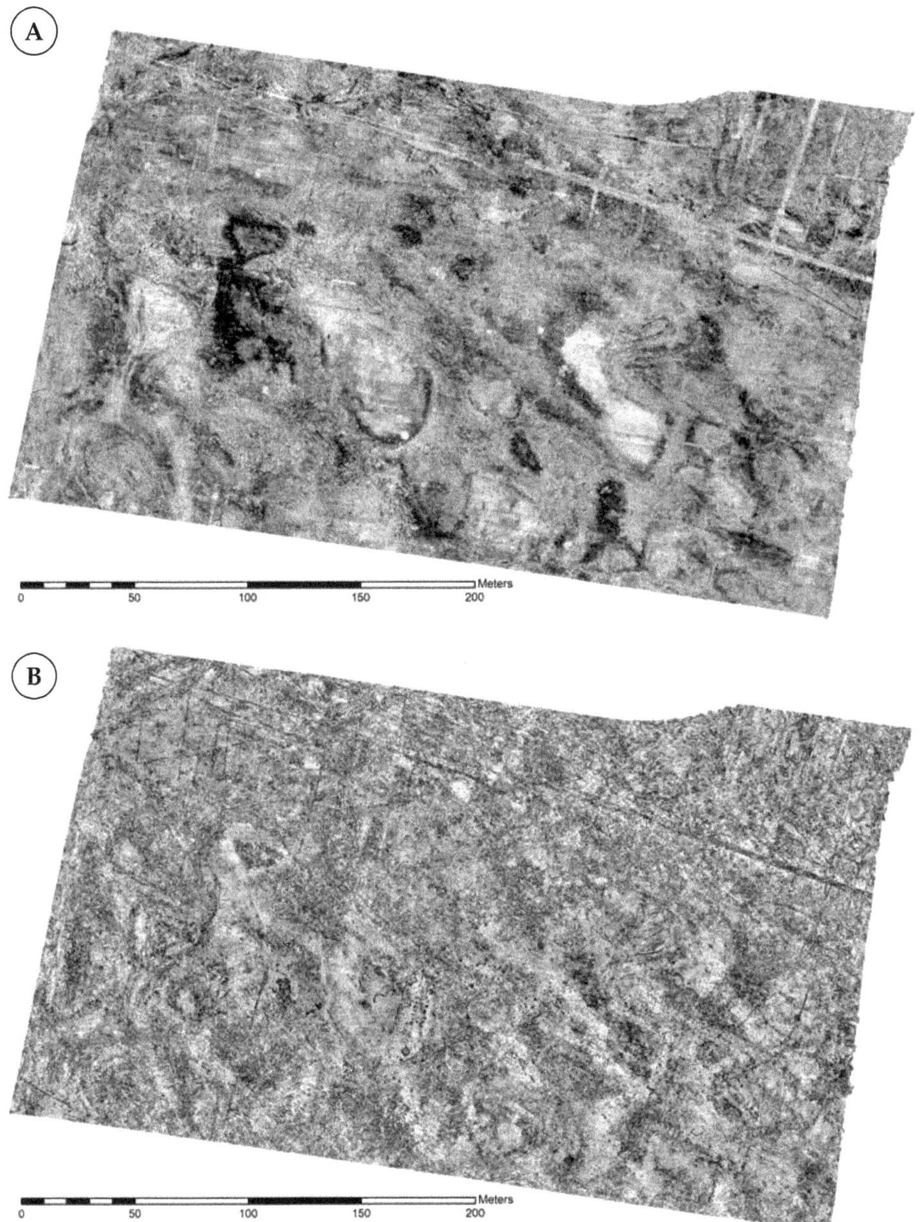

Figure 2. 7.4 ha area GPR survey at Rysensten. (**A**) Conventional GPR amplitude map after Hilbert transformation of the data showing a 5 cm thick depth-slice from approximately 40–45 cm depth. (**B**) Coherence map for the same depth range. North is pointing upwards for all data figures in this article.

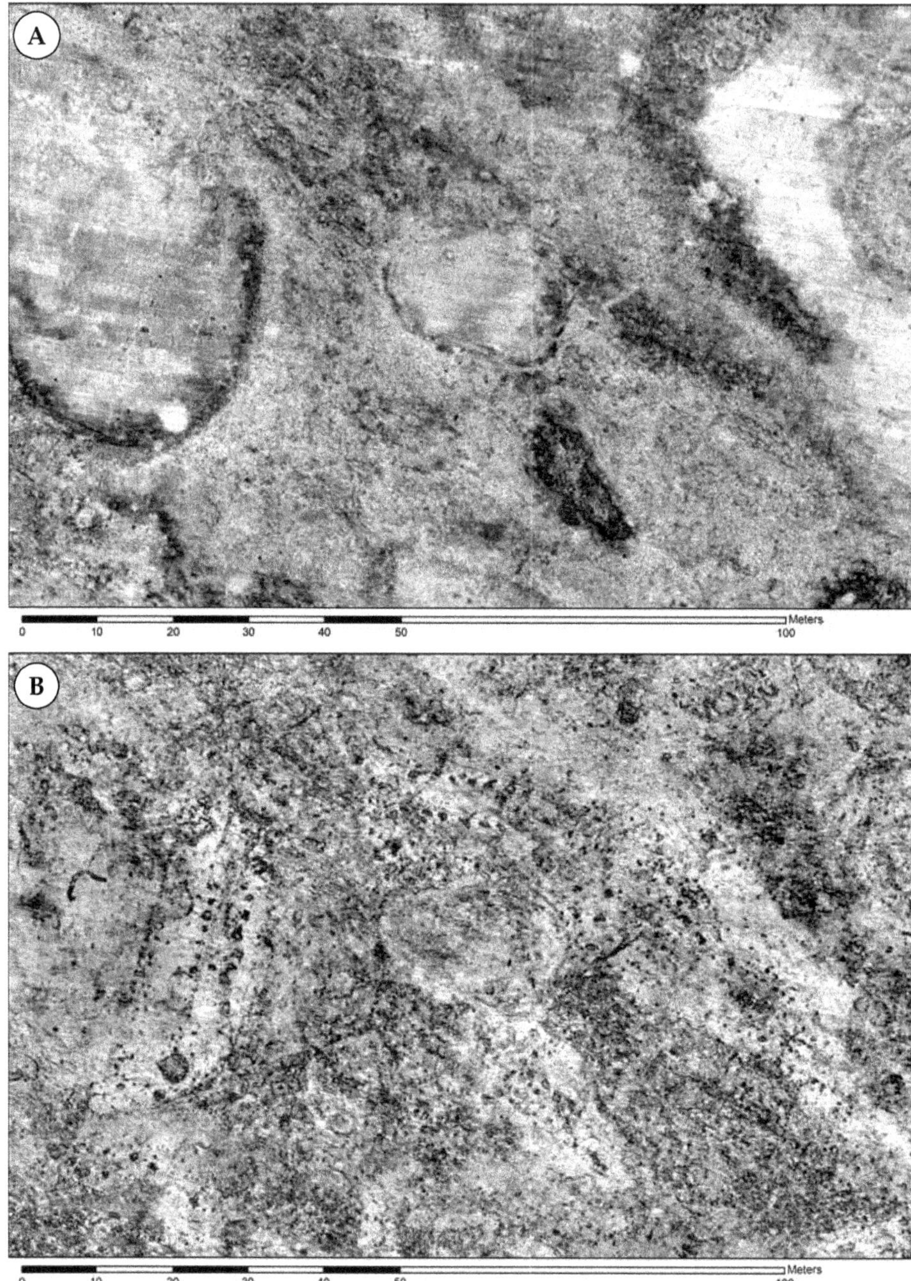

Figure 3. Detail of the same area of the Rysensten GPR data set. (**A**) Conventional GPR amplitude map after Hilbert transformation of the data showing a 5 cm thick depth-slice from approximately 40–45 cm depth. (**B**) Coherence map for the same depth range.

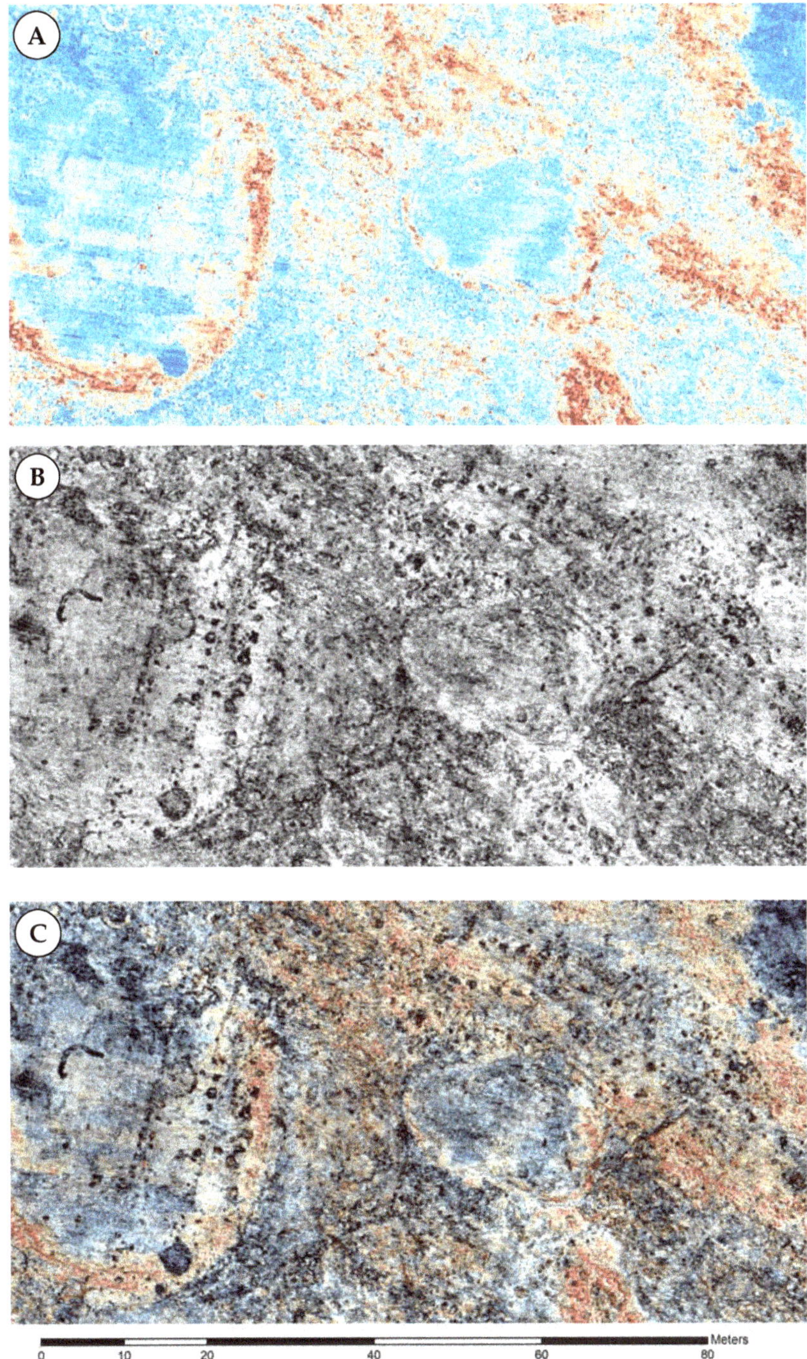

Figure 4. (**A**) Conventional GPR amplitude map after Hilbert transformation of the data showing a 5 cm thick depth-slice from approximately 40–45 cm depth using a red-white-blue seismic colour scale. (**B**) Coherence map for the same depth-range as grey-scale image. (**C**) Fusion of the two images above.

3.2. Stadil Mølleby—Denmark

The same MIRA system as the one used at Rysensten was deployed in September 2014 to survey the Viking Age settlement site of Stadil Mølleby in Denmark, which likewise had been discovered by aerial archaeology ([40], p. 137). In general, the results obtained here were very similar to those of Rysensten, consisting of building remains in the form of pits, postholes and trenches dug into sandy soil layers [4,39]. The settlement comprises a considerable number of longhouses, visible in the GPR data as parallel rows of postholes. The amplitude maps are dominated by larger patches of reflecting and absorbing soil (Figure 5A), transected by drainage trenches. The data appear affected by broad, northeast–southwest running surface disturbances, which seem to have been caused by measurement artefacts. Among the extensive, strongly reflecting and absorbing areas, it is difficult to identify the archaeologically relevant structures, mainly pits and postholes that show as small, light coloured, absorbing anomalies. Particular in areas with generally absorbing character, these anomalies become invisible or very difficult to discern.

In the coherence image (Figure 5B), the pattern of the extensive, gradually varying reflectivity of the subsurface does not show, and instead, the incoherencies or discontinuities in the GPR response caused by local soil disturbances become clearly visible. In the case of larger pits and trenches, the edge of the buried structure shows as sharply defined outline. Numerous pit-alignments indicate building remains. The larger hall buildings show as double rows of postholes (Figure 6B). In Figure 7, a fusion of the data shown in Figure 6A,B is shown. Again, a simple normal fusion with an alpha value of 0.4 for the top image (A) has been chosen, with subsequent histogram stretching applied. This approach is suited to integrate the complementary amplitude and coherence data sets for joint interpretation. It has to be kept in mind that in the case of amplitude imaging, coherence imaging, or the interpretation of fused images, the entire stack of generated slices is analysed, which permits of course for a better recognition and understanding of the anomalies and underlying structures. However, in particular, this example shows the great benefit of amplitude-independent coherence or discontinuity imaging for enhanced data analysis and interpretation possibilities.

3.3. Gjellestad Viking Ship Burial

In November 2018, the Norwegian Institute for Cultural Heritage Research (NIKU) and Østfold County—now Viken County—announced the discovery of the assumed remains of a Viking ship made by archaeological GPR prospection in an overploughed burial mound at the site of Gjellestad in Østfold County in Norway [43,44]. Using a motorised 16-channel 400 MHz MIRA system with 10.5 cm crossline trace spacing, a large area next to the monumental Jelle burial mound was investigated. Traces of eight earlier unknown burial mounds destroyed by ploughing where discovered. In one of the former mounds the buried remains of a Viking ship could clearly be identified. There have been indications that the ship's keel and bottom-most planks are likely to have been preserved in the soil. The Viking ship remains are located immediately below the topsoil at a depth of approximately 50 cm. Archaeological excavations conducted by the University Museum Oslo across the ship structure in summer 2019 have fully confirmed the findings of the GPR survey and the data interpretation of the ship shaped anomaly, as being caused by the remains of a buried Viking ship.

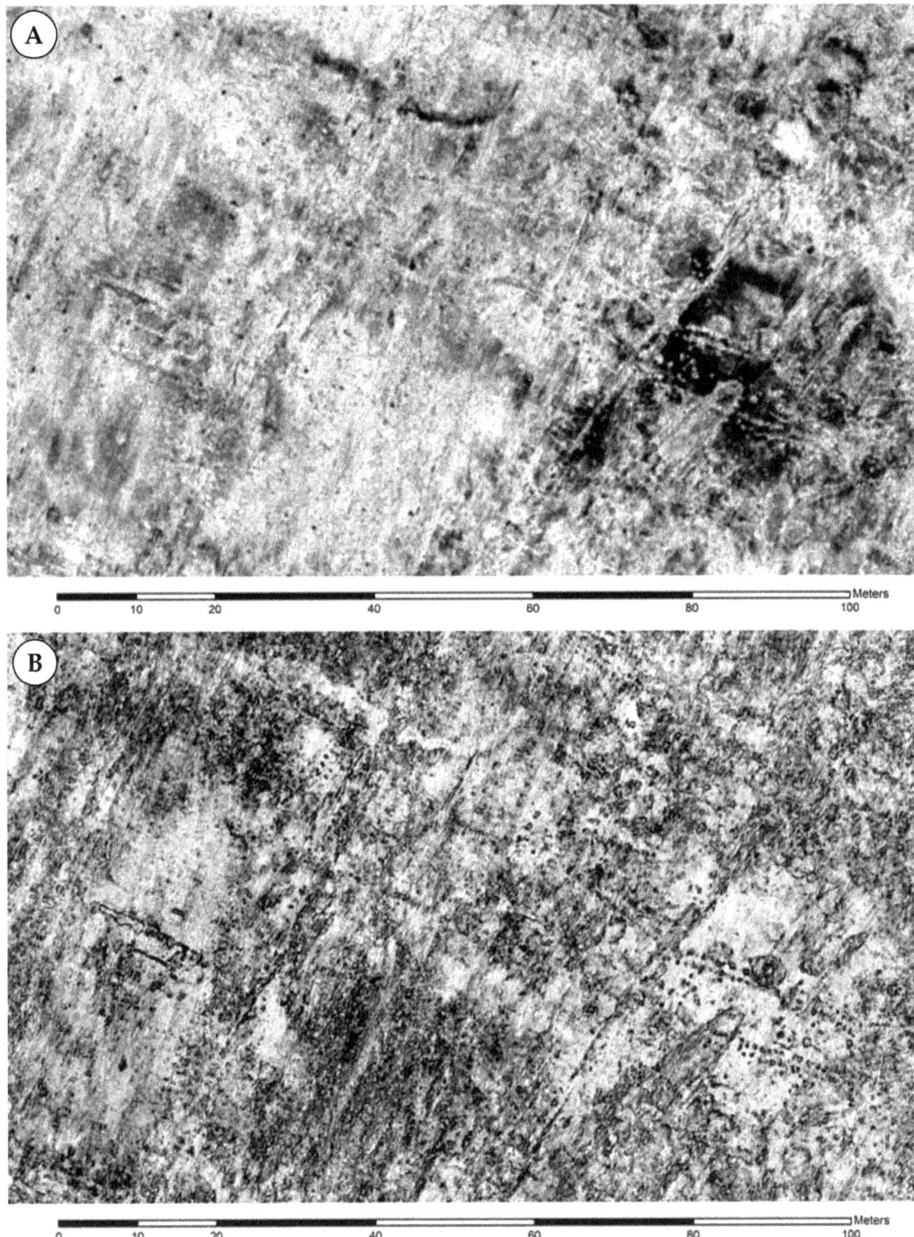

Figure 5. Stadil Mølleby: Migrated GPR depth-slices from 55–85 cm depth showing the same area. (**A**) Conventional reflection amplitude map. (**B**) Inline coherence image. Note the numerous rows of pits indicating building remains clearly visible in the coherence image.

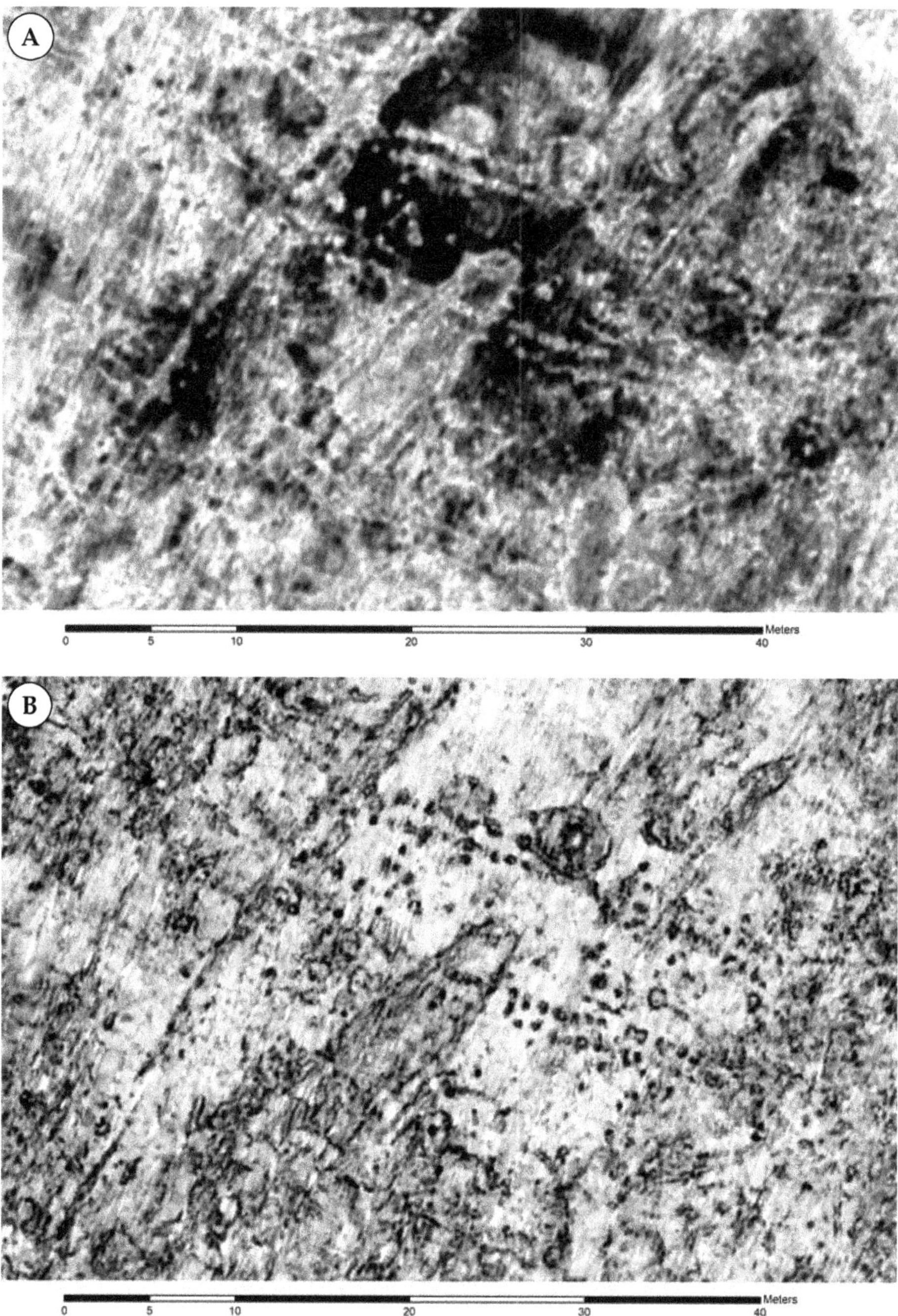

Figure 6. Stadil Mølleby (detail): Migrated GPR depth-slices from 55–85 cm depth showing the same area. (**A**) Conventional reflection amplitude map. (**B**) Inline coherence image. Note the rows of pits indicating building remains clearly visible in the coherence image.

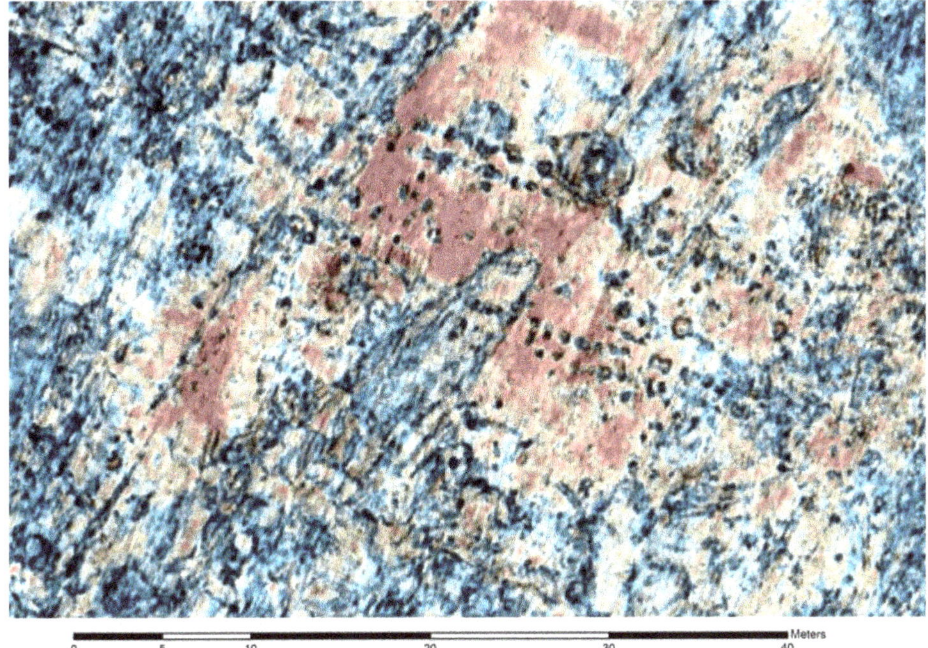

Figure 7. Stadil Mølleby (detail): Fused data image through normal fusion with an alpha value of 0.4 for the top image Figure 6A with seismic colour scale onto Figure 6B. Subsequently, histogram stretching was applied to bring out the details expressed by the black incoherence anomalies. The simple data fusion presented here is one way to merge two data images for ease of comprehension of the different information contained in each data set. The fused image preserved the crispness of the coherence anomalies, while adding the information on the higher reflective horizon marked by the red colour, into which apparently pits or postholes of a building have been dug. This fused image permits the joint simultaneous interpretation of the two complementary GPR amplitude and coherency data images.

Standard GPR amplitude maps generated after Hilbert transformation of the GPR traces show the remains of a large, north–south oriented, well-defined, circa 20 m long ship-shaped structure as absorbing the anomaly in the circular mound (Figure 8A). By inverting the grey-scale map, the absorbing structures appear dark (Figure 8C). Single-sample amplitude slices show reflective structures with a very high vertical imaging resolution. The appearance of the single-sample slices is fundamentally different to the Hilbert transformed slices, due to the alternating positive and negative amplitude values along each GPR trace (Figure 8B). The coherence images present structures in the subsurface irrespective of the reflected energy, but relative to the amount of discontinuity caused to the recorded time-series within the cross-correlation window. Thus, in this case the edge of the circular mound, the outline of the ship and internal features, the drainage trenches as well as a large number of detailed ground disturbances that do not show in the amplitude maps are visible (Figure 8D). The coherence image is highly detailed and complements the conventional data visualisation. A comparison of the archaeological interpretations based on different data visualisations is not part of this paper. However, from the examples shown, it becomes obvious that coherence imaging is a valuable possibility to reveal structural details that would be missed when using only conventional GPR data imaging.

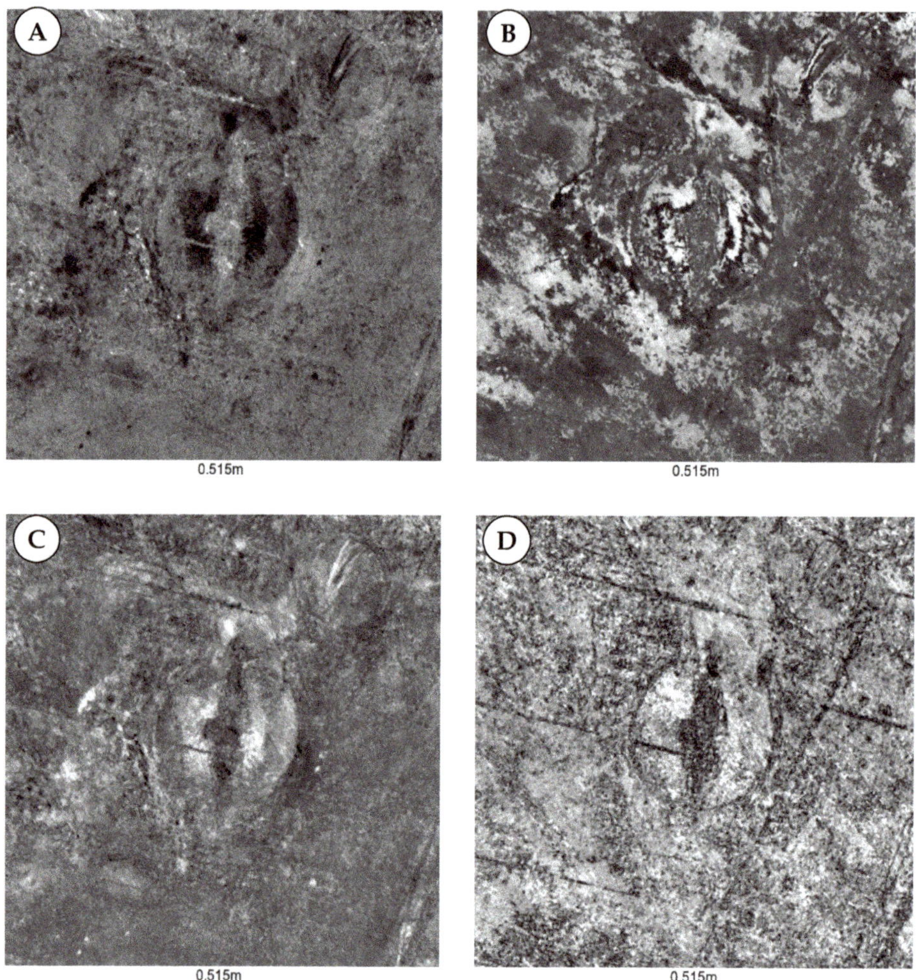

Figure 8. Four visualisations of GPR depth-slices (approx. 51.5 cm depth) from the overploughed burial mound with the remains of the Gjellestad Viking ship. North is upwards. (**A**) Standard amplitude map after Hilbert transformation. Light areas indicate absorbing structures, dark areas increased reflectivity. (**B**) Single-sample full amplitude slice. (**C**) Inverse image of (**A**). (**D**) Coherence map.

4. Discussion

In most of our trials, the *in+crossline* coherence (Figure 9e) worked better than only *inline* (Figure 9c) or only crossline (see Figure 9d) coherence, since thus incoherencies in all directions become visible. The *inline* coherence computation mainly highlights incoherencies in the direction of measurement progress. In the case of more sparsely sampled data, for instance data collected with 25 cm crossline spacing, the results of the *inline* coherence appear sharper than those of the *in+crossline* coherence due to the under-sampling in *crossline* direction.

Migrated coherence images are better defined and focused than non-migrated images. They appear as well richer in contrast, better bringing out the structures. While anomalies associated with pits appear in the unmigrated coherence images as filled structures (Figure 9f), they show as outlining edges in the migrated coherence data (Figure 9e).

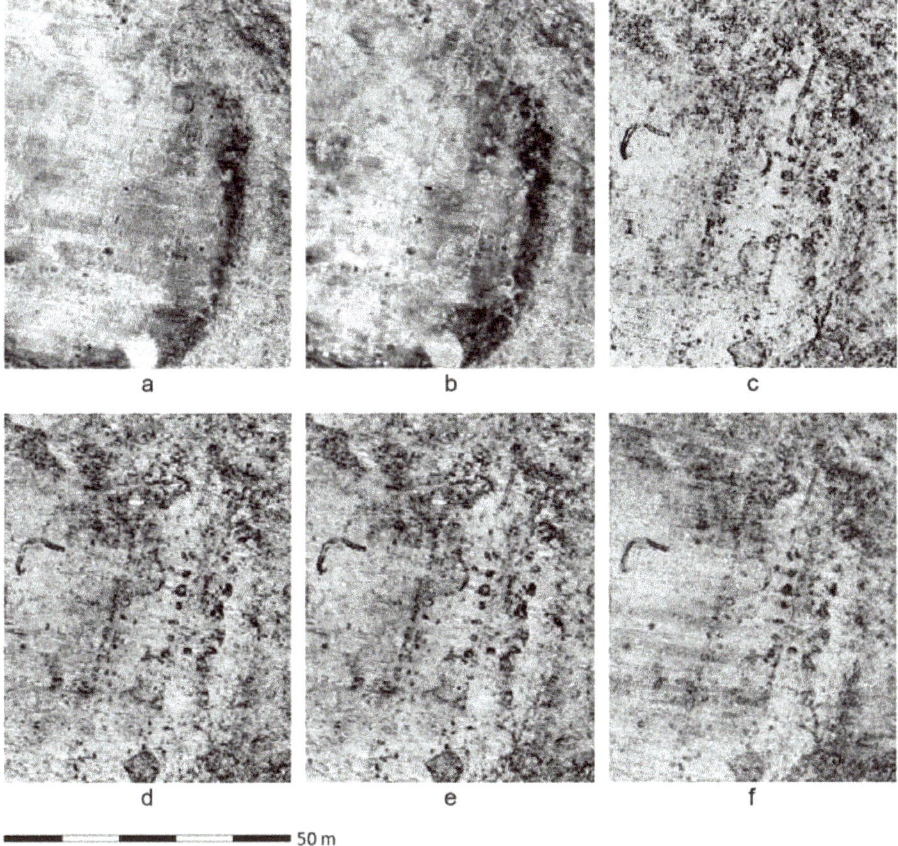

Figure 9. Depth range 40–45 cm. (a) Unmigrated amplitude values. (b) Migrated amplitude values. (c) Inline coherence migrated. (d) Crossline coherence migrated. (e) In+crossline coherence migrated. (f) In+crossline coherence unmigrated.

According to the online knowledge base [45] of the Society of Exploration Geophysicists on *Coherence*, the "analysis window chosen can have a significant effect of [*sic*] the results. Longer windows tend to mix stratigraphy, which can complicate the geologic image". Kurt Marfurt ([46], slide 54), stated that "Large vertical analysis windows can improve the resolution of vertical faults, but smears dipping faults and mixes stratigraphic features".

We have made tests regarding the analysis window defined by wavelength and the number of traces involved in the computation of the GPR trace coherence (Figure 10). The wavelength over which the correlation is computed, sensibly should be between one and two wavelengths of the mean transmitted pulse (Figure 10c,d). The wavelength of a 400 MHz pulse is 25 cm. Using less than one wavelength causes an increase in noise (Figure 10a,b). Using more than two wavelengths leads to decreased contrasts and reduced focusing (Figure 10e,f). One to two wave trains correspond to approximately the wavelength of the GPR pulse transmitted by the GPR antenna as well. In the simple approach presented here for coherence imaging based on trace cross-correlations, it is not really meaningful to compute single-sample coherence maps since the coherence computation requires a certain length. Using more than the four directly neighbouring traces does not result in improved images, but increases computation time and decreases the image contrast with increasing number of traces involved.

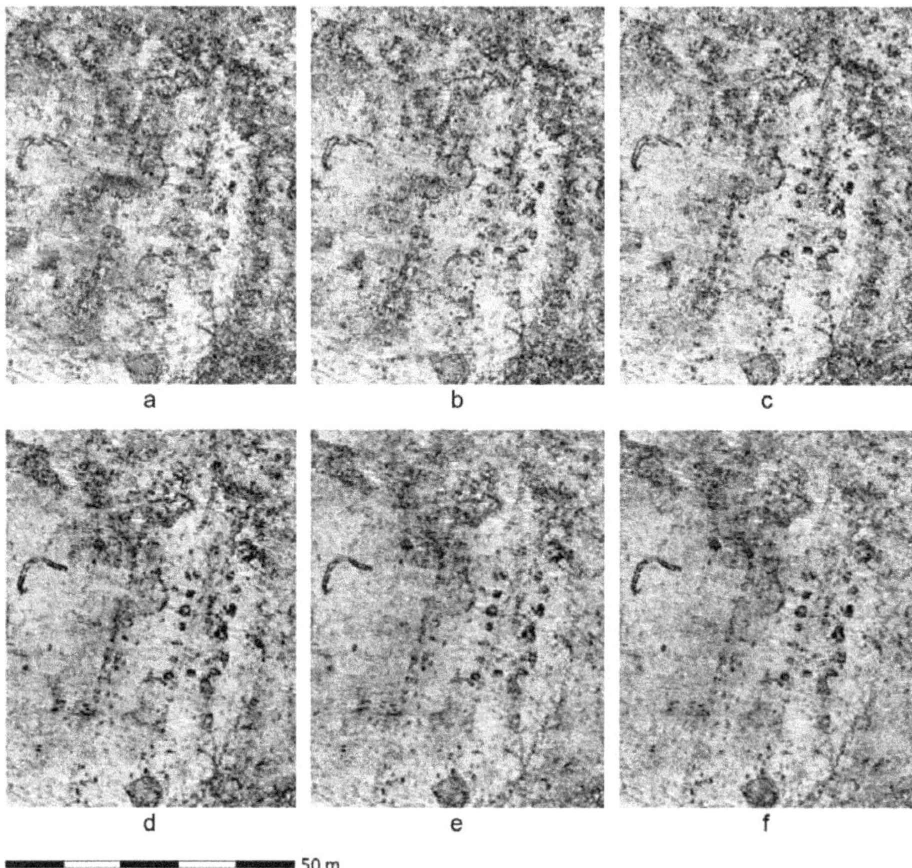

Figure 10. Coherence computation using different cross-correlation lengths: (**a**) ¼ wavelength, (**b**) ½ wavelength, (**c**) 1 wavelength, (**d**) 2 wavelengths, (**e**) 3 wavelengths, (**f**) 4 wavelengths.

The application of an average-trace removal filter is commonly used for the enhanced imaging of amplitude data. Average-trace removal applied to coherence data can have a strong effect on the resulting depth-slice images. Depending on the distance along which the average-trace is determined, strikingly different outcomes can be achieved (Figure 11). It is thought that the removal of the average-trace in coherence data corresponds a spatial high-pass, or dewow filter that accentuates short-wavelength incoherencies. The larger the effect, the shorter the distance for which the average trace is computed. In the case of our test data, a length of 10 to 20 m for the computation of the average-trace resulted in the clearest structural imaging of small discontinuities in the ground (Figure 11d,e).

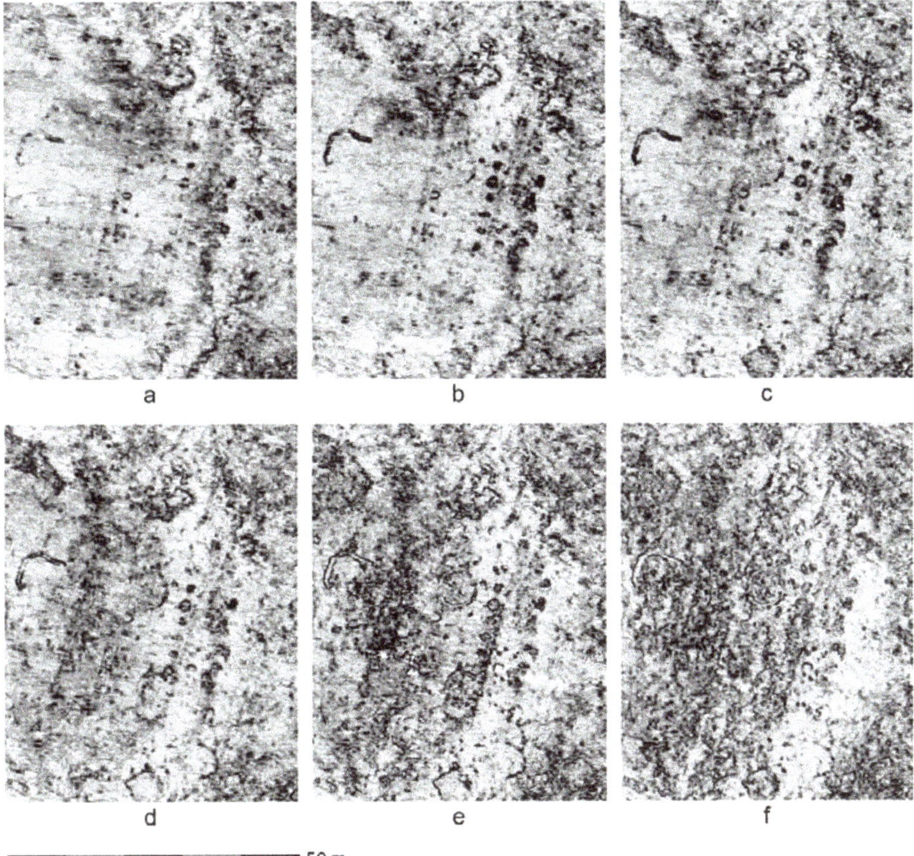

Figure 11. Coherence data after subtraction of the average trace (AT) determined over different distance ranges: (**a**) no AT filter applied, (**b**) with 80 m AT filter, (**c**) with 40 m AT filter, (**d**) with 20 m AT filter, (**e**) with 10 m AT filter, (**f**) with 5 m AT filter.

Under suitable soil conditions and regarding present archaeological remains, coherence imaging of GPR data appears to be a powerful complement to conventional amplitude mapping. Our tests have shown that in particular Scandinavian Iron Age sites, the prevalent ground conditions can be well suited for this approach. Attempts to use coherence mapping for the imaging of GPR data acquired at Roman sites have so far not been similarly successful. We assume that gains made by coherence imaging are smaller here since the amplitude maps already clearly show to a large extent the anomalies caused by the buried remains of mainly stone built architecture. Further testing is necessary to better evaluate the potential of coherence mapping for other sites and geological backgrounds. Our understanding is that single phase, or not too complex archaeological sites embedded in geologically rather homogeneously stratified soils should be well suited. Kurt Marfurt ([46], slide 54), has summarised coherence imaging for seismic processing and interpretation, which is completely transferable to GPR data visualisation. He states that "coherence

- is an excellent tool for delineating geological boundaries (faults, lateral stratigraphic contacts, etc.),
- allows accelerated evaluation of large data sets,
- provides quantitative estimate of fault/fracture presence,

- often enhances stratigraphic information that is otherwise difficult to extract,
- algorithms are local—faults that have drag, are poorly migrated, or separate two similar reflectors, or otherwise do not appear locally to be discontinuous, will not show up on coherence volumes".

We can only confirm the observations made by Marfurt [46], slide 25, who noted that the parameters affecting the quality of the coherency imaging results are static corrections and time zero, velocities, data processing filters, and the migration algorithm used. Furthermore, the method used to compute the coherence can considerably enhance the result.

5. Conclusions and Outlook

We have shown that poststack seismic geometrical attribute analysis, specifically coherence analysis, can successfully be applied for the efficient generation of novel useful visualisations of extensive high-resolution GPR data sets collected for archaeological prospection purposes. The here presented and tested coherence algorithm, based on simple trace cross-correlation, is rather primitive compared to the more advanced approaches developed by Gersztenkorn and Marfurt [36], Randen et al. [38] and Marfurt [37]. Thus, there exists still considerable room for methodological improvements of the method with regard to its application to GPR data imaging. Nevertheless, the presented examples show that coherence attribute mapping has the potential to considerably enhance GPR data visualisations and therefore offers new avenues for improved data interpretation, provided that ultra-dense GPR data covering large areas is available, since coarsely sampled data undermines the potential of this method. GPR trace coherence images should be generated routinely alongside conventional amplitude maps to permit for enhanced data interpretations. Aside from geometrical attributes, there exists a number of other seismic attributes of interest when it comes to the development of new promising approaches to GPR data analysis and interpretation. The theoretical groundwork has been laid by seismic processing wizards Satinder Chopra and Kurt Marfurt, among others. It is up to us to pick these low hanging fruits.

Author Contributions: Conceptualization, I.T.; methodology, I.T. and A.H.; software, A.H.; validation, A.H. and I.T.; formal analysis, A.H. and I.T.; investigation, I.T. and A.H.; writing—original draft preparation, I.T.; writing—review and editing, I.T. and A.H.; visualization, A.H. All authors have read and agreed to the published version of the manuscript.

Funding: This research was funded by the Ludwig Boltzmann Institute for Archaeological Prospection and Virtual Archaeology (LBI ArchPro). The LBI ArchPro (archpro.lbg.ac.at) is based on an international cooperation of the Ludwig Boltzmann Gesellschaft (A), Amt der Niederösterreichischen Landesregierung (A), University of Vienna (A), TU Wien (A), ZAMG—Central Institute for Meteorology and Geodynamics (A), Airborne Technologies (A), 7reasons (A), RGZM Mainz—Römisch-Germanisches Zentralmuseum Mainz (D), LWL—Federal state archaeology of Westphalia-Lippe (D), NIKU—Norwegian Institute for Cultural Heritage Research (N), Vestfold fylkeskommune–Kulturarv (N), and *Arc*Tron^{3D} GmbH (D). The research projects at Rysensten and Stadil Mølleby were co-funded by Holstebro Museum (Dk).

Acknowledgments: The fieldwork at Rysensten and Stadil Mølleby was conducted by Roland Filzwieser, Manuel Gabler, Erich Nau and Petra Schneidhofer. The authors thank Lis Helles Olesen for kind permission to publish the data examples from Denmark. The Gjellestad data was collected by Erich Nau and Lars Gustavsen from the Norwegian Institute for Cultural Heritage Research (NIKU) in a survey conducted by NIKU on behalf of Østfold County, now Viken County. The authors thank Sigrid Mannsåker Gundersen and Viken County Council as well as Knut Paasche from NIKU for kind permission to publish the Gjellestad data. Roland Filzwieser has kindly proofread the manuscript. Three anonymous reviewers helped to improve the manuscript.

Conflicts of Interest: The authors declare no conflict of interest.

Abbreviations

The following abbreviations are used in this manuscript:

AT	Average-Trace
LBI ArchPro	Ludwig Boltzmann Institute for Archaeological Prospection and Virtual Archaeology
GPR	Ground-Penetrating Radar
GST	Gradient Structure Tensor

References

1. Conyers, L. *Ground-Penetrating Radar for Archaeology*, 3rd ed.; Rowman and Littlefield Publishers, Alta Mira Press: Lanham, MD, USA, 2013.
2. Leckebusch, J. Ground-penetrating Radar: A Modern Three-dimensional Prospection Method. *Archaeol. Prospect.* **2003**, *10*, 213–240. [CrossRef]
3. Trinks, I.; Hinterleitner, A.; Neubauer, W.; Nau, E.; Löcker, K.; Wallner, M.; Gabler, M.; Filzwieser, R.; Wilding, J.; Schiel, H.; et al. Large-area high-resolution ground-penetrating radar measurements for archaeological prospection. *Archaeol. Prospect.* **2018**, *25*, 171–195. [CrossRef]
4. Filzwieser, R.; Olesen, L.H.; Neubauer, W.; Trinks, I.; Schlosser Mauritsen, E.; Schneidhofer, P.; Nau, E.; Gabler, M. Large-scale geophysical archaeological prospection pilot study at Viking Age and medieval sites in west Jutland, Denmark. *Archaeol. Prospect.* **2017**, *24*, 373–393. [CrossRef]
5. Goodman, D.; Nishimura, Y.; Rodgers, J.D. GPR Time Slices in Archaeological Prospection. *Archaeol. Prospect.* **1995**, *2*, 85–89.
6. Daniels, D.J. Ground penetrating radar. In *Encyclopedia of RF and Microwave Engineering*; Chang, K., Ed.; Wiley-Interscience: Hoboken, NJ, USA, 2005.
7. Trinks, I.; Johansson, B.; Gustafsson, J.; Emilsson, J.; Friborg, J.; Gustafsson, C.; Nissen, J.; Hinterleitner, A. Efficient, large-scale archaeological prospection using a true three-dimensional ground-penetrating radar array system. *Archaeol. Prospect.* **2010**, *17*, 175–186, doi:10.1002/arp. [CrossRef]
8. Fisher, S.C.; Stewart, R.R.; Jol, H.M. Ground Penetrating Radar (GPR) Data Enhancement Using Seismic Techniques. *J. Environ. Eng. Geophys.* **1996**, *1*, 89–157. [CrossRef]
9. Chopra, S.; Marfurt, K.J. Seismic attributes—A historical perspective. *Geophysics* **2005**, *70*, 3SO–28SO. [CrossRef]
10. Chopra, S.; Marfurt, K.J. Seismic attributes—A promising aid for geological prediction. *CSEG Recorder* **2006**, *31*, 115–126.
11. Randen, T.; Sønneland, L. Atlas of 3D Seismic Attributes. In *Mathematical Methods and Modelling in Hydrocarbon Exploration and Production. Mathematics in Industry*; Iske, A., Randen, T., Eds.; Springer: Berlin/Heidelberg, Germany, 2005.
12. Sheriff, R.E. *Encyclopedic Dictionary of Applied Geophysics*, 4th ed.; Society of Exploration Geophysicists: Houston, TX, USA, 2002.
13. Eastwood, J. Introduction—The attribute explosion. *Lead. Edge* **2002**, *21*, 994–995. [CrossRef]
14. Taner, M.T.; Koehler, F.; Sheriff, R.E. Complex seismic trace analysis. *Geophysics* **1979**, *44*, 1041–1063. [CrossRef]
15. Tronicke, J.; Böninger, J. GPR attribute analysis: There is more than amplitudes. *First Break* **2013**, *31*, 103–108.
16. Wikipedia Contributors. Seismic Attribute—Wikipedia, The Free Encyclopedia. 2018. Available online: https://en.wikipedia.org/wiki/Seismic_attribute (accessed on 15 May 2020).
17. Chopra, S.; Marfurt, K.J. *Seismic Attributes for Prospect ID and Reservoir Characterization*; SEG/EAGE, 2007. Available online: https://library.seg.org/doi/abs/10.1190/1.9781560801900 (accessed on 15 May 2020).
18. Marfurt, K.J. Attribute-Assisted Seismic Processing & Interpretation. Available online: http://mcee.ou.edu/aaspi/index.html (accessed on 15 May 2020).
19. Chopra, S. Adding the coherence dimension to 3D seismic data. *CSEG Rec.* **2001**, *26*, 5–8.
20. Nissen, S.; Interpretive Aspects of Seismic Coherence and Related Multi-trace Attributes; Reservoir Characteristics of Morrow/Incised-Valley Fill Plays Workshop & Morrow Incised-Valley Core Workshop. Kansas Geological Surveys, Wichita, Kansas, February 2000. Available online: http://www.kgs.ku.edu/Workshops/IVF2000/nissan-ivf/tocnav1.html (accessed on 15 May 2020).
21. Lyatsky, H. The Meaning of Anomaly. *Explor. Geophys.* **2004**, *29*, 50–51.
22. Grasmueck, M. 3-D ground-penetrating radar applied to fracture imaging in gneiss. *Geophysics* **1996**, *61*, 1050–1064. [CrossRef]
23. Henryk, M.; Golebiowski, T. Analysis of GPR Trace Attributes and Spectra for LNAPL Contaminated Ground. In Proceedings of the Near Surface 2006—12th EAGE European Meeting of Environmental and Engineering Geophysics, Helsinki, Finland, 4–6 September 2006.
24. Böninger, U. Attributes and their potential to analyze and interpret 3D GPR data. Ph.D. Thesis, Potsdam University, Potsdam, Germany, 2010.

25. Böninger, U.; Tronicke, J. Integrated data analysis at an archaeological site: A case study using 3D GPR, magnetic, and high-resolution topographic data. *Geophysics* **2010**, *75*, B169–B176. [CrossRef]
26. Böninger, U.; Tronicke, J. Improving the interpretability of 3D GPR data using target-specific attributes: Application to tomb-detection. *J. Archaeol. Sci.* **2010**, *37*, 360–367. [CrossRef]
27. Böninger, U.; Tronicke, J. Subsurface Utility Extraction and Characterization: Combining GPR Symmetry and Polarization Attributes. *IEEE Trans. Geosci. Remote. Sens.* **2012**, *50*, 736–746. [CrossRef]
28. Zhao, W.; Forte, E.; Pipan, M.; Tian, G. Ground Penetrating Radar (GPR) attribute analysis for archaeological prospection. *J. Appl. Geophys.* **2013**, *97*, 107–117. [CrossRef]
29. Zhao, W.; Tian, G.; Wang, B.; Forte, E.; Pipan, M.; Lin, J.; Shi, Z.; Li, X. 2D and 3D imaging of a buried prehistoric canoe using GPR attributes: A case study. *Near Surf. Geophys.* **2013**, *11*, 457–464. [CrossRef]
30. Zhao, W.; Forte, E.; Pipan, M. Texture Attribute Analysis of GPR Data for Archaeological Prospection. *Pure Appl. Geophys.* **2016**, *173*, 2737–2751. [CrossRef]
31. Morris, I.; Abdel-Jaber, H.; Glisic, B. Quantitative Attribute Analyses with Ground Penetrating Radar for Infrastructure Assessments and Structural Health Monitoring. *Sensors* **2019**, *19*, 1637. [CrossRef] [PubMed]
32. Kington, J. Semblance, coherence, and other discontinuity attributes. *Lead. Edge* **2015**, *34*, 1426–1544. [CrossRef]
33. Kington, J. Semblance, Coherence, and other Discontinuity Attributes. Available online: https://github.com/seg/tutorials-2015/blob/master/1512_Semblance_coherence_and_discontinuity/Discontinuity_tutorial.ipynb (accessed on 15 May 2020).
34. Bahorich, M.; Farmer, S. 3-D seismic discontinuity for faults and stratigraphic features: The coherence cube. *Lead. Edge* **1995**, *14*, 1053–1058. [CrossRef]
35. Marfurt, K.; Kirlin, R.L.; Farmer, S.L.; Bahorich, M.S. 3-D seismic attributes using a semblance-based coherency algorithm. *Geophysics* **1998**, *63*, 1150–1165. [CrossRef]
36. Gersztenkorn, A.; Marfurt, K.J. Eigenstructure-based coherence computations as an aid to 3-D structural and stratigraphic mapping. *Geophysics* **1999**, *64*, 1468–1479. [CrossRef]
37. Marfurt, K.J. Robust estimates of 3D reflector dip and azimuth. *Geophysics* **2006**, *71*, 29–40. [CrossRef]
38. Randen, T.; Monsen, E.; Signer, C.; Abrahamsen, A.; Hansen, J.O.; Sæter, T.; Schlaf, J. Three-dimensional texture attributes for seismic data analysis. In *Proceedings of the 70th Annual International Meeting*; Expanded Abstracts; SEG: Tulsa, OK, USA, 2000; pp. 668–671.
39. Filzwieser, R.; Neubauer, W.; Trinks, I.; Olesen, L.; Verhoeven, G.; Mauritsen, E.; Schneidhofer, P.; Nau, E.; Gabler, M. Geofysiske undersøgelser og luftfotoarkæologi i Vestjylland. In *Luftfotoarkæologi 2*; Olesen, L., Mauritsen, E., Broch, M., Eds.; De Kulturhistoriske Museer i Holstebro Kommune: Copenhagen, Denmark, 2019; pp. 163–177.
40. Olesen, L.; Mauritsen, E. *Luftfotoarkæologi i Danmark*; Holstebro Museum: Holstebro, Denmark, 2015; pp. 140–142.
41. Verhoeven, G.; Nowak, M.; Rebecca, N. Pixel-level image fusion for archaeological interpretative mapping. In Proceedings of the 8th International Congress on Archaeology, Computer Graphics, Cultural Heritage and Innovation 'ARQUEOLÓGICA 2.0', Valencia, Spain, 5–7 September 2016; pp. 404–407.
42. Filzwieser, R.; Olesen, L.; Verhoeven, G.; Mauritsen, E.; Neubauer, W.; Trinks, I.; Nowak, M.; Nowak, R.; Schneidhofer, P.; Nau, E.; et al. Integration of Complementary Archaeological Prospection Data from a Late Iron Age Settlement at Vesterager—Denmark. *J. Archaeol. Method Theory* **2018**, *25*, 313–333, doi:10.1007/s10816-017-9338-y. [CrossRef]
43. NIKU. The Gjellestad Ship. Available online: https://www.niku.no/en/prosjekter/jellestadskipet/ (accessed on 15 May 2020).
44. Gustavsen, L.; Gjesvold, P.; Gundersen, S.; Hinterleitner, A.; Nau, E.; Paasche, K. Gjellestad—A newly discovered central place in South-East Norway. *Antiquity* **2020**, in press.
45. SEG Wiki. Coherence. Available online: https://wiki.seg.org/wiki/Coherence (accessed on 15 May 2020).
46. Marfurt, K. *3D Seismic Attributes for Prospect Identification and Reservoir Characterization—Introduction*; AASPI, University of Oklahoma: Norman, OK, USA, 2016.

 *remote sensing*

Article

High-Resolution Coherency Functionals for Improving the Velocity Analysis of Ground-Penetrating Radar Data

Eusebio Stucchi [1,2], Adriano Ribolini [1,2,*] and Andrea Tognarelli [1,2]

[1] Department of Earth Sciences, University of Pisa, 56126 Pisa, Italy; eusebio.stucchi@unipi.it (E.S.); andrea.tognarelli@unipi.it (A.T.)
[2] CISUP, Centro per l'Integrazione della Strumentazione dell'Università di Pisa, 56126 Pisa, Italy
* Correspondence: adriano.ribolini@unipi.it

Received: 18 May 2020; Accepted: 1 July 2020; Published: 4 July 2020

Abstract: We aim at verifying whether the use of high-resolution coherency functionals could improve the signal-to-noise ratio (S/N) of Ground-Penetrating Radar data by introducing a variable and precisely picked velocity field in the migration process. After carrying out tests on synthetic data to schematically simulate the problem, assessing the types of functionals most suitable for GPR data analysis, we estimated a varying velocity field relative to a real dataset. This dataset was acquired in an archaeological area where an excavation after a GPR survey made it possible to define the position, type, and composition of the detected targets. Two functionals, the Complex Matched Coherency Measure and the Complex Matched Analysis, turned out to be effective in computing coherency maps characterized by high-resolution and strong noise rejection, where velocity picking can be done with high precision. By using the 2D velocity field thus obtained, migration algorithms performed better than in the case of constant or 1D velocity field, with satisfactory collapsing of the diffracted events and moving of the reflected energy in the correct position. The varying velocity field was estimated on different lines and used to migrate all the GPR profiles composing the survey covering the entire archaeological area. The time slices built with the migrated profiles resulted in a higher S/N than those obtained from non-migrated or migrated at constant velocity GPR profiles. The improvements are inherent to the resolution, continuity, and energy content of linear reflective areas. On the basis of our experience, we can state that the use of high-resolution coherency functionals leads to migrated GPR profiles with a high-grade of hyperbolas focusing. These profiles favor better imaging of the targets of interest, thereby allowing for a more reliable interpretation.

Keywords: Ground-Penetrating Radar; velocity analysis; coherency functionals; GPR data processing; GPR data migration

---

1. Introduction

The Ground-Penetrating Radar (GPR) method is based on electromagnetic wave (EM) propagation and its response to changes in the EM properties of the subsurface. An EM impulse (GPR wave) generated by a transmitting antenna (transmitter, TX) propagates in the subsurface and is reflected by an interface or "scatters" on a discrete object. The energy back-reflected/scattered to the surface is recorded by a receiving antenna (receiver, RX) [1,2]. The estimation of the propagation velocity field of GPR waves (hereinafter velocity field) is an important issue to obtain a reliable and focused data migration, a crucial processing step to achieve a correct final interpretation [2–5]. This is particularly true in complex subsurface contexts, where the existence of numerous and different targets (e.g., walls, voids, pavements, fractures, metal rebars, burials) interspersed in heterogeneous materials may generate several reflections and diffraction patterns. In such cases, adopting a constant velocity for migration may produce unsatisfactory results, potentially reducing the data interpretation. The subsurface contexts whose imaging may benefit from an accurate velocity field estimation are thereby numerous, from archeology to civil engineering and structural geology settings. Velocity analysis on GPR data is usually carried out using an approximate approach in which a synthetic hyperbola of known velocity is imposed on hyperbolic-shaped real data, i.e., a real diffraction evident in the GPR profile. By a visual fitting of the curvature of the synthetic hyperbola with the observed real diffraction, the GPR wave velocity is determined. Alternatively, the semblance functional [6] on a data subset extracted from the radar profile is used. These operations are performed on a point or in a selected area of the GPR profile, where the signal-to-noise ratio (S/N) of the diffractions is high. Recently, automatic tools of hyperbola detection have been introduced assisting the operators in data analysis and processing [7,8]. However, avoiding false identifications may require some manpower effort and expertise. The resulting velocity field is constant or at most varying only with depth (1D). The estimated velocity is then used for data migration, thus collapsing the diffractions and moving the observed events in the correct position on the migrated section. When a basic processing sequence is adopted the estimated velocity is frequently used just for a vertical time-to-depth conversion. This way of proceeding disregards many factors that can actually result in an inaccurate velocity field estimation, thereby not favoring the production of an optimal migrated or depth converted section. The most important of these factors is that the propagation velocity of the electromagnetic radiation can vary both vertically and laterally along the survey profile. Although in many case studies the GPR investigated medium can be approximately considered homogeneous, there are a lot of settings where this assumption is not applicable a priori, in particular where a survey covers extensive areas and the subsurface may vary for instance in terms of sediment grain-size, material density/compaction, water content, target composition, voids occurrence among the other. Another relevant issue worth mentioning is the form of the recorded event. It is commonly assumed to be an ideal point-source hyperbolic diffraction, but frequently the scatterer is not point-like but has a flat or slightly inclined top part. Thereby the recorded event can appear like a misshapen hyperbola, composed by a planar/slightly inclined reflection and one or two lateral arms. This leads to an incorrect velocity estimation caused by difficulties in fitting a synthetic hyperbola. In this case, and also when a sharp lateral target termination is present in the subsurface (e.g., the edge of wall, interruption of stratigraphic layers caused by burials) [9], it should be more appropriate to use only one arm (left or right) of the fitting hyperbola to avoid introducing significant errors in the estimated velocity.

In this study, we propose the application to GPR data of the methodology based on high-coherency functionals computation. The computation of signal coherency (the similarity of the signal between adjacent traces) according to these functionals has long been proven to be effective in seismic exploration to estimate reflection event parameters in a common midpoint gather (CMP), even in cases of low S/N [10–18]. Moreover, this methodology offers a higher resolution of the estimated parameters with respect to the semblance functional [6] that is frequently used in various applications, including GPR velocity analysis [3]. The technique consists of the computation of high-resolution coherency maps by means of functionals that have been adapted from a pre-stack (source and receiver location some offset apart) to a post-stack (source and receiver coincident) velocity analysis. Applying this technique to GPR data may offer an additional processing tool capable of improving the correct location and focusing of subsurface objects in GPR profiles thanks to a more accurate velocity field estimation. We apply this technique first to a synthetic radar profile to test the effectiveness of the proposed method and evidence some possible issues potentially manifesting when real data are processed. Then we estimate the velocity field on a real data set composed of 41 parallel GPR profiles acquired in an archaeological site (Badia Pozzeveri, Lucca, Italy) and migrate the data in time with a Kichhoff algorithm. The archeological site was used because there is the possibility of ground-truthing the results of GPR data processing thanks to the evidence of excavations in the same area previously covered by the survey [9]. The migrated GPR lines were used to build some time-slices, depicting spatial coherencies of reflected energy consistent with archaeological targets [19–25].

2. Materials and Methods

2.1. Site Description and GPR Survey

The real GPR data of this study come from the archaeological site of Badia Pozzeveri (43°49′20″N, 18°38′39″E), located about 10 km south-east of the city of Lucca (north-west Italy). Here, an 11th-century church was believed to be larger than it is today, and buried structures were expected [26]. A GPR survey was conducted with a radar system of the IDS Georadar Company [27] equipped with a monostatic transmitter and receiver that operate at 600 MHz (nominal peak frequency). A common offset acquisition of 41 parallel survey lines 50 cm-spaced was performed covering an area of about 660 m^2. The in-line trace spacing of 1.2 cm was controlled by an odometer wheel. The configuration for acquisition provided 1024 samples in a time window of 100 ns (Figure 1). Previous GPR processing revealed the existence of walls buried at variable depths in the modern churchyard together with other reflective areas potentially related to anthropogenic features (e.g., tombs, burials, channels) [9]. These results found a good match with the excavation results undertaken after the GPR survey. Indeed, the excavation operation made it possible to discover the foundations and parts of the structures of the church, as well as several medieval and modern tomb features. Finally, in the area of the modern churchyard, the excavation revealed a channel for the drainage of rainwater, contained in a funnel-shaped section and originating near the entrance of the church.

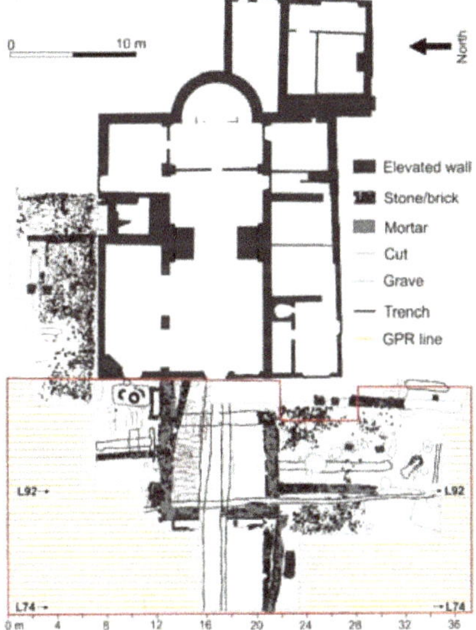

Figure 1. The archaeological site of Badia Pozzeveri. The plan of the elevated structures and of the unearthed archaeological findings is reported. The red box corresponds to the area surveyed with the GPR, and the yellow lines are the traces of the GPR profiles (the codes correspond to GPR profiles mentioned in the text). Although the GPR profiles entirely cover the area, they are interrupted to make the archaeological symbols more readable.

2.2. High Coherency Functionals Rationale

The travel time equation of electromagnetic waves recorded at the surface from a point diffractor in a homogeneous medium is given by:

$$t = \sqrt{t_0{}^2 + 4(x - x_0)^2 / V^2}$$

where t is travel time, x is the horizontal coordinate of the transmitting and receiving antennae, (x_0,z_0) the coordinates of the diffractor with the vertical axis z pointing downward, $t_0 = 2\,z_0/V$ the two-way travel time for $x = x_0$ and V the velocity of the electromagnetic wave in the medium.

A common way to estimate the velocity V is to use the semblance functional (Cs) that measures the ratio between the signal energy and the total energy in a given time window that follows a predefined hyperbolic trajectory within a user-defined aperture or number of summed traces:

$$C_s = \frac{1}{M}\frac{\sum_{t=t_0-T/2}^{t=t_0+T/2}\left(\sum_{i=1}^{M} d_i(t_0, V_s)\right)^2}{\sum_{t=t_0-T/2}^{t=t_0+T/2}\sum_{i=1}^{M} d_i^2(t_0, V_s)}$$

where (t_0,Vs) are the parameters of the hyperbola, T is the window length in ns, M is the number of traces considered, d_i are the data of the i_{th} trace. The semblance is known to be a robust coherency measure against noise [6]. However, the semblance coherency functional is only able to compute low-resolution panels, limiting the precision and accuracy of the estimated subsurface velocity.

In signal processing, and in particular in seismic exploration, a wide variety of tools has been developed to detect a signal that is buried in the noise [13,15–17]. Such tools exploit the signal coherency among different traces gathered in common midpoint gathers (CMP). Even in the CMP case, the travel time equation is hyperbolic, and thereby the techniques that have been developed can easily be adapted to the GPR velocity estimation. Due to the high number and variety of existing tools found in literature, it is not possible to comment on them all in detail here. Therefore, we only give a brief description of those, in addition to the semblance, we have found more promising in the tests we performed on GPR data in terms of resolution, noise rejection and computational time.

The first one we consider is the Complex Matched Analysis (Ccm) functional proposed by Spagnolini et al. [28]. It makes use of an approximate estimation of the wavelet to obtain a more efficient rejection of the random noise and is able to better discriminate between interfering events. The Ccm is given by:

$$C_{cm} = \frac{1}{M}\frac{\sum_{t=t_0-T/2}^{t=t_0+T/2}\left|\sum_{i=1}^{M} D_i(t_0, V_s)\right|^2}{\sum_{t=t_0-T/2}^{t=t_0+T/2}\sum_{i=1}^{M}\left|D_i(t_0, V_s)\right|^2}$$

where (t_0,Vs), T, and M have the same meaning as before and D_i is the i_{th} trace data filtered in the Hilbert domain by an estimated or known wavelet. Its efficacy has been proved on seismic data also by Tognarelli et al. [17,29]; however, we expect a resolution similar to the estimated wavelet that is employed and, ultimately, to the semblance.

The second functional we consider is given by the Complex Matched Coherency Measure ($Ccmcm$) algorithm, originally described by Key and Smithson [11]. It exploits the high resolution that can be achieved using the eigenvalue analysis of the data covariance matrix computed in a short time window that encompasses the signal. Considering the first eigenvalue as due to the signal and the others as due to uncorrelated random noise, the S/N can be computed as:

$$S/N = \frac{\lambda_1 - \sum_{i=2}^{M}\frac{\lambda_i}{M-1}}{\sum_{i=2}^{M}\frac{\lambda_i}{M-1}}$$

where λ_i are the eigenvalues of the data covariance matrix in decreasing order of magnitude and M is the number of traces used. The computed S/N re-scaled in the [0 1] interval gives the $Ccmcm$ measure. The resolution that is achieved with this functional is very high. However, the combined use of a complex matched filter and the eigenstructure analysis of the data covariance matrix allows for better discrimination of interfering or crossing events [17].

For this reason, a third functional has been introduced, given by a combination of the $Ccmcm$ and Ccm algorithms, trying to benefit from the positive properties of both. In other words, the $Ccmcm$ has been integrated with the *Ccm* by Grandi et al. [14,15], thus combining the robustness inherited by the *Ccm* with the resolution capabilities of *Ccmcm*. Therefore, the new functional is given by:

$$C_{cmKS} = C_{cmcm}C_{cm}$$

Overall, we tested more than 15 functionals, built on the basis of the original description given in the cited references and also obtained through their combination. Our choice has represented the best trade-off between the required characteristics of a signal coherency estimator (i.e., high resolution and noise rejection capability) and allowed us to pick the velocity spectra on GPR data with good reliability.

3. Results

3.1. Velocity Analysis of the Synthetic Data

We tested the proposed methodology on a synthetic data set built by using the 2D forward modelling software GPRSIM 3.0 by Geophysical Archaeometry Laboratory [30] based on ray-tracing techniques [31]. Via a discrete grid, we built a ground model (Figure 2) composed by two regions of different relative permittivity (ε_r), resulting in a velocity of approximately 12.2 cm/ns and 6.7 cm/ns on the left and right sides, respectively. Inside each region, two iron objects of different sizes are placed at different depths. We run a GPR simulation according to the same parameters (antenna frequency, temporal window, sampling frequency in the spatial and temporal domains) adopted in the GPR survey composing the real dataset (see later in the text). Furthermore, and consistently with the constructive parameter of the employed GPR, we simulated a transmitter radiating GPR waves over a 170° angle (symmetrical half aperture of 85° with respect to the vertical). The impulse shape was set as Ricker wavelet, resulting in the computed radar profile shown in Figure 3. Note that in the asymptotic hyperbolic arms far away from the top of the events, some arrivals are not computed. This is due to the discretization of the grid or of the starting angles of the rays, and some arrivals are missed even if smaller discretizations than the ones used are set. However, we can consider this problem as a more severe test that mimics the absence of the signal in the velocity analysis because of missing traces. As it will be evident in the computation of the coherency spectra, the proposed methodology is solid enough to overcome this issue if it is not too severe.

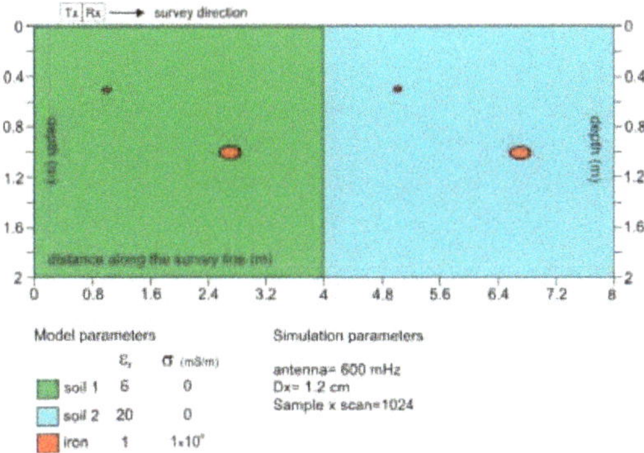

Figure 2. The synthetic model adopted to simulate the appearance of diffraction hyperbolas in GPR profiles. Subsurface EM characteristics and simulation parameters are indicated. Surface locations of the objects are labelled A, B, C, and D. Tx: transmitter; Rx: receiver.

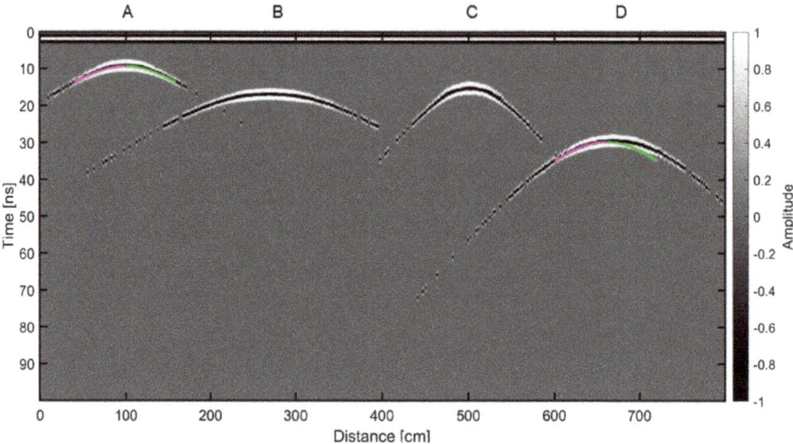

Figure 3. The synthetic GPR section showing the events generated by the buried objects of the model in Figure 2. Two fitting hyperbolas with their left (magenta) and right (green) arms are also displayed. While in A the whole hyperbola matches quite well the observed diffraction, in D only the magenta arm shows a good fit. This is because only the left arm of the diffracted event is used in the velocity analysis. Including also the right arm would result in an incorrect velocity estimation due to the fact that the object in D, for its dimension, is not a point diffractor, and consequently the shape of the observed event is not a true hyperbola.

The velocity analysis using the previously-described functionals has been carried out on the events of the GPR profile in Figure 3 with the parameters illustrated in Table 1. As examples, coherency maps computed for the positions A and D are displayed in Figure 4a,b, along with the picked events marked by a red circle. Each of these (t_0, Vs) pairs corresponds to the hyperbolic trajectory imposed on the GPR profile at the analysis location (hyperbola with magenta and green arms, Figure 3).

Table 1. Parameters used in the velocity analysis on synthetic data by the coherency functionals.

n	64	**Number of Time Samples of the Velocity Analysis Window**
win_ap	101	Aperture of the velocity analysis window (number of traces)
v	4–20	Velocity interval scanned by velocity analysis (cm/ns)
dv	0.1	Increment in the velocity scanned (cm/ns)
dt_0	0.098	Time increment in the functional computation (ns)
noise	0.1	Standard deviation of the added white noise to avoid functional singularity
wavelet	600 MHz	Maximum of the Ricker wavelet frequency spectrum

As it can be observed in Figure 4a,b, the resolution of *Ccmcm* and *CcmKS* functionals is much higher than the one given by the semblance and by *Ccm*, and consequently, the velocity picking is much easier and accurate. It is well worth noting that the velocity analysis described has been performed using only the portion of the data to the left of the apex (the data pertaining to the magenta hyperbola arm). The reason for this choice is that the full hyperbola does fit an event in the radargram if it originates from an object that can actually be considered a point diffractor. The object in A is sufficiently small to fulfill this requirement, as testified by the fitting of the green right arm also to the data. Instead, the dimensions of the object in D do not satisfy this point-like requirement and generate an event that differs from a classic hyperbola. Also, including in the velocity analysis the data in the correspondence of the green arm, we obtain a less accurate coherency map, and the velocity picking becomes inevitably more difficult or even impossible. For the event in A, the manual picking allows us to estimate a velocity of 11.8 cm/ns at 9.2 ns, and for the event in D a velocity of 6.6 cm/ns at 29.5 ns using the *Ccmcm* functional. These values are in good agreement with the background velocities given above (12.2 cm/ns and 6.7 cm/ns respectively).

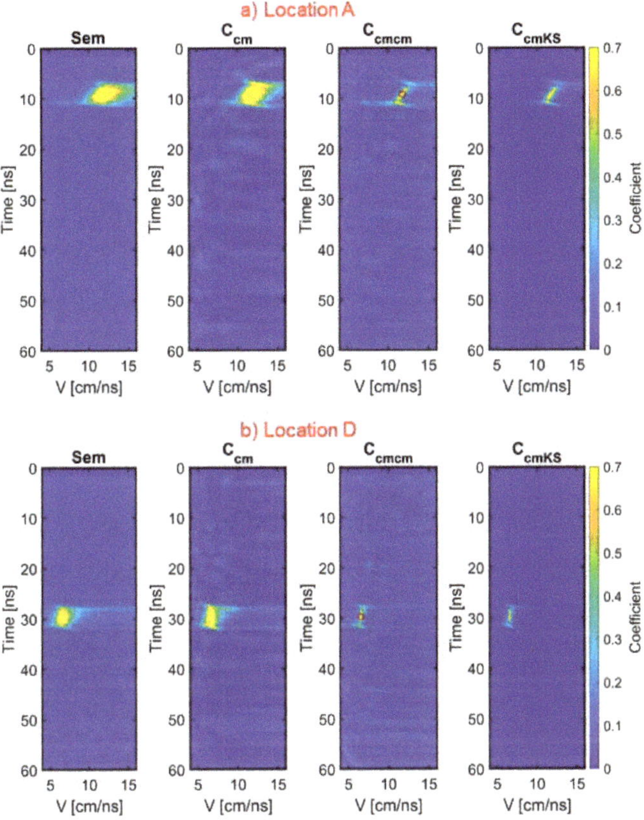

Figure 4. (**a**) Map of the coherency measure in location A; (**b**) map of the coherency measure in location D. These maps are obtained by the following functionals: Semblance (*Cs*), Complex Matched Analysis (*Ccm*), Complex Matched Coherency Measure (*Ccmcm*), and *CcmKS*, the combination of *Ccm* and *Ccmcm*. The higher velocity resolution of the functionals that make use of the data covariance matrix is clearly observed. The red circles correspond to the picked (t_0, Vs) pairs that characterize the hyperbolic trajectories plotted in Figure 3. Colour scale and gain are identical for all functionals.

The picked (t_0, Vs) pairs are used to time migrate the radar profile with a Kirchhoff algorithm [32]. Figure 5 displays in frames (a) and (b) the outcomes obtained using a constant velocity field built with the lower (6.7 cm/ns) and higher (12.2 cm/ns) velocity, respectively, and in frame (c) the outcomes when the input to migration is the varying velocity model built with the picked values (in A, B, C, and D). It is clear that only in the latter case do we have a satisfactory result from migration, and that in (a) and (b) only a portion of the radar profile is correctly migrated.

From this application of high-coherency functionals to synthetic data, it is evident that the migrated GPR profile can benefit from an accurate velocity analysis carried out both across time in a determined location and along the whole profile where hyperbolic diffractions are present.

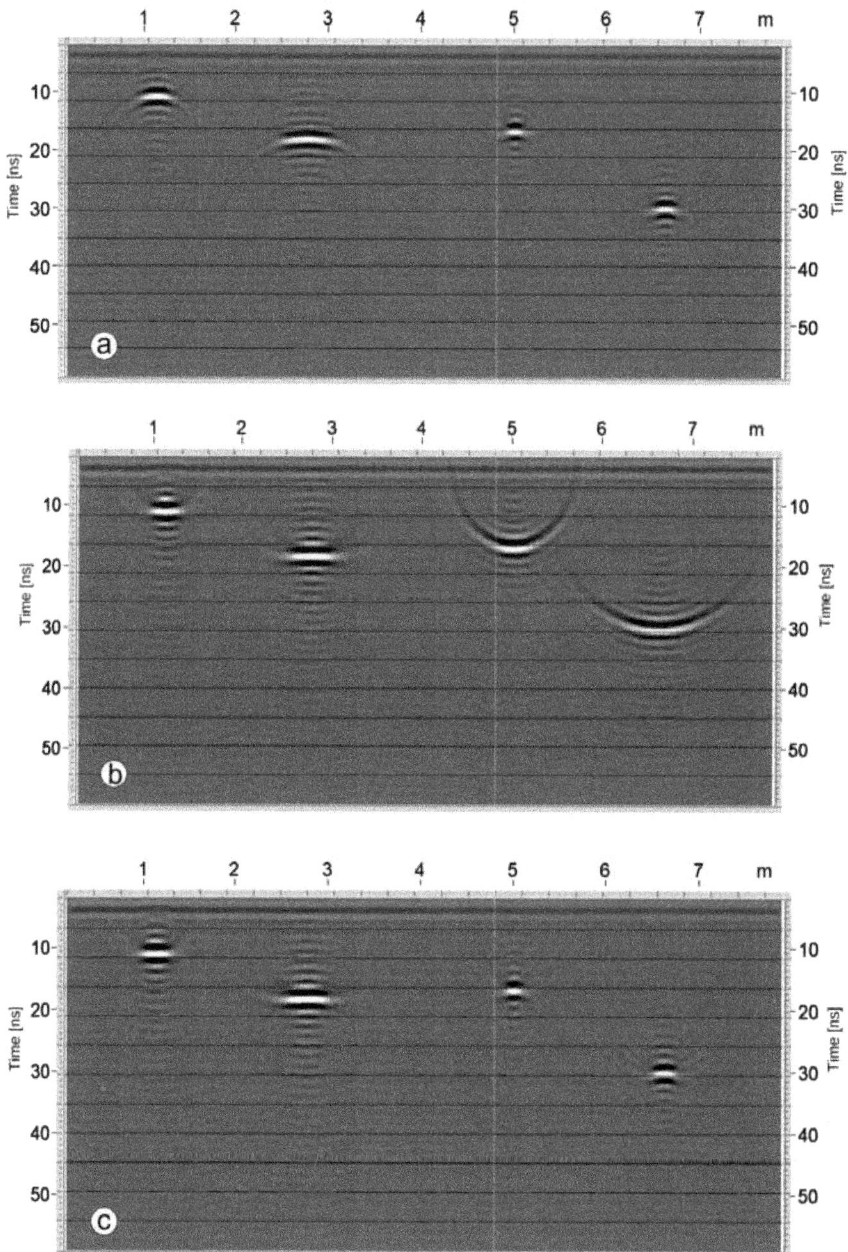

Figure 5. Migrated synthetic section using: (**a**) a constant velocity of 6.7 cm/ns; (**b**) 12.2 cm/ns; (**c**) a varying velocity field obtained by the velocity analysis in A, B, C and D. When the migration velocity is not correct, we obtain in (**a**) under-migrated data, A and B positions, and in (**b**) over-migrated data, C and D positions. Only in (**c**) are the events correctly migrated.

3.2. Velocity Analysis on the Real Data

We performed the velocity analysis using the previously described functionals on data that have undergone the following processing steps to increment the S/N ratio and to reduce the reverberations: (1) background removal; (2) a band-pass filter in the range of 80-100-600-700 MHz to select only the frequencies related to the transmitted electromagnetic energy; (3) an automatic gain control on a time window of 10 ns to recover the low amplitudes of the signal at a later time; (4) a predictive deconvolution (prediction lag 2 ns, filter length 4 ns) to reduce the reverberations observed on the radargram, thus facilitating the velocity analysis; (5) a more selective band-pass filter in the range of 80-120-450-550 MHz.

Coherency maps of GPR data can only be significant in correspondence with the diffracted energy that has a hyperbolic trajectory. Due to the time required to select the locations and to compute the coherency maps, we performed the velocity analysis only on some selected GPR profiles (L-lines) of the data set where the diffraction events are more evident. Figure 6a,b show a portion of the L74 and L92 lines with imposed the estimated trajectories after the velocity picking on the coherency panels. The different colors of the left and right arms of the hyperbolas allow for easier identification of the data used in the analysis. Indeed, after some attempts, we chose to employ a 64-sample time window around the right arm for the L74 line, and an identical window around the left arm for the L92 line. In Figure 7a–d we show the coherency panels used to pick the (t_0, Vs) pairs and the values actually picked with red asterisks in positions E and F of L74, and G and H of L92. The velocity analysis in point E highlights the difficulties that may be found to match real diffraction patterns on GPR data. We selected the right arm (the use of the left one resulted in a too high velocity and an over-migrated section) and performed the analysis every 6 cm, eventually choosing the best velocity function on the basis of hyperbola fitting and migrated results. This made it possible to make a good compromise regarding the issue of possible uncertainties in velocity estimation due to the finite dimensions of the subsurface targets [8]. We made use of both the *Ccmcm* and the semblance functionals since the contribution of each of them was significant in the velocity analysis phase. Semblance was helpful for giving a rough indication of the regions in the (t_0, Vs) map where the signal concentrates, while *Ccmcm* was useful for its higher resolution, which allows us to select more precise values and discard artefacts introduced by the semblance algorithm.

In Figure 7 we also show the *CcmKS* functional that is generally less noisy than the *Ccmcm* due to the *Ccm* weighting, but at least in case E was less effective. A possible explanation can be given by looking at the Semblance. The sample-by-sample weights introduced by the *Ccm* in the *CcmKS*, which are similar to the semblance, filter out the coherency observed around 22 ns at 10 cm/ns that was detected by the *Ccmcm* functional. Note that having more than a single coherency measure available (the ones displayed plus the *Ccm* which is very similar to the Semblance and therefore not shown) helps to better discriminate the (t_0, Vs) pairs related to signal, leading to an improved velocity analysis. The parameters used for the real data case are indicated in Table 2. The velocity values obtained are used to build a velocity field and to migrate with a Kirchhoff algorithm the GPR profiles (L-lines) without imposing limits on the dip angle and on the available frequencies. This permits on the one hand to collapse all the hyperbola arms and on the other hand to preserve the maximum possible resolution allowed by the data. The migrated GPR profiles in Figure 6a,b are displayed in Figure 8a,b respectively, for an analysis of the results. As can be observed, along both profiles the diffracted energy is fairly collapsed in correspondence with the apex region (only point E appears slightly under-migrated), allowing us to better define the discontinuities in the subsurface.

This is more evident in the close-up of Figure 9a,b pertaining to the L92 line before and after migration, where the arrows indicate some interesting locations. It can be observed that after migration, the dimension of the objects illuminated by the electromagnetic energy is laterally well-delimited, thus giving a more realistic description of the size and shape of the buried targets. Moreover, thanks to the multi-channel (multi-trace) filtering process operated by the migration operator, random noise is reduced after migration.

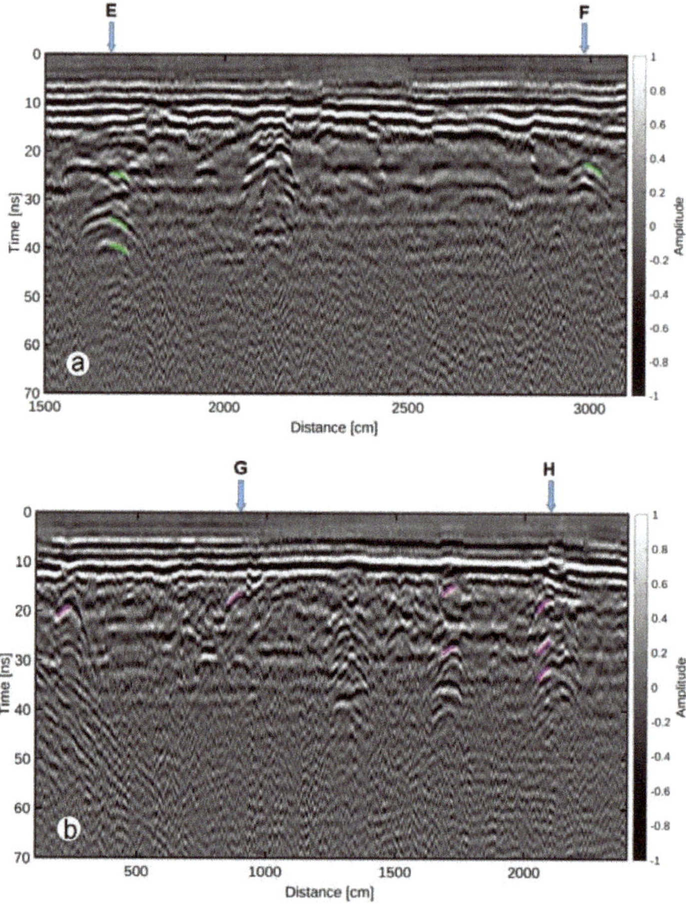

Figure 6. (**a**) GPR section of the L74 Line containing the right arm (plotted in green) of the hyperbolae whose (t_0,Vs) parameters are picked in the coherency panels of Figure 7a for the location E, and Figure 7b for the location F; we decided to fit the right arm of the misshapen hyperbola at 34 ns in E because it appears more regular. (**b**) GPR section of the L92 Line. In this case, the left arms (plotted in magenta) of the hyperbolae are displayed. Their (t_0,Vs) parameters are picked on the coherency panels of Figure 7c for location G and Figure 7d for location H. The other magenta half-hyperbolae on the L92 Line correspond to additional positions where the velocity analysis has been carried out.

Table 2. Parameters used in the velocity analysis on real data by the different functionals.

n	64	**Number of Time Samples of the Velocity Analysis Window**
win_ap	81	Aperture of the velocity analysis window (number of traces)
v	4–20	Velocity interval scanned by velocity analysis (cm/ns)
dv	0.1	Increment in the velocity scanned (cm/ns)
dt_0	0.098	Time increment in the functional computation (ns)
wavelet	400 MHz	Maximum of the Ricker wavelet frequency spectrum

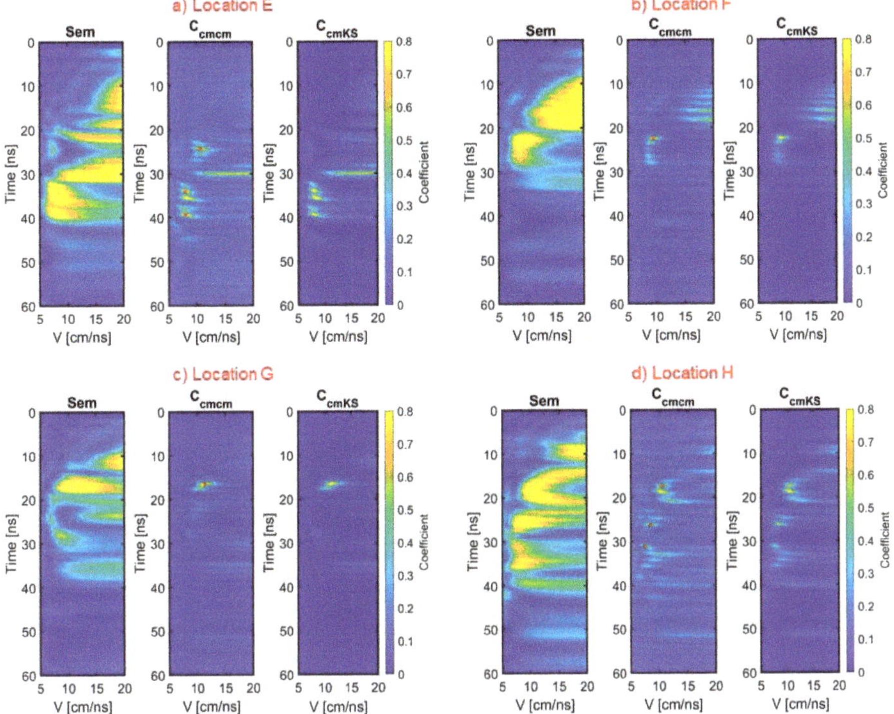

Figure 7. Coherency maps of the Semblance (*Cs*), Complex Matched Coherency Measure (*Ccmcm*) and *CcmKS* functionals computed in locations E and F of Line 74 (**a,b** respectively), and G and H of Line 92 (**c,d** respectively). The red asterisks indicate the (t_0,*Vs*) pairs picked and corresponding to the parameters of the hyperbolae shown in Figure 6a,b. The colour scale and gain are identical for all functionals.

3.3. The Effect of Varying Velocity Field on Time-slices

The migrated GPR lines were used to build time-slices (amplitude maps) parallel to the surface and cut at different times. It is well known that this kind of representation is largely used to correlate spatially coherent reflected energy (i.e., reflections) with targets at different depths. Testing whether the high-coherency functionals increase the readability of reflective areas in a time-slice representation is thereby crucial. To do this, we gridded and sliced three GPR datasets: (i) unmigrated data, (ii) data migrated with a constant velocity (10 cm/ns), and (iii) data migrated according to the varying velocity field reconstructed via the high-coherency functionals described above. The value of 10 cm/ns, used for the constant velocity migration, comes from the average of the picked velocity values on the coherency maps relative to several GPR profiles. In building the time-slices, we took care to keep slicing and gridding parameter constant (slice thickness: 4 ns; the square amplitude of amplitude values; cell dimension: 0.05 m; search area: circular; search radius: 1 m), as well as the interpolation statistical method (i.e., inverse weighted distance).

All these operations were performed with the GPRSlice software via Geophysical Archaeometry Laboratory [30]. For the sake of comparison, we illustrate only three time-slices in the temporal interval of 26–36 ns, centered at 28, 31, 34 ns and 4 ns thick (Figure 10). Instead of the same scale of reflection strength (amplitude), we used a relative normalization for each time-slice because this allows us to better imagine the subsurface targets and make the comparison more effective.

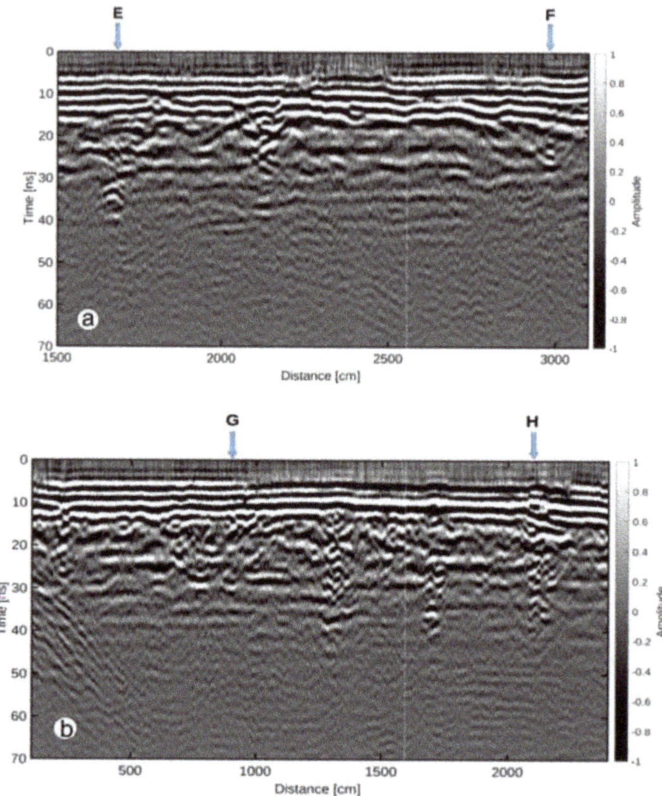

Figure 8. (a) The time-migrated section using the 2D picked velocity field of the L74 Line; (b) the time-migrated section using the 2D picked velocity field of the L92 line. The diffraction events shown satisfactory collapse at their apex, and the true geometry of the subsurface objects is better-imaged.

It is evident that the spatially-coherent reflective areas are clearer and of higher resolution in the time-slices built using the GPR lines migrated with a variable velocity field reconstructed via the high-resolution functionals (Figure 10). Moreover, the reflection intensity appears to have increased, and spot reflections, associated with local subsurface conditions (e.g., localized increases in density due to the presence of boulders/blocks concentrations or fragments of man-made remnants), are reduced. The time-slices built with GPR profiles migrated with a constant velocity (10 cm/ns) show content in amplitude higher than those built with un-migrated GPR profiles. However, in some sectors, it appears that the migration process has not favored the continuity of the reflections even if compared with those depicted in the un-migrated time-slices. This is an effect of the non-optimal velocity used in the migration that hampers the focusing of the reflected energy at the correct position.

As can be seen from the comparison with the excavation results (Figure 1), the linear-shaped reflections are consistent with the archaeological findings unearthed in the investigation area. There is a clear correspondence both with masonry structures composed of stones/bricks and cuts consisting in channels for rainwater drainage. Relative to the channel EW crossing the central part of the time-slices (especially visible in the interval 29–33 ns), the backscattered energy from the filling material (coarse gravel and boulders) has returned in the GPR profiles hyperbolic diffractive figures, although with an irregular form [9]. Note that we selected only three time-slices ranging between 26–36 ns, and some of the archeological structures may be not evident in Figure 10 because they are shallower or deeper.

Indeed, the evident NS-oriented wall developed between 22 and 30 m of horizontal distance is at about 5–10 ns.

The test of high-resolution functionals on such a kind of data was quite demanding. This is because the structure of the buried walls is composed laterally by interlocked decimetric boulders (interconnected boulders and open voids) and internally by boulders dispersed in an abundant mortar. In GPR profiles, the tops of these structures correspond to a planar reflection, while the boulders placed on the sides and corners of the wall to arms of hyperbolas (i.e., half hyperbolas) (see the black arrow pointing to 22 ns and 1700 cm of distance in Figure 9a). Moreover, the boulders in the mortar-dominant core generated small hyperbolic diffractions with interfering arms [9]. While the latter is difficult to model and manage even with a variable velocity field, the approach based on high-resolution functionals has clearly made possible a more than satisfactory focusing of the principal diffraction hyperbolas, and therefore a better definition of the wall structures in the time-slices.

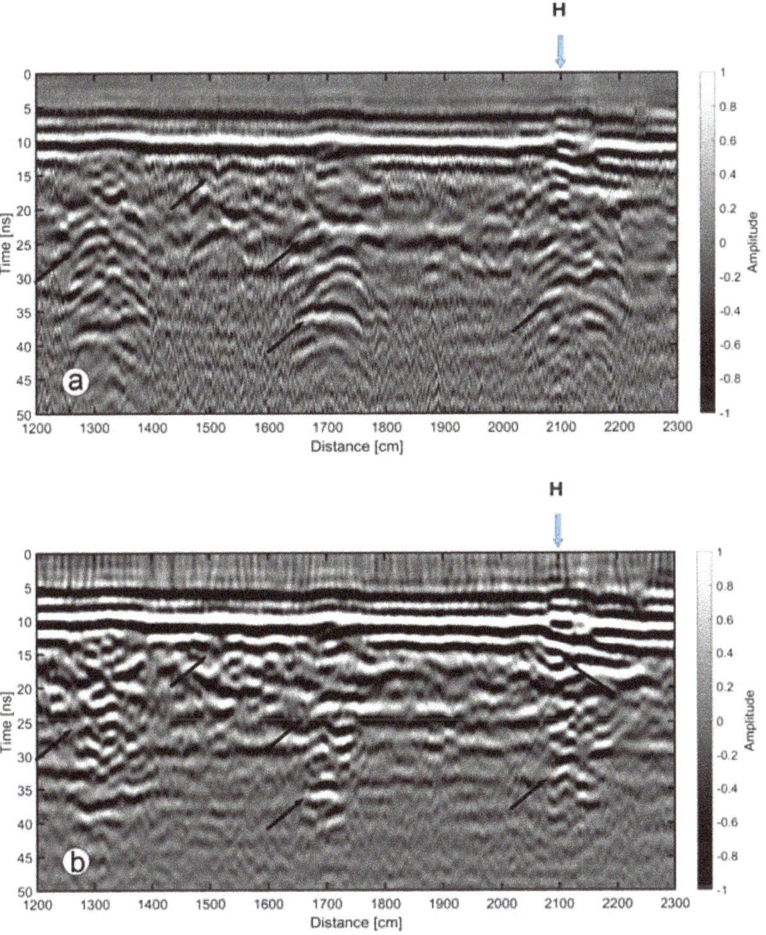

Figure 9. (a) A close-up of the L92 section; (b) a close-up of the L92 migrated section. Migration has the effect of delineating with higher accuracy the geometry of the buried objects. The black arrows indicate some points in the section where the migration outcomes are particularly evident.

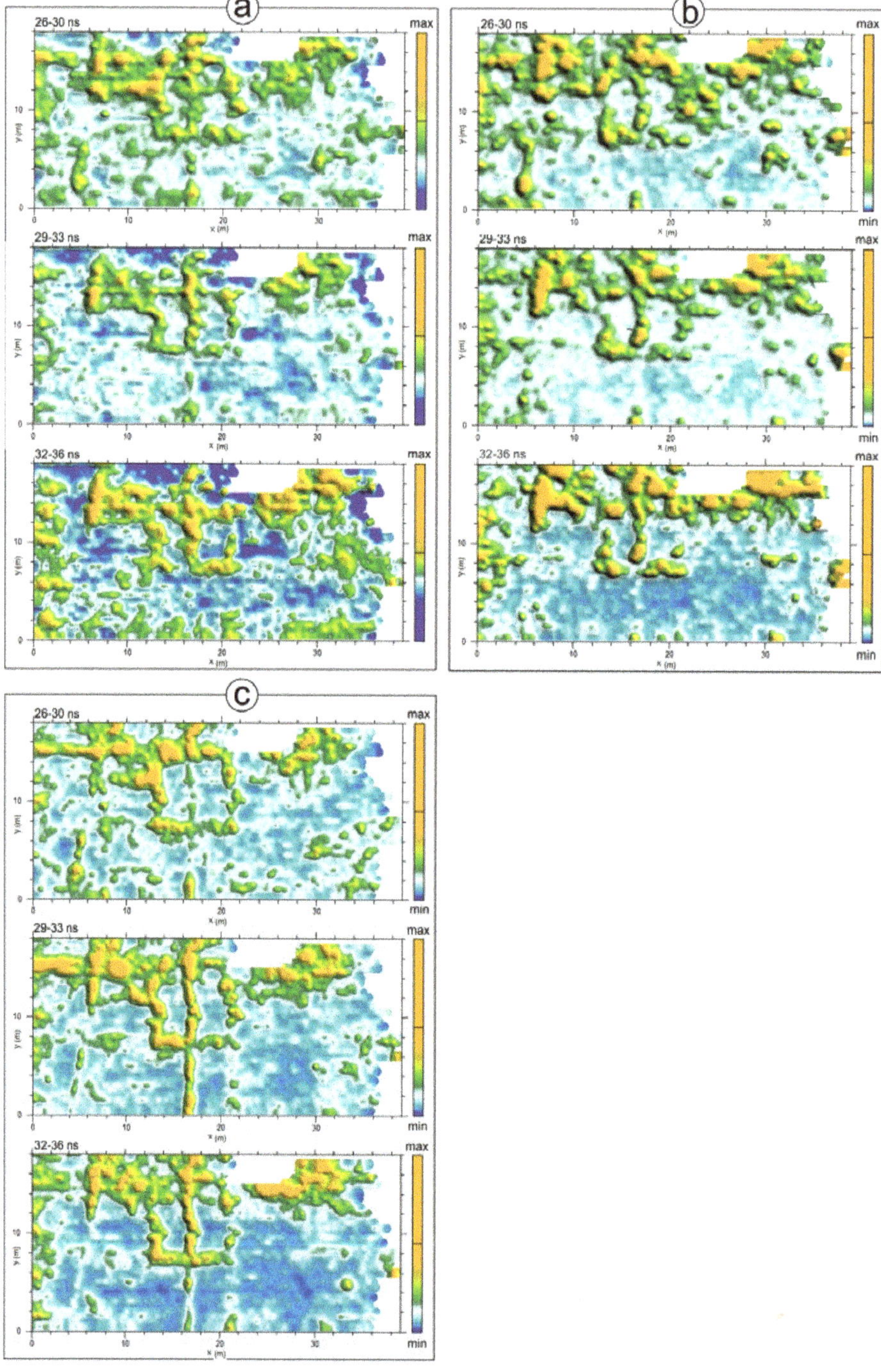

Figure 10. Time-slices generated from (**a**) unmigrated data, (**b**) data migrated with a constant velocity field (10 cm/ns), and (**c**) data migrated according to the varying velocity field.

4. Conclusions

The aim of this paper was to verify whether the use of high-resolution coherency functionals could improve the quality in terms of S/N ratio and interpretability of GPR data by introducing a variable velocity field in the migration process.

After carrying out tests on synthetic data to schematically simulate the problem, assessing the types of functionals most suitable for GPR data analysis, a varying velocity field relative to a real dataset was calculated. This dataset was acquired in an archaeological area for which the position of the targets was already available because these have been unearthed after the GPR survey.

The best functionals turned out to be the *Ccmcm* and the *CcmKS*, because they are able to compute coherency maps characterized by high resolution and strong noise rejection that make it possible to better define the time-velocity pairs of the observed diffracted events. In the velocity analysis, particular attention was paid to checking the fitting between the reconstructed diffraction hyperbola arms and those generated by buried structures, which do not consist of ideal scattering points but targets with a certain lateral extension as in most real cases. In the selected data, we did not include the top planar part of the event, limiting the application of the method to the hyperbola arms, reducing the possibility of wrong velocity estimation. We observed that with the precisely-picked 2D velocity field migration algorithms perform better than in the case of constant or 1D velocity field, with satisfactory collapsing of the diffracted events and moving of the reflected energy to the correct position.

The varying velocity field was applied to all the GPR profiles composing the survey and covering the entire archaeological area. The time-slices constructed are of a higher quality than those obtained from non-migrated or migrated GPR profiles at a constant velocity. The improvements are inherent to resolution, continuity, and energy amplitude of linear reflective areas.

On the basis of our experience, we can affirm that the use of high-resolution functionals allows us to obtain migrated GPR profiles with a high grade of hyperbola focusing. On these profiles, the targets of interest are better-imaged and the interpretation can be carried out more reliably. A varying velocity field, even sparsely sampled like the one we have built, can allow for more accurate positioning in time of the objects of interest with respect to the constant or 1D velocity model routinely used, or when no data migration is applied. This could be relatively important when studying an archaeological site or a geological setting, cases in which a slightly incorrect positioning of targets/layers is acceptable. However, accurate reconstruction of the target geometry and positioning becomes crucial when it comes to studying other topics, e.g., those related to civil engineering and micro-structural settings, where even modest velocity variations can lead to reconstruction errors which may be quite relevant.

The main drawbacks we experienced were the time required by the functional computation (approximately 5 min on an i7 laptop for each location) and the time required to choose the locations for performing the velocity analysis. Concerning this last issue, one suggestion could be to perform the velocity analysis on a whole line (or the more promising portions of a line) and then inspect the velocity volume (time, velocity, inline distance) to select the best coherency maps on which to perform the velocity analysis. This can be done at a very high computational cost but can save some manpower time. An alternative option could be exploiting automatic tools of hyperbola detection to generate a dataset of locations on which to perform the high-resolution velocity analysis. By applying the functionals only to this limited number of points, selected along the GPR profiles, allows us to limit computational costs.

Notwithstanding these drawbacks, we believe that in cases where an accurate and detailed image of the subsurface is desired, this methodology is worth the effort.

Author Contributions: Conceptualization, E.S. and A.R.; methodology, E.S. and A.T.; software, E.S., A.R. and A.T.; validation, E.S. and A.R.; formal analysis, E.S., A.R. and A.T.; investigation, E.S. and A.R.; resources, E.S., A.R. and A.T.; data curation, E.S. and A.R.; writing—original draft preparation, E.S. and A.R.; writing—review and editing, E.S., A.R. and A.T.; funding acquisition, A.R. All authors have read and agreed to the published version of the manuscript.

Funding: GPR data acquisition was funded by the CA.RI.LU grant: "New high resolution Georadar techniques for archeological contexts analysis" (Leader. A. Ribolini).

Acknowledgments: We thank the Landmark Graphics Corporation for the use of the ProMax software employed in the seismic data processing. We thank F. Coschino (University of Pisa) and A. Fornaciari (University of Pisa) for providing the archeological data. M. Bini (University of Pisa) and I. Isola (INGV, Pisa) cooperated in the GPR data acquisition. The comments of four anonymous reviewers were helpful to improve the manuscript.

Conflicts of Interest: The authors declare no conflict of interest.

References

1. Davis, J.L.; Annan, A.P. Ground Penetrating Radar for high resolution mapping of soil and rock stratigraphy. *Geophys. Prospect.* **1989**, *37*, 531–551. [CrossRef]
2. Annan, A.P. Electromagnetic principles of ground penetrating radar. In *Ground Penetrating Radar: Theory and Applications*; Jol, H.M., Ed.; Elsevier: Amsterdam, The Netherlands, 2009; pp. 3–40.
3. Cassidy, N.J. Ground penetrating radar data processing, modelling and analysis. In *Ground Penetrating Radar: Theory and Applications*; Jol, H.M., Ed.; Elsevier: Amsterdam, The Netherlands, 2009; pp. 141–176.
4. Persico, R. *Introduction to Ground Penetrating Radar. Inverse Scattering and Data Processing*; IEEE Press Wiley: Hoboken, NJ, USA, 2014; p. 400.
5. Allroggen, N.; Tronicke, J.; Delock, M.; Böniger, U. Topographic migration of 2D and 3D ground-penetrating radar data considering variable velocities. *Near Surf. Geophys.* **2015**, *13*, 253–259. [CrossRef]
6. Neidell, N.S.; Taner, M.T. Semblance and other coherency measures for multichannel data. *Geophysics* **1971**, *36*, 482–497. [CrossRef]
7. Maas, C.; Schmalzl, J. Using pattern recognition to automatically localize reflection hyperbolas in data from ground penetrating radar. *Comput. Geosci.* **2013**, *58*, 116–125. [CrossRef]
8. Laurence, M.; Persico, R.; Matera, L.; Lambot, S. Automated Detection of Reflection Hyperbolas in Complex GPR Images with no a priori knowledge on the medium. *IEEE Trans. Geosci. Remote Sens.* **2016**, *54*, 580–596.
9. Ribolini, A.; Bini, M.; Isola, I.; Coschino, F.; Baroni, C.; Salvatore, M.C.; Zanchetta, G.; Fornaciari, A. GPR versus Geoarchaeological Findings in a Complex Archaeological Site (Badia Pozzeveri, Italy). *Arch. Prosp.* **2017**, *24*, 141–156. [CrossRef]
10. Sguazzero, P.; Vesnaver, A. A comparative analysis of algorithms for seismic velocity estimation. In *Deconvolution and Inversion*; Bernabini, M., Rocca, F., Treitel, S., Worthington, M., Eds.; Blackwell: London, UK, 1987; pp. 267–286.
11. Key, S.C.; Smithson, S.B. New approach to seismic-reflection event detection and velocity determination. *Geophysics* **1990**, *55*, 1057–1069. [CrossRef]
12. Grion, S.; Mazzotti, A.; Spagnolini, U. Joint estimation of AVO and kinematics parameters. *Geophys. Prospect.* **1998**, *46*, 405–422. [CrossRef]
13. Sacchi, M.D. A bootstrap procedure for high-resolution velocity analysis. *Geophysics* **1998**, *63*, 1716–1725. [CrossRef]
14. Grandi, A.; Stucchi, E.; Mazzotti., A. Multicomponent velocity analysis by means of covariance measures and complex matched filters. In Proceedings of the 74th Annual International Meeting, Denver, CO, USA, 10–14 October 2004; Society of Exploration Geophysicists: Huston, TX, USA, 2004; pp. 2415–2418.
15. Grandi, A.; Mazzotti, A.; Stucchi, E. Multicomponent velocity analysis with quaternions. *Geophys. Prospect.* **2007**, *55*, 761–777. [CrossRef]
16. Abbad, B.; Ursin, B. High-resolution bootstrapped differential semblance. *Geophysics* **2012**, *77*, U39–U47. [CrossRef]
17. Tognarelli, A.; Stucchi, E.; Ravasio, A.; Mazzotti, A. High-resolution coherency functionals for velocity analysis: An application for subbasalt seismic exploration. *Geophysics* **2013**, *78*, U53–U63. [CrossRef]
18. Gong, X.; Wang, S.; Zhang, T. Velocity analysis using high-resolution semblance based on sparse hyperbolic Radon transform. *J. Appl. Geophys.* **2016**, *134*, 146–152. [CrossRef]
19. Goodman, D.; Nishimura, Y.; Rogers, J.D. GPR timeslices in archaeological prospection. *Archaeol. Prospect.* **1995**, *2*, 85–89.
20. Nuzzo, L.; Leucci, G.; Negri, S.; Carrozzo, M.T.; Quarta, T. Application of 3D visualization techniques in the analysis of GPR data for archaeology. *Ann. Geophys.* **2002**, *45*, 321–337.

21. Leckebush, J. Ground-penetrating radar: A modern three-dimensional prospection method. *Archaeol. Prospect.* **2003**, *10*, 213–240. [CrossRef]

22. Grasmueck, M.; Weger, R.; Horstmeyer, H. Three-dimensional ground-penetrating radar imaging of sedimentary structures, fractures, and archaeological features at submeter resolution. *Geology* **2004**, *32*, 933–936. [CrossRef]

23. Soldovieri, F.; Orlando, L. Novel tomographic based approach and processing strategies for GPR measurements using multifrequency antennas. *J. Cult. Herit.* **2009**, *10*, 83–92. [CrossRef]

24. Bini, M.; Fornaciari, A.; Ribolini, A.; Bianchi, A.; Sartini, S.; Coschino, F. Medieval phases of settlement at Benabbio castle, Apennine Mountains, Italy: Evidence from ground penetrating radar survey. *J. Archaeol. Sci.* **2010**, *37*, 3059–3067. [CrossRef]

25. Tinelli, C.; Ribolini, A.; Bianucci, G.; Bini, M.; Landini, W. Ground penetrating radar and palaeontology: The detection of sirenian fossil bones under a sunflower field in Tuscany (Italy). *Comptes Rendus Palevol* **2012**, *11*, 445–454. [CrossRef]

26. Fornaciari, A. Badia Pozzeveri. Dalla terra alla storia. *LUK* **2014**, *20*, 30–36.

27. IDS GeoRadar. Available online: https://idsgeoradar.com/ (accessed on 12 May 2020).

28. Spagnolini, U.; Maciotta, L.; Manni, A. Velocity analysis by truncated singular value decomposition. In Proceedings of the 63rd Annual International Meeting, Washington, DC, USA, 26–30 September 1993; Society of Exploration Geophysicists: Huston, TX, USA; pp. 677–680.

29. Tognarelli, A.; Aleardi, M.; Mazzotti, A. Estimation of a high resolution P-wave velocity model of the seabed layers by means of global and a gradient-based FWI. In Proceedings of the Second Applied Shallow Marine Geophysics Conference, Barcelon, Spain, 4–8 September 2016; European Association of Geosciences and Engineering. [CrossRef]

30. Geophysical Archaeometry Laboratory. Available online: https://www.gpr-survey.com/ (accessed on 12 May 2020).

31. Goodman, D. Ground-penetrating radar simulation in engineering and archaeology. *Geophysics* **1994**, *59*, 224–232. [CrossRef]

32. Yilmaz, O. *Seismic Data Analysis: I—Processing, Inversion, and Interpretation of Seismic Data*; Doherty, S.M., Ed.; Society of Exploration Geophysics: Tulsa, OK, USA, 2001; p. 2065.

Article

Convolutional Neural Network with Spatial-Variant Convolution Kernel

Yongpeng Dai, Tian Jin *, Yongkun Song, Shilong Sun and Chen Wu

College of Electronic Science and Technology, National University of Defense Technology, Changsha 410073, China; dai_yongpeng@nudt.edu.cn (Y.D.); yongkunsong@nudt.edu.cn (Y.S.); sunshilong@nudt.edu.cn (S.S.); wuchen@nudt.edu.cn (C.W.)
* Correspondence: tianjin@nudt.edu.cn

Received: 23 July 2020; Accepted: 25 August 2020; Published: 30 August 2020

Abstract: Radar images suffer from the impact of sidelobes. Several sidelobe-suppressing methods including the convolutional neural network (CNN)-based one has been proposed. However, the point spread function (PSF) in the radar images is sometimes spatially variant and affects the performance of the CNN. We propose the spatial-variant convolutional neural network (SV-CNN) aimed at this problem. It will also perform well in other conditions when there are spatially variant features. The convolutional kernels of the CNN can detect motifs with some distinctive features and are invariant to the local position of the motifs. This makes the convolutional neural networks widely used in image processing fields such as image recognition, handwriting recognition, image super-resolution, and semantic segmentation. They also perform well in radar image enhancement. However, the local position invariant character might not be good for radar image enhancement, when features of motifs (also known as the point spread function in the radar imaging field) vary with the positions. In this paper, we proposed an SV-CNN with spatial-variant convolution kernels (SV-CK). Its function is illustrated through a special application of enhancing the radar images. After being trained using radar images with position-codings as the samples, the SV-CNN can enhance the radar images. Because the SV-CNN reads information of the local position contained in the position-coding, it performs better than the conventional CNN. The advance of the proposed SV-CNN is tested using both simulated and real radar images.

Keywords: spatial-variant convolution neural network (SV-CNN); spatial-variant convolution kernel (SV-CK); radar image enhancing; MIMO radar; neural networks; imaging radar

---

1. Introduction

Convolutional neural networks (CNNs) make good use of the convolution kernels in the first several layers to detect distinctive local motifs and construct feature maps. The convolution kernel functions are similar to the filter banks. All the units in a feature map share the same filter bank. As a result, the total number of layer nodes in a CNN is drastically reduced. For this reason, the CNN has been widely used and has become the state-of-the-art image processing method [1]. In [2–5], the researchers used the CNN to perform the whole image and handwriting recognition task. Authors of [6–8] introduced the advantages of CNN in the applications of the edge and keypoint detection. Researchers in [9,10] made a further step; they replaced all the fully connected layers with convolution layers and used the modified CNNs to classify each pixel of an image and then semantic segmentation could be made. In [11–14], the researchers removed both the pool layers and the fully connected layers of a CNN and obtained a pixel-to-pixel convolutional neural network, which can evaluate the super-resolution task of an image. In [15–18], the researchers approved that CNN can also be used to deal with the complex-valued data, and to enhance the radar images. The purpose of the enhancement is to sharpen the main lobes of the radar image and suppress the sidelobes.

The function of the convolution kernels in these CNNs is detecting the local motifs of the images, which are assumed to be spatially invariant. However, features of motifs in images are sometimes related to the motifs' positions, such as the radar images, especially those of the near-field radar systems. The point spread function (PSF) of a near-field radar system is spatially varying. Accordingly, the features of the motifs in those images are spatially varying. Thus, it is more reasonable to consider a spatial-variant convolutional neural network (SV-CNN) which consists of spatial-variant convolution kernels (SV-CK) to deal with the radar images.

The local position information is sometimes important, such as when evaluating the language translation [19], the photo classification [20], and the point cloud classification [21]. The local position information plays a more important role in radar image processing, because of the spatially varying PSF in the radar images. The local position information can be extracted by several kinds of layers. The most commonly used ones are the fully connected layers; the local position information can be learned by the network cells. However, when the network is used to convert one image to another, at least the connection number equals to the product of the input number and output number are needed, which will make the network hardly acceptable for the current devices. Besides, the network will lack generation ability. The self-attention layers tackle spatial awareness well [19]. However, a huge number of connections are still needed. Thus these layers are commonly used in the translation and the picture recognition field [20]. The graph convolution layers are with spatial awareness and are widely used for dealing with the point, such as the protein and gene recognition [21] as well as figure classification [22]. All the above neural networks are not capable of converting a big size image to another while taking the spatial information into account.

In this paper, we propose a special SV-CNN with SV-CKs. It can take the spatial information into account and hardly increase the size of the network compared to a conventional CNN. In this paper, the proposed SV-CNN is used to suppress the sidelobes of a multi-input–multi-output (MIMO) imaging radar. (In our condition, the target is near the radar, and the PSF is spatially varying.) The images of a MIMO radar suffer from sidelobes because of the limitations on the signal bandwidth and the total aperture length of the MIMO array. The sidelobes can be considered as false peaks and severely impact the quality of the radar images. A lot of research has been proposed to suppress the sidelobes such as the coherence factor (CF) algorithm [23,24], the sparsity-driven methods [25–27], and the deconvolution methods [28]. However, the CF algorithm suppresses the weak targets as it suppresses the sidelobes. The sparsity-driven methods are faced with a heavy computation burden and the lack of robustness. The deconvolution method also suffers from a heavy computation burden. Besides, it needs a precise PSF of the imaging system. However, for the MIMO radar, it is a spatially variant one. Recently, research has focused on the CNN to suppress the sidelobes [29]. However, as illustrated, the PSF of the MIMO radar is a spatially variant one while the convolution kernels of the CNN are spatially invariant ones. So, the proposed SV-CNN has better performance in this task. The SV-CNN is with spatial awareness, so it performs better than the conventional CNN. Besides, it performs better than the conventional sidelobe-suppressing algorithms.

The rest of this paper is organized as follows: Section 2 illustrates the MIMO radar imaging and the spatial-variant characters of the MIMO radar image. Section 3 illustrates the structure of the SV-CNN and the training method. Section 4 validates the enhancement ability of the SV-CNN, and some of its features through simulation. In Section 5, the advantage of SV-CNN is verified using experimental data. Finally, Section 6 gives the conclusion.

2. Spatial-Variant Characters of MIMO Radar Image

MIMO imaging radars are widely used for its low complicity and more degrees of freedom. They are well suited for the near-field imaging applications, such as the through-wall radar, the security inspection radar, and the ground-penetrating radar. A real implementation of a MIMO radar is illustrated in the experiment part. However, due to the spatial variance of the radiation pattern of the MIMO radar antenna array, the PSF of the MIMO imaging radar systems are spatial-variant.

Correspondingly, the features/shapes of the motifs in MIMO radar images are spatial-variant. So, it is more reasonable to use an SV-CNN to enhance the MIMO radar images. In this section, the MIMO radar imaging procedure and the spatial-variant feature of the motifs/PSF in radar images are introduced.

2.1. MIMO Radar Imaging

Wide-frequency band MIMO radar with a two-dimensional antenna array has the capability of obtaining three-dimensional radar images. Radar devices transmit the radar signal through each of the transmitting antennae and record the echo signal from the target using each receiving antenna. The transmitted wide-frequency band radar signal can be expressed by (1), and its corresponding echo signal from the target can be expressed by (2).

$$s(t) = \int W(f)e^{-j2\pi ft}df \tag{1}$$

$$s_{echo}(t) = \sigma s(t - \tau) \tag{2}$$

where $W(f)$ is the amplitude–frequency function of the radar signal, and it is usually a square window function on a certain frequency band. σ is the radar cross-section of the target and t represents the time.

τ is the time interval between the transmitted signal and the received signal and is defined as (3). **r** represents the position of the target, $\mathbf{r_t}$ and $\mathbf{r_r}$ represent the positions of the transmitting antenna and the receiving antenna of the radar, respectively. c is the velocity of light.

$$\tau = \frac{\|\mathbf{r} - \mathbf{r_t}\|_2 + \|\mathbf{r} - \mathbf{r_r}\|_2}{c} \tag{3}$$

Back-projection (BP) algorithm is a commonly used radar imaging algorithm. Its main procedure is to back-project and accumulate the echo signal to a matrix that corresponds to the imaging scene. Considering a MIMO imaging radar with N_T transmitting antennas and N_R receiving antennas, the intensity of each pixel **r** can be written as

$$I(\mathbf{r}) = \sum_{i=1}^{N_T} \sum_{j=1}^{N_R} s_{i,j} \left(\frac{\|\mathbf{r} - \mathbf{r_i}\|_2 + \|\mathbf{r} - \mathbf{r_j}\|_2}{c} \right) \tag{4}$$

where $\mathbf{r_i}$ and $\mathbf{r_j}$ represent the position of the ith transmitting antenna and the jth receiving antenna respectively. $s_{i,j}(t)$ is the range compressed echo signal which is transmitted by the ith transmitting antenna and received by the jth receiving antenna. When the position vector **r** erodes all the pixels (or voxels) in the imaging scene, a radar image is obtained. The intensity of each pixel indicates the reflected power of the corresponding position in the imaging scene.

The substance of the radar detecting is the sampling of the imaging scene. An ideal radar image is the radar cross-section (RCS) distribution map of the imaging scene. However, for the real radar systems, the wideband signal offers high range resolution of the radar system. However, according to the principle of Fourier Transform, when recovering the signal according to the echo with a limited bandwidth, sidelobes occur on the range profile. Correspondingly, the function of the antennas is the spatial sampling of the echo signal, and the sidelobes will also occur on the azimuth profile because of the limitation on the total aperture length of the antenna array. The sidelobes severely impact the quality of the radar images and form false peaks. Thus, they should be removed.

2.2. Spatial Variance Motifs in the Radar Image

The output of the imaging system for an input point source is called the point spread function (PSF). The radar image can be considered as the convolution of PSF and the RCS distribution map of the imaging scene. For a point target at the position of $\mathbf{r_0}$, its echo radar signal can be represented as

(5). Thus, the corresponding radar image which is also the PSF of the radar system at the position $\mathbf{r}_0$ can be expressed as (6).

$$S_{echo}(t) = \sigma \int W(f)e^{-j2\pi f(t-\frac{\|\mathbf{r}_0-\mathbf{r}_t\|_2+\|\mathbf{r}_0-\mathbf{r}_r\|_2}{c})} df \tag{5}$$

$$I(r) = \sum_{i=1}^{N_T}\sum_{j=1}^{N_R}\left\{\sigma \int W(f)e^{-j2\pi f(\frac{\|\mathbf{r}-\mathbf{r}_i\|_2+\|\mathbf{r}-\mathbf{r}_j\|_2}{c}-\frac{\|\mathbf{r}_0-\mathbf{r}_i\|_2+\|\mathbf{r}_0-\mathbf{r}_j\|_2}{c})} df\right\} \tag{6}$$

As we can see in (6), there will be cross-terms of $\mathbf{r}$ and $\mathbf{r}_0$ after simplification. So, the shape of the PSF varies with the change of position $\mathbf{r}_0$. Accordingly, the shape of the main lobe in the radar image for a point target depends on its position $\mathbf{r}_0$. It can be illustrated in the following figure. Figure 1a shows an imaging scene with nine point targets. Figure 1b is the corresponding radar image. The shapes of the main lobes of the nine point targets are different due to the difference in their positions.

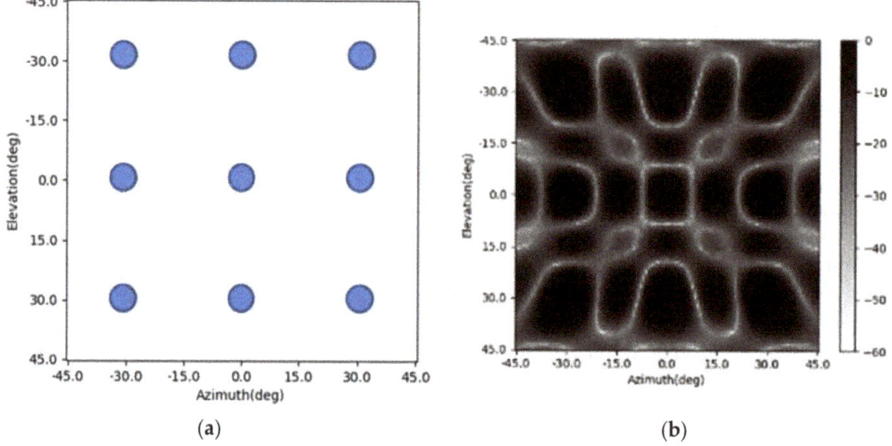

Figure 1. Imaging scene and imaging results; (**a**) imaging scene; (**b**) original radar image.

3. Conventional Radar Image-Enhancing CNN and SV-CNN

The proposed SV-CK is an evolution of the conventional convolution kernel. The proposed SV-CK contains two parts: a basic data kernel and a position kernel. The position kernel reads information of local position from the position coding embedded into the input sample and controls the data kernel. Thus, the SV-CK has spatial-variant characters.

In this section, the conventional radar image-enhancing CNN in [18] is introduced firstly. Then the SV-CK and SV-CNN are proposed. The enhancing results of the two neural networks are compared in Section 4 to show the advantage of the proposed one.

3.1. Conventional Radar Image-Enhancing CNN

In [18], a convolutional neural network is constructed and trained to enhance the radar image. After enhancement, the main lobes of targets in the radar image are sharpened, and the sidelobes in the radar image are suppressed. Its structure is shown in Figure 2.

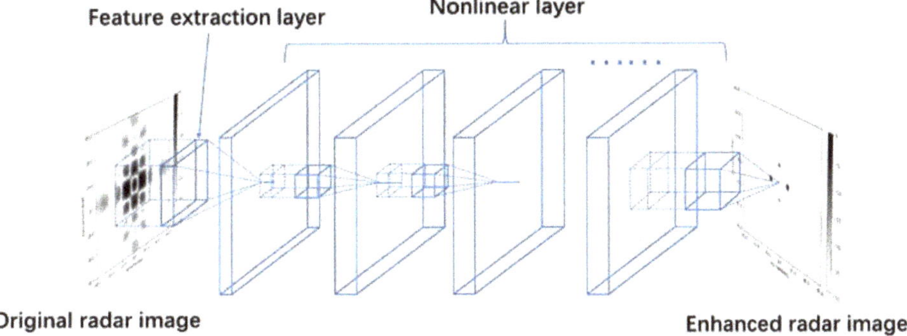

Figure 2. Structure of the conventional radar image-enhancing CNN.

In this paper, a CNN with a similar structure of [18] is constructed to serve as the comparison of the SV-CNN. The CNN is composed of several convolution layers with the rectified linear unit (ReLU) as activation. What is special is that its first layer contains some two-channel convolution kernels. They can take the complex-valued two-dimensional radar image as the input samples (one channel for the real part and the other for the imaginary part). The last convolution layer comprises only one kernel and is not followed by a ReLU. Its output is the final enhanced radar image. The enhancing results can be seen in the next section. The function of this neural network is to sharpen the main lobes and suppress the sidelobes of the radar image. In this way, the radar image is enhanced. However, while enhancing the radar image, the whole CNN can be seen as a spatial-invariant weighted function. The result of each pixel in the enhanced radar image can be seen as the convolution result of the pixels around it and the spatial-invariant weighted function. As discussed earlier, it is more reasonable to use a CNN with spatial-variant convolution kernels to enhance the radar image because of the spatial-variant feature of the motifs in the radar image.

3.2. SV-CK and Position-Coding

As discussed above, a CNN with spatial-variant convolution kernels will give better results when enhancing a MIMO radar image. So, in this section, the SV-CNN is proposed.

The SV-CK and its input sample are shown in the lower half and upper half of Figure 3, respectively. Two parts constitute the SV-CK, which are the data kernel and the position kernel. After being trained, the data kernel reads features from the data channels of the input sample while the position kernel reads the corresponding position information from the position-coding. The position kernel has an influence on the data kernel, and determines what feature it reads and outputs. Thus, an SV-CK is obtained.

The structure of the input samples is shown in the upper half of Figure 3. As discussed before, the first two channels are the input data (complex-valued radar image) and the rest of the four channels are the position-coding. The essences of the position-coding are tensor meshes which indicate the position of each pixel in the data channels. They bring position information into the SV-CNN while training. Consequently, the spatial-variant features can be extracted by the SV-CKs. As shown in the figure, two channels are used to indicate position on one dimension, and the remaining two are used to indicate position on the other dimension. For the two channels of position-coding on the same dimension, the corresponding two elements are complementary, and their sums equal to a constant. Just as shown in Figure 3, the third and fourth channels of the input are the position-coding on the horizontal dimension. The values of the elements on one channel linearly increase, while the values of the elements on the other channel decrease gradually. On the upper right corner of Figure 3, one of the position-coding channels is zoomed in to illustrate its structure more clearly. Specifically, if there is only one channel indicating the position in one direction, the convolution kernels might take it as a constant weighted function of the features, and consequently be confused. The advantage of using two channels of complementary position-coding is indicated in Section 4.

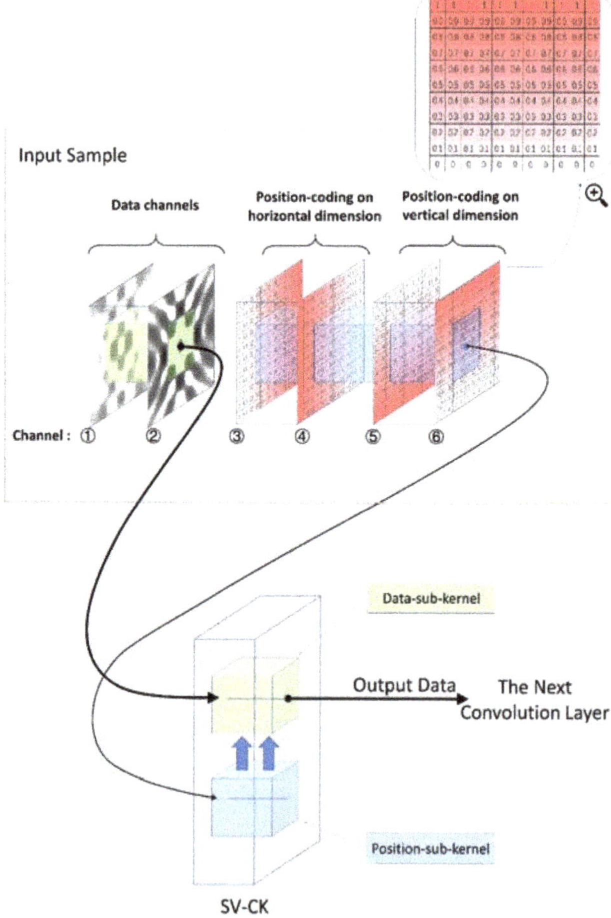

Figure 3. Structure of input sample with position-coding and the SV-CK.

3.3. SV-CNN

The structure of the SV-CNN is illustrated in Figure 4. There are several convolution layers with SV-CK. The comparison of the CNN and the SV-CNN is shown in Figure 5. During training, the input samples are organized as illustrated in Section 3.2, and the labels are the amplitude of ideal radar image with sharp main lobes and no sidelobe. After being trained, the network takes complex-valued radar images with four-channel position-coding as the input and outputs amplitude of enhanced radar images with sharp main lobes and low sidelobes.

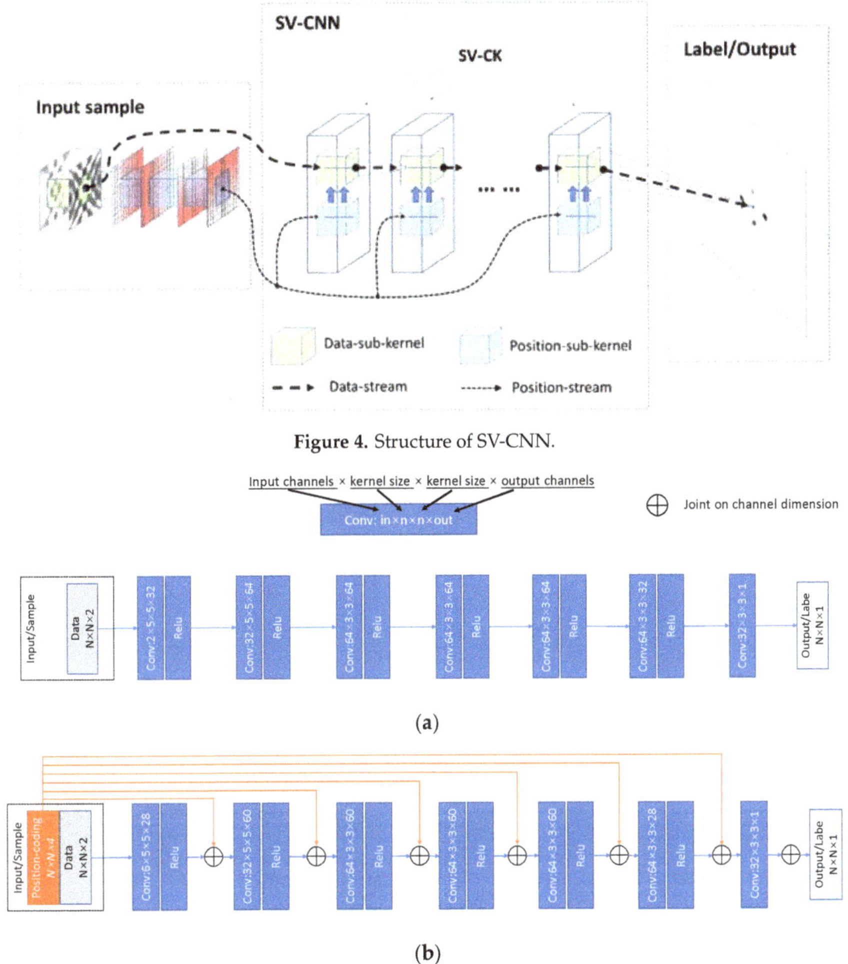

Figure 4. Structure of SV-CNN.

Figure 5. Structure and parameters of (**a**) CNN; (**b**) SV-CNN.

As shown in Figures 4 and 5b, there are two streams in the SV-CNN. (1) The first one is the data stream. The data kernel extracts the features from the input sample or the output of the former layer and outputs feature maps to the next layer until the final output (the enhanced radar image) is obtained. (2) The second one is the position-coding stream. Position-codings are transmitted to position kernels in each layer through skip connection. It helps the position kernels to read the information of the local position and guides them to control the corresponding data kernel. The input data are covered by four channels of position-codings. The convolution kernels process the data and the position-coding together, and the ReLU after each of the convolutional layers will totally mix them up. The parameters which determine the relationships between the position-codings and the data are contained in the convolution kernels and can be optimized through training. Thus, an SV-CNN is obtained and its superiority is tested in the next section.

4. Implementation and Simulation

In this section, some features of the SV-CNN are tested. Firstly, in part A, the training procedure of the SV-CNN and its enhancing results are given. Then, in part B, the training loss and the testing loss of the SV-CNN are compared to those of a CNN with a similar structure to show the superiority of the SV-CNN. In part C, the guided backpropagation method is used to discover the influential information in the input samples. The results showed that the position-coding does offer useful information for the SV-CNN. Finally, in part D, the guided backpropagation method is used to show the importance of position-coding on each layer. Then, the position-codings which make no difference to the final result are cut off to reduce the complexity of the SV-CNN.

4.1. Structure of the SV-CNN and the Training Procedure

A seven-layer SV-CNN was proposed to enhance the radar image. An illustration of the structure is shown in Figure 4 and its specific structure and the compared CNN are shown in Figure 5. The input samples are organized as illustrated in Section 3.2. There are two channels for data and four channels for the position-coding. The four position-coding channels are inset into the input of each layer through the skip connection as illustrated in Figures 4 and 5b.

The samples and the labels are valued while training the neural network. While generating the samples, three steps are taken. Firstly, simulate several point scatterers in the imaging scene. Then, simulate the radar signal and the corresponding echo of these point scatterers. Finally, calculate the original radar images as the input samples. As for the labels, just project the RCS of each simulated point on the RCS distribution maps whose value of each pixel equals to the RCS of the point target at the corresponding position. If the pixel does not correspond to a point scatterer, then its value is zero. A total of 22,000 sample label pairs like this were simulated; 20,000 of them were randomly chosen to train the neural network and 2000 of them were chosen to test it.

The training procedure is evaluated on a PC with a CPU of I7, two pieces of 16 GB RAM, and a GPU of 2080 Ti. The networks are established using the PyTorch. After being trained, the SV-CNN is given the ability to enhance the radar image. The original radar images and the enhanced ones are shown in Figure 6. Besides, the enhancing results of the SV-CNN are compared to the enhancing result of the CNN trained using the same procedure. The structure of the compared CNN is illustrated in Figure 5a. Two imaging scenes were simulated as shown in Figure 6a,b. The original radar images are shown in Figure 6c,d. The enhancing results of the CNN are shown in Figure 6e,f, and the enhancing results of the SV-CNN are shown in Figure 6g,h. As we can see in the figure, after enhanced, the main lobes of motifs in the radar image are sharpened and the sidelobes are suppressed. However, there are false peaks in the enhancing result of the CNN. The false peaks might result in false alarms in the detection procedure, which is unacceptable. So, the SV-CNN performs better while enhancing radar images. The detailed performance of the networks is summarized in Table 1.

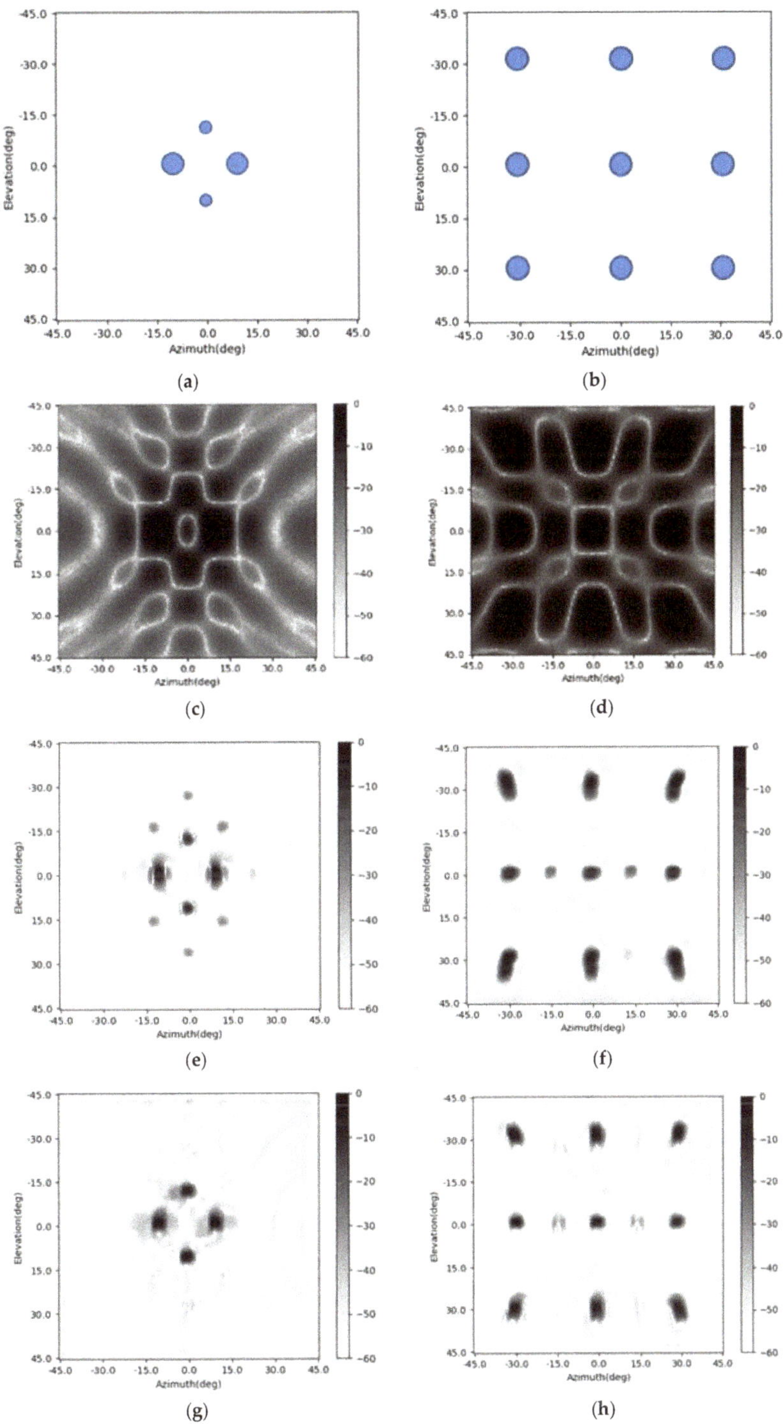

Figure 6. Enhancing result of the CNN and the SV-CNN (**a**,**b**) imaging scenes; (**c**,**d**) original radar images; (**e**,**f**) enhancing results of the CNN; (**g**,**h**) enhancing result of the SV-CNN.

Table 1. MSLL comparison of the above networks.

Networks	Scene	4 Targets	9 Targets
Original image		−9.51 dB	−7.24 dB
CNN		−22.3 dB	−18.7 dB
SV-CNN with 2-channel Position-codings		−22.2 dB	−8.6 dB
SV-CNN with 4-channel complementary position-codings		−31.8 dB	−32.4 dB
SV-CNN with 4-channel orthogonal position-codings		−42.2 dB	−25.7 dB
Simplified SV-CNN with 4-channel complementary position-codings		−38.5 dB	−33.3 dB
Simplified SV-CNN with 4-channel orthogonal position-codings		−43.8 dB	−46.9 dB

4.2. Comparison of CNN and SV-CNNs Trained Using Samples with Different Forms of Position-Coding

In this part, the training loss and the testing loss of the proposed SV-CNN are compared to the other three networks. (1) The conventional CNN; (2) the SV-CNN trained using samples with two position-coding channels. Its input samples are with the similar structure shown in Figure 3, however, only the third and the fifth channels serve as the position-coding, and each channel indicates the position on one direction; the fourth and the sixth channels are deleted. (3) SV-CNN trained using samples with four orthogonal position-coding channels. The four complementary position-coding channels are processed by the sinusoidal function, and then they became orthogonal to each other, while training with the same parameters. The training loss and the testing loss of these CNNs are shown in Figure 7a,b, respectively.

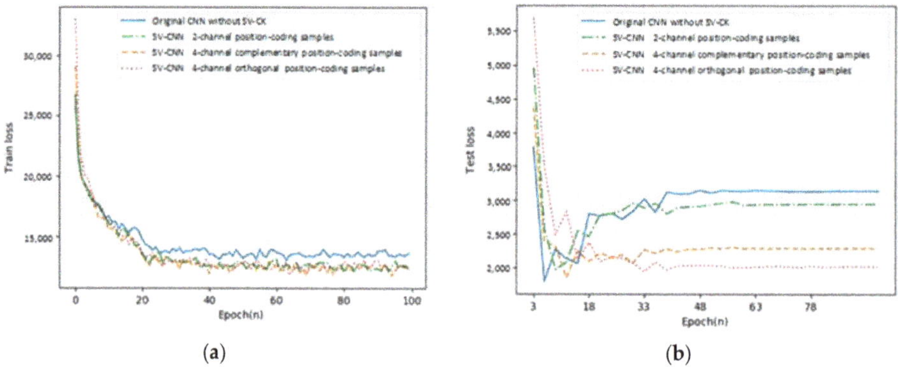

Figure 7. Training loss and testing loss of the CNN and SV-CNN trained using samples with different forms of position-coding; (a) training loss; (b) testing loss.

As shown in the figure, both the training loss and the testing loss of the SV-CNN trained using samples with four position-coding channels (both complementary and orthogonal) decrease the fastest, and after the networks stabilize, both the training loss and the testing loss of the SV-CNN trained using samples with four position-coding channels are lower.

Besides, the enhancing results of the four CNNs are shown in Figures 6 and 8. The imaging scenes are set as Figure 6a,b. Their corresponding imaging results are shown in Figure 6c,d. The enhancing results of the CNN are shown in Figure 6e,f. The enhancing results of the SV-CNN trained using samples with four complementary position-coding channels are shown in Figure 6g,h. The enhancing results of the SV-CNN trained using samples with two position-coding channels are shown in Figure 8a,b, and those of the SV-CNN trained using four orthogonal position-coding channels are shown in Figure 8c,d. The dynamic ranges of all these figures are assigned as 60 dB. As shown in the figures, the false peaks occur in the enhancing results of the conventional CNN and the SV-CNN trained using samples with two position-coding channels. What is even worse is that the enhancing results of the SV-CNN trained using samples with two position-coding channels become asymmetric due to the asymmetric position-coding values. Both the SV-CNNs trained using samples with four position-coding channels have better effects on suppressing the sidelobes.

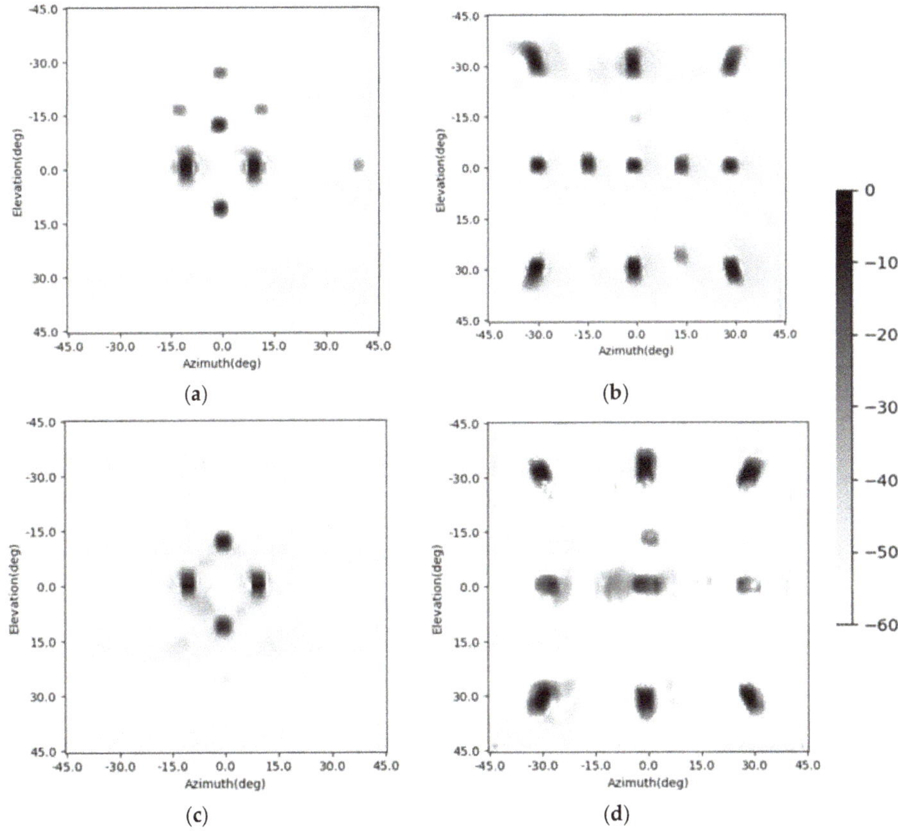

Figure 8. Enhancing results; (**a**,**b**) enhancing results of the SV-CNN trained using samples with two position-coding channels; (**c**,**d**) enhancing results of the SV-CNN trained using samples with four orthogonal position-coding channels.

4.3. Testing the Function of SV-CK and Position-Coding

In this part, the guided backpropagation method is used on the SV-CNN to show if the SV-CKs extract local position information from the position-coding. The guided backpropagation is a combination of the deconvolution method and the backpropagation method [29–31]. It is an efficient way of visualizing what concepts in the graph have been learned by the neural network. An imaging scene shown as Figure 6b is simulated. There are nine point targets in the simulated imaging scene, and the corresponding radar image is as shown in Figure 6d. The guided backpropagation method is used to show what concepts the network took while enhancing this sample. The influential concepts on each channel of this sample are shown in Figure 8. The results on two data channels are shown in Figure 9a,b. The reception field and important features that the network used to enhance the radar image can be seen in these figures. The guided backpropagation results on the four position-coding channels are shown in Figure 9c–f. The influential features on the horizontal dimension are shown in Figure 9c,d, and the influential features on the vertical dimension are shown in Figure 9e,f. As shown in these figures, the tendency of values read from the two complementary position-codings on the same dimension is contrary. This tendency is in accord with the tendency of the position-codings. This phenomenon confirms that the position information contained in the position-coding is extracted by the network.

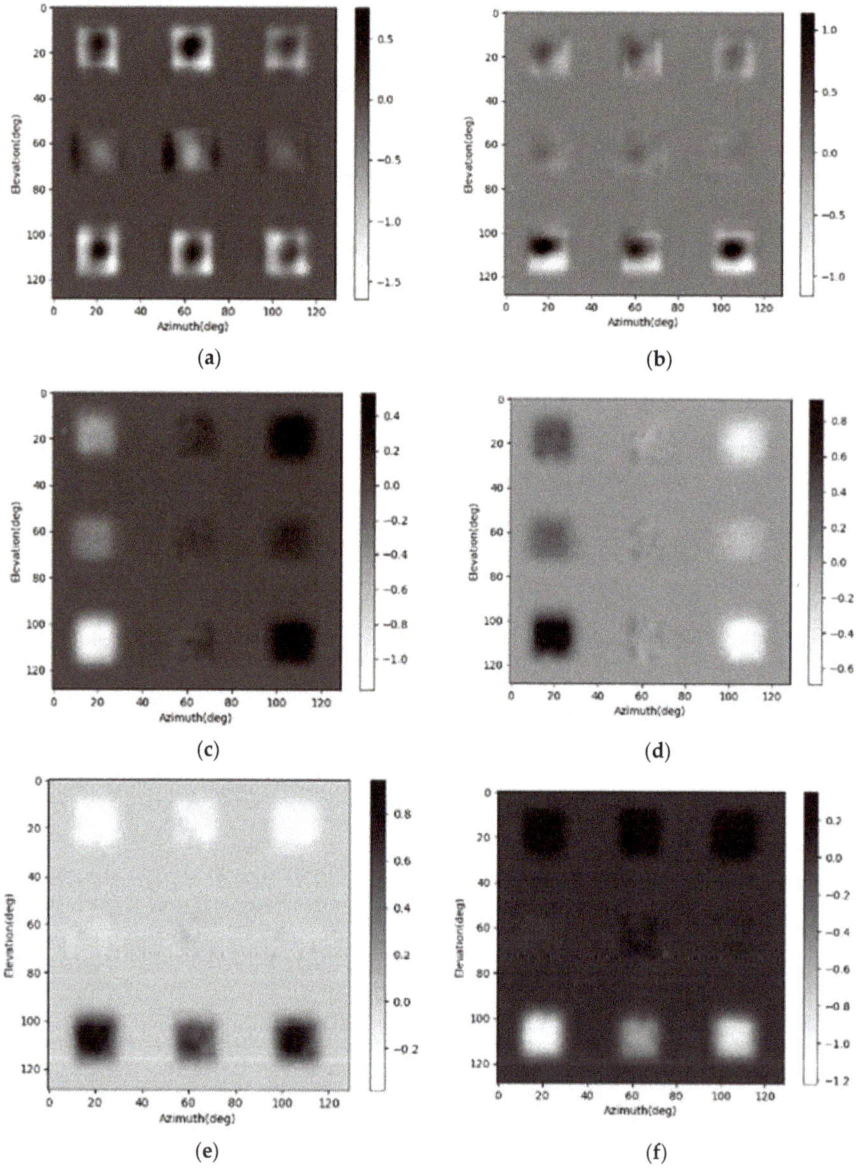

Figure 9. Guided backpropagation results on each channel of the sample. (**a**) Results on the 1st channel; (**b**) results on the 2nd channel; (**c**) results on the 3rd channel; (**d**) results on the 4th channel; (**e**) results on the 5th channel; (**f**) results on the 6th channel.

4.4. Simplification of the SV-CNN

As shown in Figures 4 and 5b, the four position-coding channels are inset into the input of each layer through the skip connection. These skip connections increase the complexity of the SV-CNN. In this part, the guided backpropagation method is used to evaluate the weightiness of the position-codings inset into each layer, and cut off the unnecessary ones. The guided backpropagation method is used to show what the SV-CNN extracts from the input of each layer. If there is no obvious spatial-variant feature extracted from the position-codings of one layer, then they are not necessary.

The guided backpropagation results on the position-codings of each layer are calculated, and those of the first, the sixth, and the seventh layers are shown in Figures 9–11, respectively. As we can see, for the simulated nine point scatterers, the guided backpropagation results on the position-codings of the first layer vary tremendously. The backpropagation results on the position-codings of the sixth layer only vary slightly. As for the seventh layer, the backpropagation results have hardly any variety. The results mean that, for the first and the sixth layer, the spatial-variant features in the input data are extracted by the SV-CK, while the SV-CKs in the seventh layer do not extract these spatial-variant features. Thus, the corresponding skip connection from the position-codings to the seventh layer is unnecessary and can be cut off. In this way, the SV-CNN is simplified. Then, using the same training method in Section 3.1 to train the network, the training loss and testing loss of the simplified SV-CNN are shown and compared in Figure 12. As we can see in the figure, after simplification, there is no significant degradation on both the training loss and the testing loss. Besides, after simplification, its training loss and testing loss are still lower than those of the conventional CNN, and they may sometimes be even lower than those of the SV-CNN that has not been simplified.

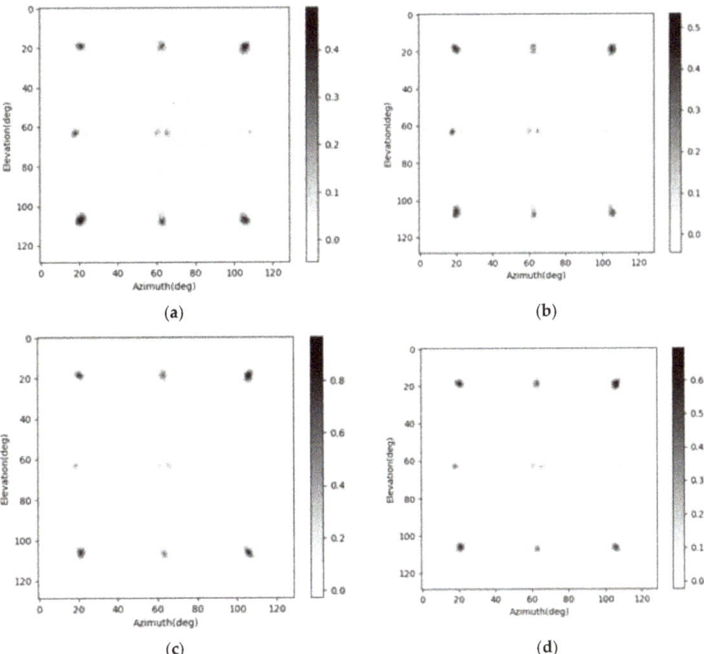

Figure 10. Guided backpropagation on position-coding of the 6th layer. (**a**) Results on the 1st position-coding channel; (**b**) results on the 2nd position-coding channel; (**c**) results on the 3rd position-coding channel; (**d**) results on the 4th position-coding channel.

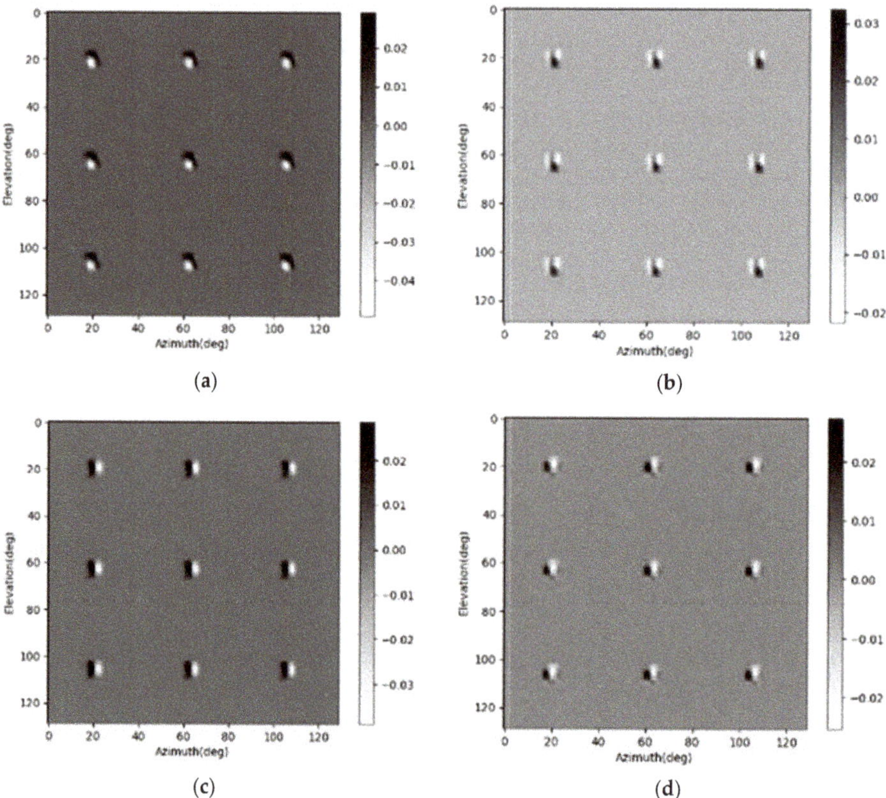

Figure 11. Guided backpropagation on position-coding of the 7th layer. (**a**) Results on the 1st position-coding channel; (**b**) results on the 2nd position-coding channel; (**c**) results on the 3rd position-coding channel; (**d**) results on the 4th position-coding channel.

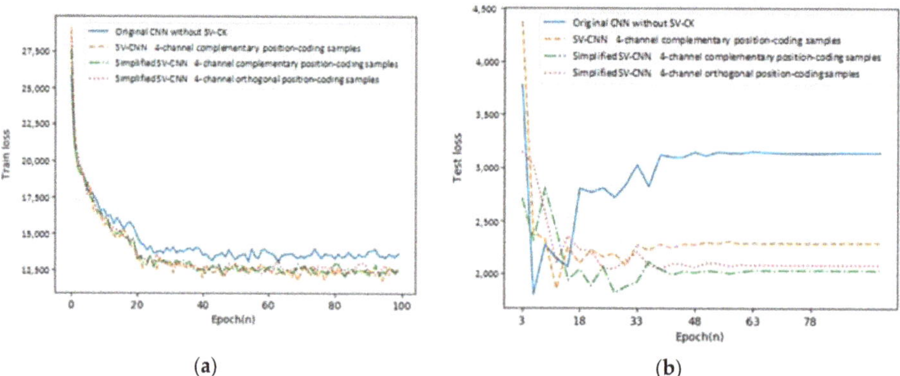

Figure 12. Training loss and testing loss of the CNN, the SV-CNN, and the simplified SV-CNNs; (**a**) training loss; (**b**) testing loss.

The enhancing results of the simplified SV-CNN are shown in Figure 13. The imaging scene is the same as those illustrated in Figure 6. The upper row is the enhancing result of the simplified SV-CNN trained using samples with four complementary position-coding channels, while the lower row is the enhancing result of the simplified SV-CNN trained using samples with four orthogonal position-coding channels. The dynamic ranges of all the figures are set to 60 dB. Compared to the enhancing results shown in Figures 6 and 7, there is no obvious degradation in the performance of the simplified SV-CNN. The performance of the simplified SV-CNN is still higher than that of the CNN. Besides, the simplified SV-CNN trained using samples with four orthogonal position-coding channels sometimes performs better in sidelobe suppression (Figure 13d).

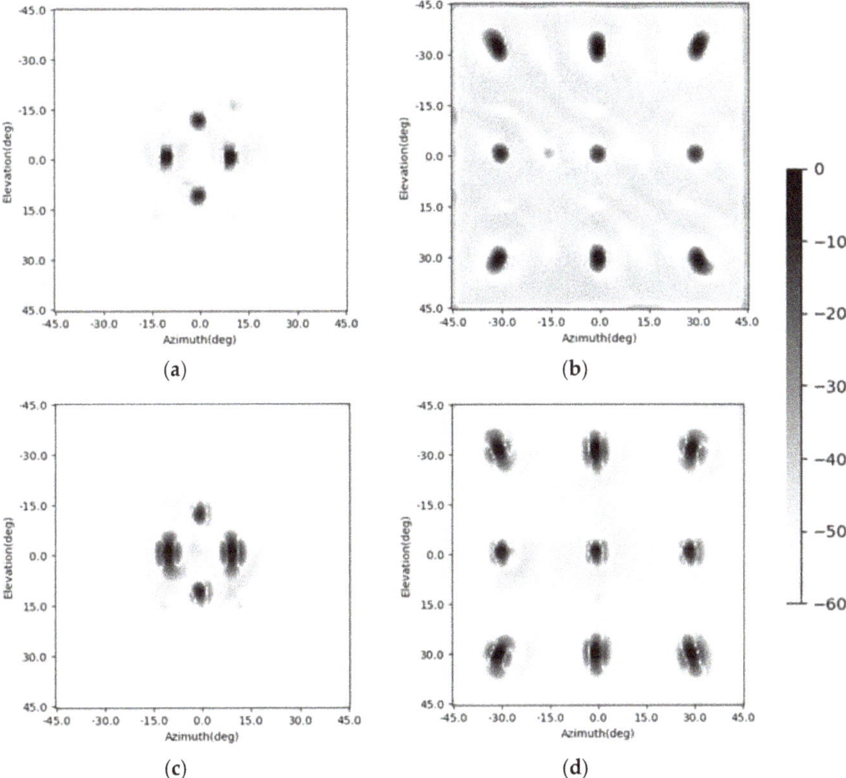

Figure 13. Enhancing results of the simplified SV-CNNs; (**a,b**) enhancing results of the simplified SV-CNN trained using samples with four-channel complementary position-coding; (**c,d**) enhancing results of the simplified SV-CNN trained using samples with four-channel orthogonal position-coding.

The SV-CNNs are trained and analyzed in this simulation part. It can be seen from the simulation results that compared to the CNN and the SV-CNN trained using samples with two position-coding channels, the SV-CNN trained using samples with four position-coding channels performs better in enhancing radar images. The SV-CNN trained using samples with both four complementary position-coding channels and four orthogonal position-coding channels has good performance in enhancing radar images. Besides, the performance of the simplified SV-CNN does not degrade obviously, and is sometimes even better. The maximum sidelobe levels (MSLLs) of the enhancing results of the above networks are listed in Table 1. The results also support the conclusion that the networks with four channels of position-codings perform better. For the original radar images, there is more than one point target in both of the images. So, their sidelobes are accumulated and higher than the theoretical value of −13.2 dB.

4.5. Comparison to Other Existing Methods

In this part, the enhancing results of the four-target scene in Section 4.1 are reviewed, and another three existing methods are added into the comparison. This time, the Gaussian noise is added to the simulated signal and makes the signal-to-noise ratio 15 dB. The original radar image, the enhancing results of the CF algorithm [23,24,32,33], the results of the orthogonal matching pursuit (OMP) sparsity driven method [27], and the Lucy–Richardson deconvolution algorithm [28] are shown in Figure 14a–e, respectively, and the results of the CNN and the SV-CNN are shown in Figure 14f,g, respectively. Besides, the MSLLs of these methods are listed in Table 2.

It can be seen from the results that all of these methods can suppress the sidelobes and improve the quality of the images. In the original, the sidelobes of these four points overlap, and this makes the MSLL higher than the test theoretical value of −13.2 dB. The Lucy–Richardson deconvolution and the CF algorithm can suppress the sidelobes to −28.39 dB and −27.75 dB, respectively. The performance of the OMP algorithm strongly depends on the parameter of the sparsity. If it is set as 4 and equals the number of the targets, there will be no sidelobes. However, in real-world detection, it cannot be foreseen. So, a result of the OMP with the sparsity of 10 is also given. With the interference of noise, there are sidelobes with MSLL of −22.85 dB. In the results of the CNN and the SV-CNN, the MSLL is −25.11 dB and −42.71 dB, respectively. According to the above results, the SV-CNN offers better results, especially when compared to the conventional CNN.

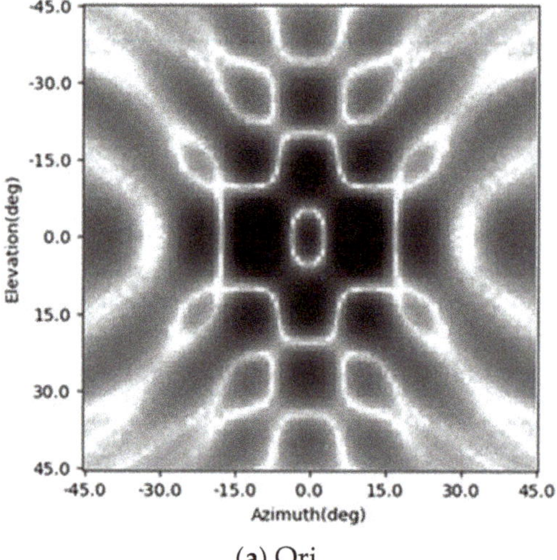

(a) Ori

Figure 14. *Cont.*

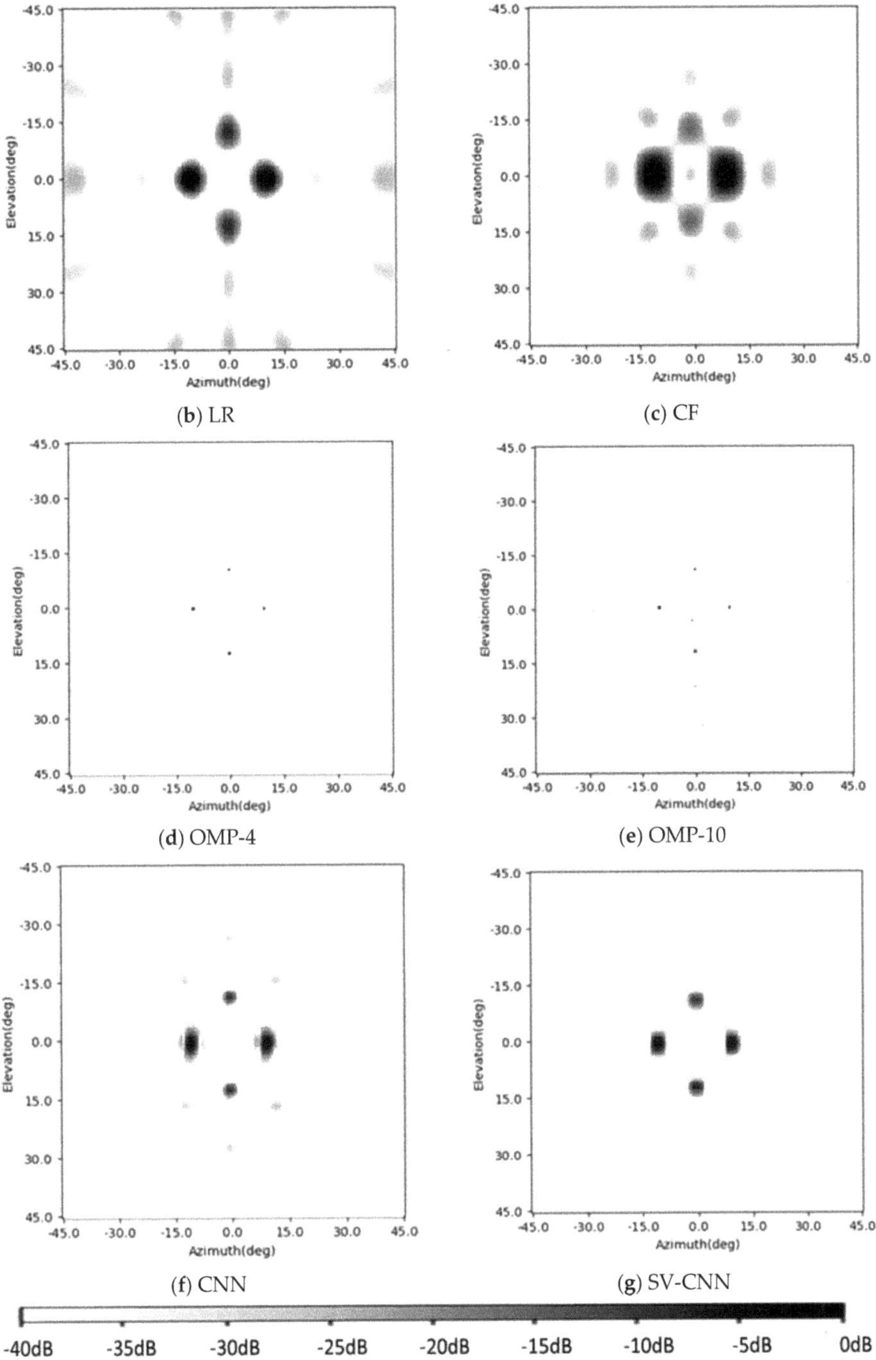

Figure 14. Enhancing results of different methods; (**a**) the original image; (**b**) the enhancing result of the Lucy–Richardson algorithm; (**c**) the result of the CF algorithm; (**d**) the enhancing result of the OMP with the sparsity of 4; (**e**) the enhancing result of the OMP with the sparsity of 10; (**f**) the enhancing result of the CNN; (**g**) the enhancing result of the SV-CNN.

Table 2. MSLL comparison of the above different methods on simulated images.

Methods	Original Image	Lucy–Richardson Deconv	CF	OMP_4
MSLL	−9.48 dB	−28.39 dB	−27.75 dB	−180 dB
Methods	**OMP_10**	**CNN**	**SV-CNN**	
MSLL	−22.85 dB	−25.11 dB	**−42.71 dB**	

5. Experiment

In this experiment part, a radar device is implemented to obtain the radar image of the imaging scene and test the proposed SV-CNN. The MIMO radar is shown in Figure 15. The small units fixed on the plane are its antenna. Eight of them are transmitting antennas and 10 of them are receiving antennas. During operation, the transmitting antennas take turns to transmit L/S band signal with 600 MHz bandwidth, and the receiving antennas take turns to receive the echo signal. It takes 0.1 s to switch all the antenna channels. Then, the radar images of the imaging scene can be obtained through the BP algorithm illustrated in Section 2.

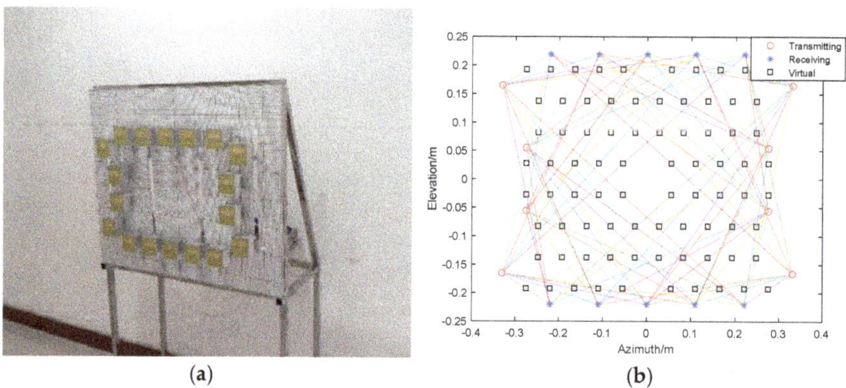

(a)   (b)

Figure 15. The experimental MIMO radar; (**a**) the photo; (**b**) the topology of the antenna array.

5.1. Performance for Point Targets

In this part, the imaging and enhancing results of two corner reflectors are given. The imaging scene in this experiment is shown in Figure 16a. Two corner reflectors are placed in front of the radar system. The point scatterers are 3 m away from the center of the radar antenna array. The original radar image of the imaging scene is shown in Figure 16b, and the enhancing results of the Lucy–Richardson deconvolution and the CF algorithm are shown in Figure 16c,d, respectively. The enhancing results of the OMP with the sparsity of 2 and 5 are shown in Figure 16e,f, respectively. Finally, the enhancing results of the conventional CNN and the simplified SV-CNN trained using samples with four position-coding channels are shown in Figure 16g,h, respectively. The MSLLs of these results are listed in Table 3.

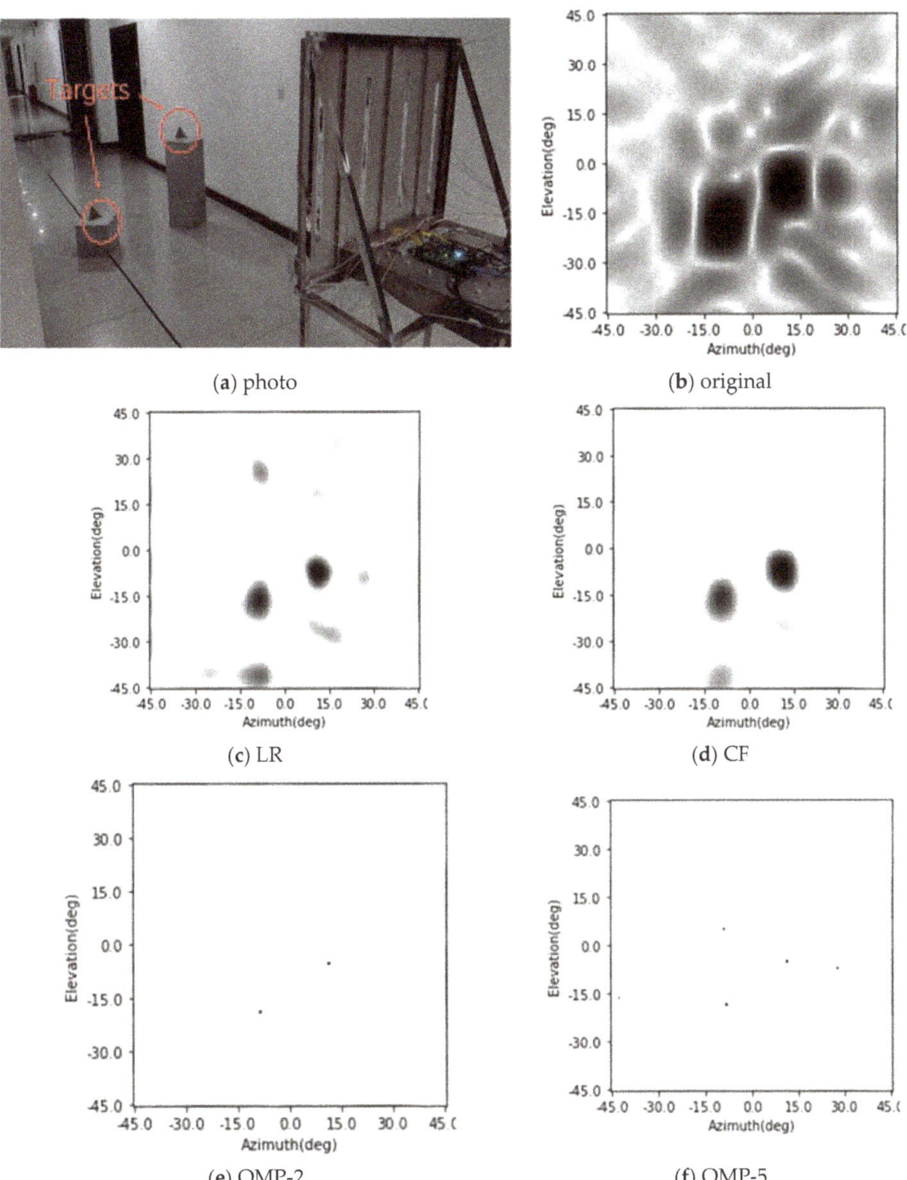

(**a**) photo   (**b**) original

(**c**) LR   (**d**) CF

(**e**) OMP-2   (**f**) OMP-5

Figure 16. *Cont.*

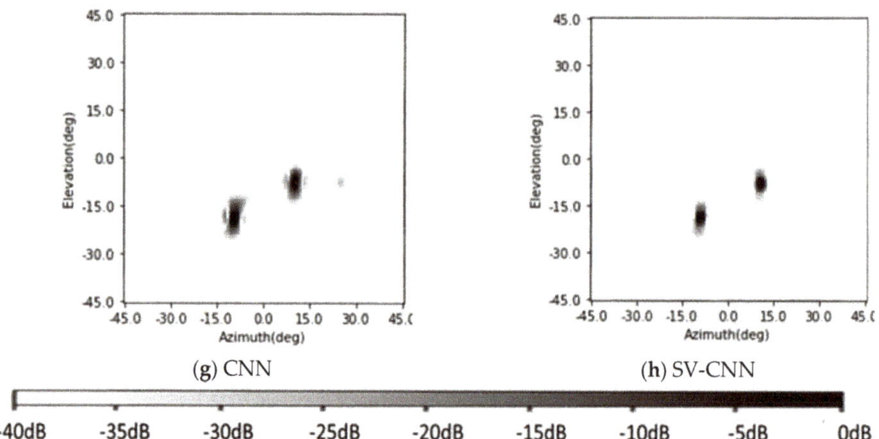

Figure 16. Imaging scene and radar images. (**a**) Imaging scene; (**b**) the original image; (**c**) the enhancing result of Lucy–Richardson algorithm; (**d**) the result of the CF algorithm; (**e**) the enhancing result of the OMP with the sparsity of 2; (**f**) the enhancing result of the OMP with the sparsity of 4; (**g**) the enhancing result of the CNN; (**h**) the enhancing result of the SV-CNN.

Table 3. MSLL comparison of the above different methods on real recorded images.

Methods	Original Image	Lucy–Richardson Deconv	CF	OMP_2
MSLL	−8.87 dB	−15.46 dB	−25.32 dB	−180 dB
Methods	OMP_5	CNN	SV-CNN	
MSLL	−7.4 dB	−26.95 dB	**−34.84 dB**	

It can be seen in the results that the sidelobes in the real recorded radar image are a little higher than in the simulation one, because the scatterers used are slightly expanded ones. Compared to the simulation results, the performance of the Lucy–Richardson deconvolution slightly degrades, because of these expanding characters. The results of the OMP with the sparsity of 2 is still free of sidelobes. However, the performance of the OMP with the sparsity of 5 degrades because of the inevitable noise and the mismatch between the real system and the theoretical model. From the results, we can see that the SV-CNN also offers the best results in our occasion.

5.2. Performance for Extending Targets

In this part, the SV-CNN is used to enhance the radar images of human targets to evaluate the degradation in a real setup. The radar images of a human body are enhanced using the SV-CNN. The photo of the imaging scene is as shown in Figure 17a,b. A human stands 3 m away from the radar and with his arms opening and dropping, respectively. The original radar images are shown in Figure 17c,d, respectively. The enhancing results of the Lucy–Richardson deconvolution algorithm, the CF algorithm, the OMP algorithm, the CNN, and the SV-CNN are shown in Figure 17. The power reflected by the human limbs varies strongly with the variety of the incident angle of the radar signal. So, the final enhanced radar image is obtained through accumulating the enhancing results of several continuous frames of images. Because of the accumulation, the quality of all these images is improved. As can be seen, for these extremely extended targets, the sidelobes of the Lucy–Richardson is the highest for it is originally proposed for real-valued optical images. The results of the OMP are faced with several sidelobes when dealing with the real recorded targets, as well as the CNN (at about (0°, −42°)). The sidelobes in both results of the CF and the SV-CNN are low. However, the proposed SV-CNN offers sharper main lobes. In all of these enhanced results, the actions can be easily discriminated. However, the SV-CNN offers the best results.

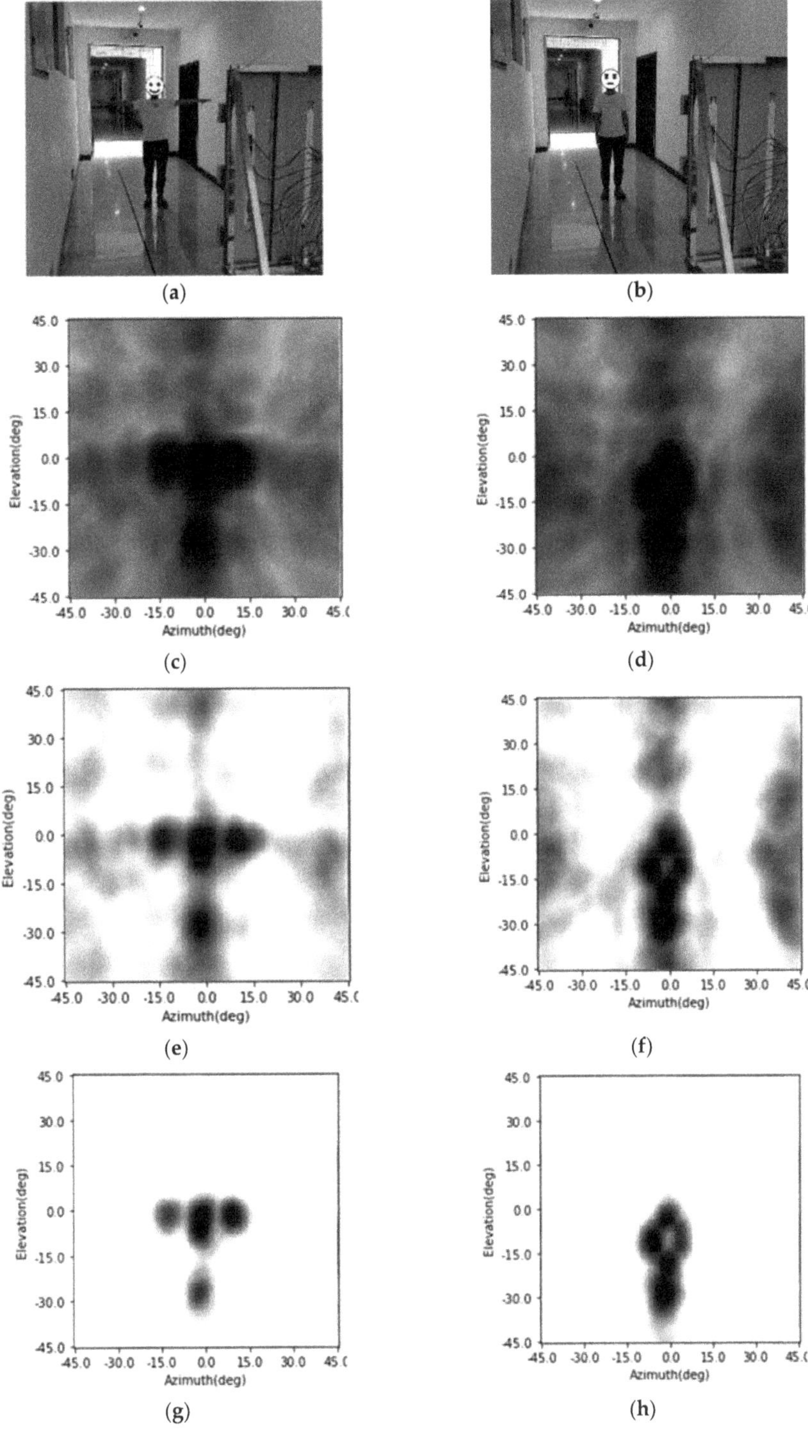

Figure 17. *Cont.*

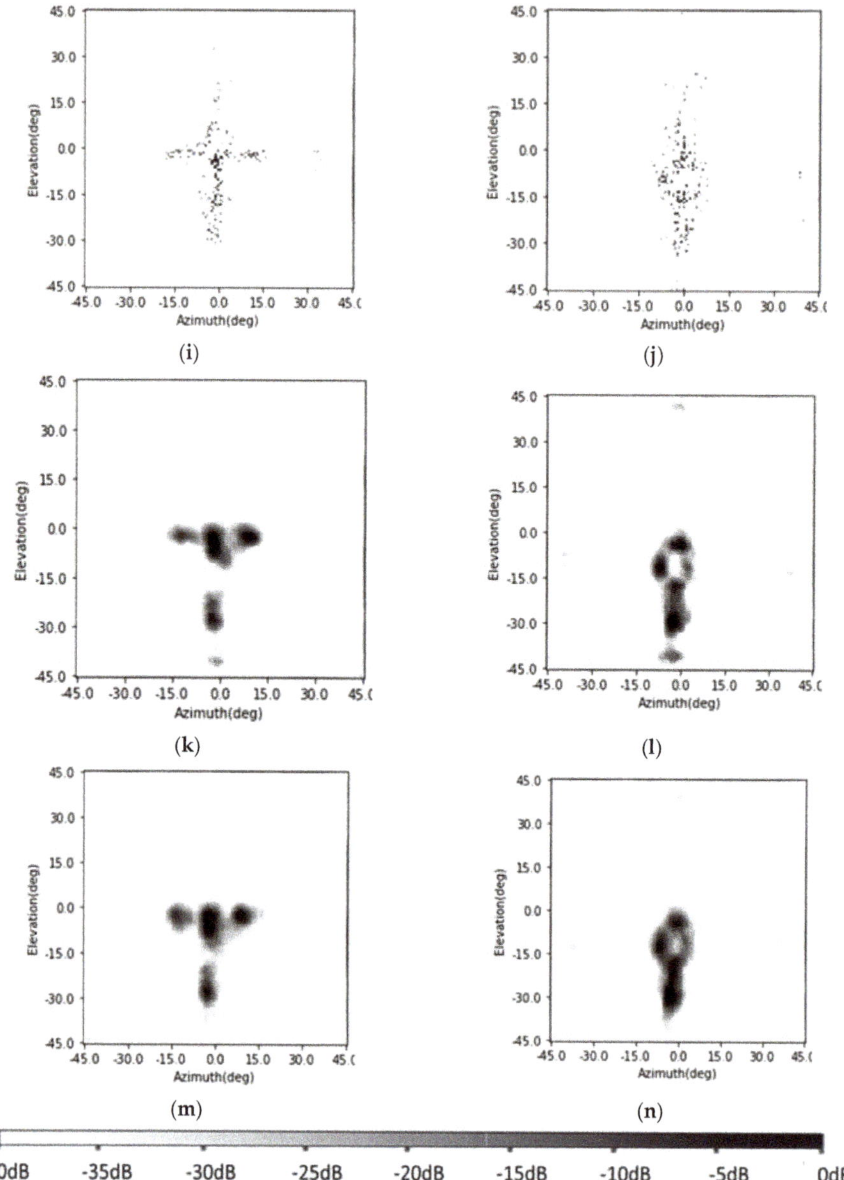

Figure 17. Targets and images; (**a,b**) imaging scene of the two actions; (**c,d**) the original radar image; (**e,f**) LR-deconv results; (**g,h**) CF results; (**i,j**) OMP results; (**k,l**) CNN results; (**m,n**) SV-CNN results.

6. Discussions

In this paper, we proposed the SV-CNN to deal with images with spatially variant features. Compared to conventional CNN, the proposed SV-CNN is with spatial awareness. Thus, the SV-CNN performs better when faced with spatially variant features.

In Sections 4 and 5, we use the proposed SV-CNN to suppress the sidelobes in MIMO radar images as an illustration. The enhancing results of the proposed SV-CNN are compared to several state-of-art methods including the CF algorithm [23,24,32,33], the OMP sparsity driven method [27], and the Lucy–Richardson deconvolution algorithm [28], as well as the conventional CNN.

Since the MIMO radar images in our condition are with spatially variant features (as illustrated in Figure 1), our proposed SV-CNN performs best among these methods. The superiority of the SV-CNN can be seen in Figures 14, 16 and 17.

Besides, it is pointed out that the SV-CNN should extract spatial information from four-channel position-codings. The enhancing results will degrade and even become asymmetric when there are only two channels of position-coding, because the network might take the two channels of position-coding as a weight function to show the degree of importance of the input samples. Simulation results in Section 4.2 support this standpoint.

7. Conclusions

In this paper, a spatial-variant convolutional neural network (SV-CNN) with spatial-variant convolution kernels (SV-CK) is proposed. While extracting features, the proposed SV-CNN can take the local position information into account compared to conventional CNNs. Thus it has better performance when the shapes of the motifs in the images depend on the local position.

The proposed SV-CNN is trained to enhance the radar images to illustrate its function. After being trained using radar images with position-codings, it can suppress the sidelobes in the radar images. The SV-CKs can extract spatial-variant features from the radar image. Thus it has better enhancing results and leaves less false peaks in the enhanced image. Simulation and experimental results showed that the SV-CNN trained using samples with four position-coding channels gives good results, even after simplification.

The proposed SV-CNN is a special CNN and is with spatial awareness. It shall have better performance in tasks with spatially variant features. In future works, we will test its performance in image segmentation and even try to use it as an imaging algorithm.

Author Contributions: Conceptualization, Y.D.; methodology, Y.D.; validation, C.W.; formal analysis S.S., Y.S.; investigation S.S., Y.S.; resources, T.J.; data curation, C.W.; writing—original draft preparation, Y.D.; writing—review and editing, S.S.; visualization, C.W.; supervision, T.J.; project administration, T.J.; funding acquisition, T.J. All authors have read and agreed to the published version of the manuscript.

Funding: This work was supported in part by the National Natural Science Foundation of China under Grant 61971430.

References

1. Rusk, N. Deep learning. *Nat. Methods* **2015**, *13*, 35. [CrossRef]
2. Mersa, O.; Etaati, F.; Masoudnia, S.; Araabi, B.N. Learning Representations from Persian Handwriting for Offline Signature Verification, a Deep Transfer Learning Approach. In Proceedings of the 4th International Conference on Pattern Recognition and Image Analysis, Tehran, Iran, 6–7 March 2019.
3. Zhong, P.; Gong, Z.; Li, S.; Schonlieb, C.B. Learning to diversify deep belief networks for hyperspectral image classification. *IEEE Trans. Geosci. Remote Sens.* **2017**, *55*, 3516–3530. [CrossRef]
4. Bi, N.; Chen, J.; Tan, J. The handwritten Chinese character recognition uses convolutional neural networks with the GoogLeNet. *Int. J. Pattern Recognit. Artif. Intell.* **2019**, *33*, 1940016. [CrossRef]
5. Simonyan, K.; Zisserman, A. Very Deep Convolutional Networks for Large-Scale Image Recognition. *arXiv* **2014**, arXiv:1409.1556.
6. Haoxiang, L.; Zhe, L.; Xiaohui, S.; Jonathan, B.; Hua, G. A convolutional neural network approach for face detection. In Proceedings of the IEEE Conference on Computer Vision and Pattern Recognition (CVPR) 2015, Boston, MA, USA, 7–12 June 2015; pp. 5325–5334.
7. Gao, J.; Li, H.; Han, Z.; Wang, S.; Hu, X. Aircraft detection in remote sensing images based on background filtering and scale prediction. *Lect. Notes Comput. Sci.* **2018**, *11012*, 604–616.

8.   Raza, S.E.A.; AbdulJabbar, K.; Jamal-Hanjani, M.; Veeriah, S.; Quesne, J.L.; Swanton, C.; Yuan, Y. Deconvolving convolution neural network for cell detection. In Proceedings of the IEEE 16th International Symposium on Biomedical Imaging, Venice, Italy, 8–11 April 2018.

9.   Shelhamer, E.; Long, J.; Darrell, T. Fully Convolutional Networks for Semantic Segmentation. *IEEE Trans. Pattern Anal. Mach. Intell.* **2017**, *39*, 640–651. [CrossRef]

10.   Qi, C.R.; Su, H.; Mo, K.; Guibas, L.J. PointNet: Deep learning on point sets for 3D classification and segmentation. In Proceedings of the 30th IEEE Conference Computer Vision and Pattern Recognition, CVPR, Honolulu, HI, USA, 21–26 July 2017; pp. 77–85.

11.   Ren, H.; El-Khamy, M.; Lee, J. CT-SRCNN: Cascade Trained and Trimmed Deep Convolutional Neural Networks for Image Super Resolution. In Proceedings of the IEEE Winter Conference Applications Computer Vision, WACV, Lake Tahoe, NV, USA, 12–15 March 2018; pp. 1423–1431.

12.   Dong, C.; Loy, C.C.; He, K.; Tang, X. Image super-resolution using deep convolutional networks. *IEEE Trans. Pattern Anal. Mach. Intell.* **2016**, *38*, 295–307. [CrossRef]

13.   Ledig, C.; Theis, L.; Huszár, F.; Caballero, J.; Cunningham, A.; Acosta, A.; Aitken, A.P.; Tejani, A.; Totz, J.; Wang, Z.; et al. Photo-realistic single image super-resolution using a generative adversarial network. *CVPR* **2017**, *2*, 4.

14.   Dong, C.; Loy, C.C.; Tang, X. Accelerating the super-resolution convolutional neural network. *Lect. Notes Comput. Sci.* **2016**, *9906*, 391–407.

15.   Liu, C.; Wang, L.G.; Nehorai, A.; Li, L.; Cui, T.J.; Teixeira, F.L. DeepNIS: Deep neural network for nonlinear electromagnetic inverse scattering. *IEEE Trans. Antennas Propag.* **2018**, *67*, 1819–1825. [CrossRef]

16.   Qin, Y.; Gao, J.; Wang, H.; Deng, B.; Li, X. Enhanced radar imaging using a complex-valued convolutional neural network. *IEEE Geosci. Remote Sens. Lett.* **2018**, *16*, 35–39.

17.   Liu, T.; Su, Y.; Huang, C. Inversion of ground penetrating radar data based on neural networks. *Remote Sens.* **2018**, *10*, 730. [CrossRef]

18.   Dai, Y.; Jin, T.; Song, Y.; Du, H.; Zhao, D. SRCNN-Based enhanced imaging for low frequency radar. *Prog. Electromagn. Res. Symp.* **2018**, *2018*, 366–370.

19.   Bahdanau, D.; Kyunghyun, C.; Bengio, Y. Neural machine translation by jointly learning to align and translate. *arXiv* **2014**, arXiv:1409.0473.

20.   Du, Y.; Yuan, C.; Li, B.; Zhao, L.; Li, Y.; Hu, W. Interaction-aware spatio-temporal pyramid attention networks for action classification. *Lect. Notes Comput. Sci.* **2018**, *11220*, 388–404.

21.   Gilmer, J.; Schoenholz, S.S.; Riley, P.F.; Vinyals, O.; Dahl, G.E. Neural message passing for quantum chemistry. *arXiv* **2017**, arXiv:1704.01212.

22.   Defferrard, M.; Bresson, X.; Vandergheynst, P. Convolutional neural networks on graphs with fast localized spectral filtering. *Adv. Neural Inf. Process. Syst.* **2016**, 3844–3852.

23.   Burkholder, R.J.; Browne, K.E. Coherence factor enhancement of through-wall radar images. *IEEE Antennas Wirel. Propag. Lett.* **2010**, *9*, 842–845. [CrossRef]

24.   Tu, X.; Zhu, G.; Hu, X.; Huang, X. Grating lobe suppression in sparse array-based ultrawideband through-wall imaging radar. *IEEE Antennas Wirel. Propag. Lett.* **2016**, *15*, 1020–1023. [CrossRef]

25.   Wang, X.; Li, G.; Liu, Y.; Amin, M.G. Two-level block matching pursuit for polarimetric through-wall radar imaging. *IEEE Trans. Geosci. Remote Sens.* **2018**, *56*, 1533–1545. [CrossRef]

26.   Amin, M.G.; Ahmad, F. Compressive sensing for through-the-wall radar imaging. *J. Electron. Imaging* **2013**, *22*, 231–250. [CrossRef]

27.   Zhu, X.; He, F.; Ye, F.; Dong, Z.; Wu, M. Sidelobe suppression with resolution maintenance for SAR images via sparse representation. *Sensors* **2018**, *18*, 1589. [CrossRef] [PubMed]

28.   Zhao, D.; Jin, T.; Dai, Y.; Song, Y.; Su, X. A three-dimensional enhanced imaging method on human body for ultra-wideband multiple-input multiple-output radar. *Electron* **2018**, *7*, 101. [CrossRef]

29.   Zeiler, M.D.; Taylor, G.W.; Fergus, R. Adaptive deconvolutional networks for mid and high level feature learning. In Proceedings of the IEEE International Conference on Computer Vision, Barcelona, Spain, 6–13 November 2011; pp. 2018–2025.

30.   Zhao, J.J.; Mathieu, M.; Goroshin, R.; Lecun, Y.J. Stacked what-where auto-encoders. *arXiv* **2015**, arXiv:1506.02351.

31.   Zefler, M.D.; Fergus, R. Visualizing and understanding convolutional networks. *arXiv* **2013**, arXiv:1311.2901v3.

32. Jiang, Y.; Qin, Y.; Wang, H.; Deng, B.; Liu, K.; Cheng, B. A side-lobe suppression method based on coherence factor for terahertz array imaging. *IEEE Access* **2018**, *6*, 5584–5588. [CrossRef]
33. Li, P.C.; Li, M.L. Adaptive imaging using the generalized coherence factor. *IEEE Trans. Ultrason. Ferroelectr. Freq. Control* **2003**, *50*, 128–141.

 *remote sensing*

Article

Wavelet Scattering Network-Based Machine Learning for Ground Penetrating Radar Imaging: Application in Pipeline Identification

Yang Jin [1],* and Yunling Duan [2]

[1] Department of Structural Engineering, Delft University of Technology, Postbus 5, 2600 AA Delft, The Netherlands
[2] Department of Hydraulic Engineering, Tsinghua University, Beijing 100084, China; yduan@tsinghua.edu.cn
* Correspondence: J.Jin-3@tudelft.nl; Tel.: +31-06-2004-8305

Received: 23 September 2020; Accepted: 5 November 2020; Published: 7 November 2020

Abstract: Automatic and efficient ground penetrating radar (GPR) data analysis remains a bottleneck, especially restricting applications in real-time monitoring systems. Deep learning approaches have good practice in automatic object identification, but their intensive data requirement has reduced their applicability. This paper developed a machine learning framework based on wavelet scattering networks to analyze GPR data for subsurface pipeline identification. Wavelet scattering network is functionally equivalent to convolutional neural networks, and its null-parameter property is intended for non-intensive datasets. A double-channel framework is designed with wavelet scattering networks followed by support vector machines to determine the existence of pipelines on vertical and horizontal traces separately. Classification accuracy rates arrive around 98% and 95% for datasets without and with noises, respectively, as well as 97% for considering surface roughness. Pipeline locations and diameters are convenient to determine from the reconstructed profiles of both simulated and practical GPR signals. However, the results of 5 cm pipelines are sensitive to noises. Nonetheless, the developed machine learning approach presents promising applicability in subsurface pipeline identification.

Keywords: ground penetrating radar; wavelet scattering network; machine learning; support vector machine; pipeline identification

1. Introduction

Ground penetrating radar (GPR) is a well established non-destructive technology for the geophysical investigation of subterranean structures and substances by propagating electromagnetic waves. It remotely monitors the underground conditions as an echo listener and then produces highly-correlated signal profiles, e.g., for applications in infrastructure maintenance [1], archaeology surveys [2], and stratum investigation [3]. Detecting finite objects, namely, pipelines [4], land mines [5], and voids [6], from the noisy background is among common utilization; these targets generate distinct hyperbolas in the recorded profiles. Location accuracy is significant in object identification as errors reach around 30 cm in mapping underground utility infrastructure in real urban environments (summarized by Šarlah et al. [7]) and improving location accuracy has aroused research concerns recently. While absorbing new technologies these years for evolution, GPR has enhanced automaticity and efficiency in broad area measurement. By contrast, automatic and time-efficient data analysis remains a bottleneck especially restricting applications in online monitoring systems [8]. Deconvolution and filtering results depend absolutely on artificial processing experience and interpretation. Various developed inversion techniques based on Maxwell's equations to migrate scattered signals back requires much integral computation [9,10], improving analysis accuracy but increasing computation burden. Therefore, automatic and desirable data processing methods have been subjected to meticulous investigations.

Artificial intelligence provides computational learning approaches for unmanned data processing. Different from static programs under explicit human instructions, machine learning algorithms acquire inner statistical properties according to sampled joint distribution [11] while ignoring the prior knowledge that is indispensable for classic GPR analysis, and then automatically extract targeted characteristics from radargrams. Various novel machine learning frameworks have contributed to the promotion of object identification performance, e.g., applying the Viola–Jones Algorithm in pattern recognition approach for locating reflection hyperbolas [12], utilizing one computer vision technique, namely, histogram of oriented gradients for landmine detection [13], and adapting region-based convolutional neural networks (CNNs) for subterranean objects recognition [14]. Deep learning including CNN is an increasingly glorious branch of machine learning, against extremely complicated problems but preserving promising learning accuracy. It constructs a considerable size of networks that have the versatile capability to learn GPR signal features. However, the favorable learning reliability relies on multitudinous training data that are difficult to collect all in GPR measurement. For instance, 84 original radargrams in [15] are insufficient for training the AlexNet CNN model containing 60 million parameters. Either extending the datasets over thousands by translation and scaling [15,16] or generating radargrams by numerical simulation [14,17,18] can complement data shortage, but these sources bring doubts on training excessive duplicates or time-consuming problems in data production, respectively. Therefore, a deep learning equivalent framework intended for non-intensive datasets is necessary to expand GPR-related learning researches.

Mallat [19] proposed a wavelet scattering convolution network with translation and rotation invariant operators based on the wavelet transform in 2012. Structurally similar to deep CNN, it decomposes input signals into multi-layer components, each layer consisting of linear and nonlinear operations. Specially, the convolution kernels are predefined by the chosen wavelet, similar to band-pass filters extracting characteristics with physical meanings, and activation functions are replaced by modulus operators that solve the covariant issue of the wavelet transform. As a result of such a model structure, the wavelet scattering network contains no parameters while keeping complicated, which indicates that data volume reduction and model reliability are available simultaneously. It has practically outperformed deep CNN in handwritten digits recognition and texture classification with an accuracy rate of up to 99.7% [20,21], and in synthetic aperture radar target recognition with an accuracy rate of up to 97.63% [22]. However, modulus operation in the wavelet scattering network eliminates location information, and, to the best of our knowledge, no existing researches apply it in automatic objects detection or GPR radargram interpretation.

In this paper, a machine learning framework consisting of wavelet scattering networks and support vector machine (SVM) is investigated for subterranean pipelines identification from GPR profiles. The 2D dataset is decomposed into 1D vertical signals and horizontal signals separately, intending to acquire pipeline coordinates by classification. Both numerically simulated and practical data are utilized to extend the applicability of the learning model. The remainder of this paper is organized as follows. Section 2 describes the data and methodology. Section 3 evaluates the applicability of the proposed learning framework in pipeline identification. Section 4 discusses the results and future improvement. Section 5 concludes the paper.

2. Materials and Methods

2.1. Data Description

To investigate the applicability of machine learning in pipeline identification, we use numerically simulated and practical datasets in this paper.

While considerable GPR data with similar subterranean attributes are required in the learning procedure to acquire internal statistical probability distribution but difficult to collect experimentally, numerical simulation provides an unlimited data production approach. An open source software, "gprMax" [23], is a favorable option to generate GPR profiles. It numerically solves Maxwell's equations by the Finite-Difference Time-Domain method [24] and offers advanced subterranean modeling, succeeding in both academic and industrial applications [25–27]. In this research, we simulated a stochastic number (range: 0–16) of cylinder pipelines (random diameters ranging from 5 to 40 cm) buried randomly inside a 2 m × 1 m subsurface domain and then produced 40 such GPR profiles (a non-intensive dataset) with downsampled data resolution 400 × 448. The underground soil has certain attributes, with the relative permittivity of 8, the conductivity of 0.02 S/m, the relative permeability of 1 and the magnetic loss of 0. Pipelines are simulated as perfect electric conductors. The Ricker waveform with the central frequency 600 MHz is created to simulate the GPR antenna for subsurface detection. Example domains with pre-buried pipelines and corresponding GPR profiles are illustrated in Figure 1.

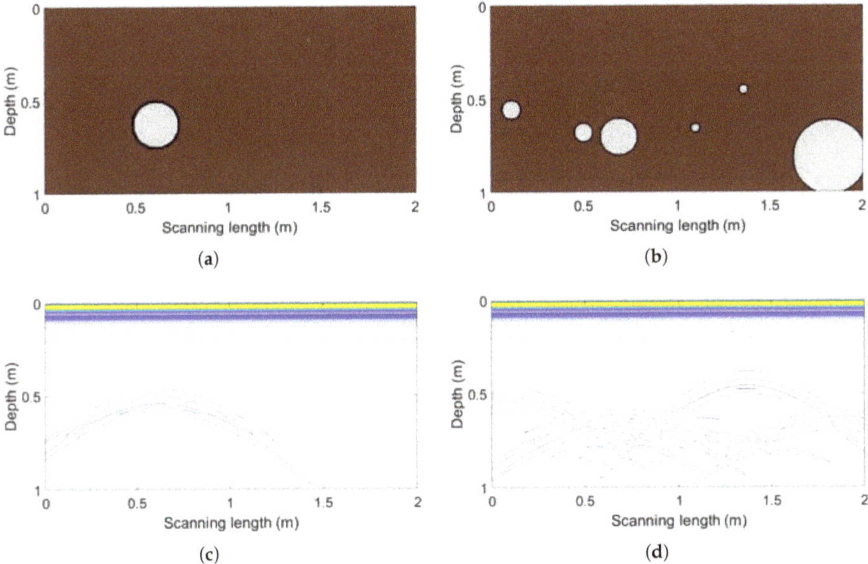

Figure 1. (**a**) Example subsurface section with only one pre-buried pipeline, where the brown color represents the soil and the white color represents the pipeline. (**b**) Example subsurface section with multiple pre-buried pipelines. (**c**) GPR signal profile corresponding to the subsurface section in panel (a). (**d**) GPR signal profile corresponding to the subsurface section in panel (b).

However, subterranean formation and measurement environment are generally idealized in numerical simulation, which overlooks the system noise, widens transmission angles of electromagnetic waves, and simplifies the material attributes. To extend the applicability of machine learning to a practical phase, three GPR profiles (one provided by the authors of [28]) from field measurements are adopted as the ultimate test set, seen in Figure 2. The first subsurface profile (Figure 2a) contains two separate pre-buried pipelines with central depths of 1 m, diameters of 0.32 m, and their interval is 2.5 m. The central frequency of the GPR antenna is 700 MHz. The second profile contains 5 × 2 double-layer concrete cylinders with central depths of 0.3 m & 0.45 m, diameters of 0.1 m and horizontal intervals of 0.4 m. The central frequency of the GPR antenna is 500 MHz. The third profile contains 3 pre-buried pipelines inside limestone areas (detailed information seen in [28]), and the detection frequency is 250 MHz.

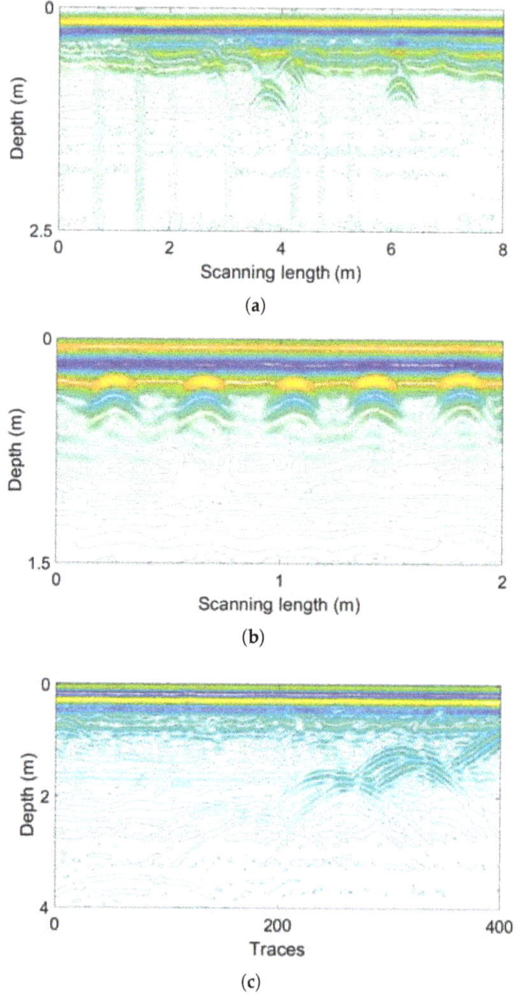

Figure 2. Field GPR profiles of (**a**) two separate pre-buried pipelines, (**b**) double-layer concrete cylinders, and (**c**) three distributed pipelines [28].

2.2. Pre-Processing

For machine understanding better the targeted characteristics, two preliminary procedures, de-"wow" and amplitude gain, are considered to preprocess the original data.

The "wow" phenomenon refers to the unavoidable occurrence of low-frequency noise components in each GPR trace, predominantly arising from inductive coupling effects or electronic saturation along the ground–air interface [29,30]. This results in zero-offset signals and affects the subsequent procedure as the low-frequency component will obscure real signals after amplitude gain. A simple but practical approach to eliminate the "wow" effects is subtracting the mean signal amplitude.

$$x^{\star}(t) = x'(t) - \frac{1}{n}\sum_{i=1}^{n} x'(t_i) \tag{1}$$

where $x'(t)$ is the original signal trace, $x^\star(t)$ represents the signal after de-"wow", and t_i $(i = 1, 2, \ldots, n)$ is the sampling time. Figure 3 shows an example of de-"wow" results.

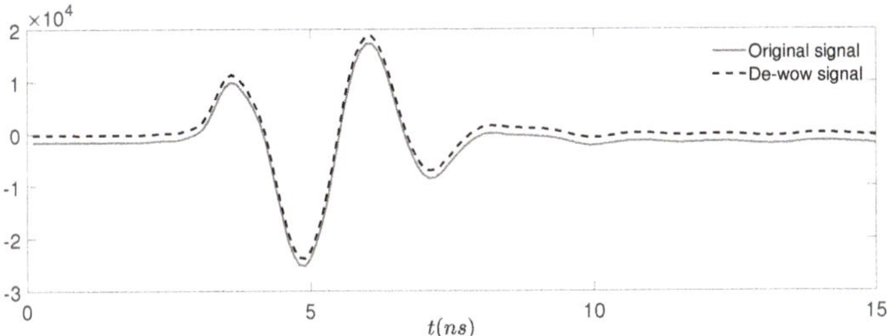

Figure 3. De-"wow" results of a representative GPR trace.

In the two-way propagation of electromagnetic waves, the energy attenuates with time due to dielectric loss and geometrical spreading consumption (heavy scattering and refraction at interfaces) [31]. Amplitude gain is indispensable for enhancing the later arrived signals to a distinguishable degree. Considering the aforementioned attenuation, exponential and linear functions are combined for amplitude gain operation.

$$F(t) = e^{2\alpha(t-t_0)} + 2\beta(t - t_0)$$
$$x(t) = F(t)x^\star(t) \tag{2}$$

where $x(t)$ represents signals after preprocessing, α is the exponential attenuation coefficient corresponding to dielectric loss, β is the linear attenuation coefficient related to geometrical spreading consumption, and t_0 is the corrected time at the ground–air interface.

The choices of α & β are based on experience. In this paper, we determined $\alpha = 0.0156\frac{N-N_0}{t-t_0}$ & $\beta = 0.12\frac{N-N_0}{t-t_0}$ for simulated GPR signals, $\alpha = 0.0112\frac{N-N_0}{t-t_0}$ & $\beta = 0$ for the first two measurement profiles, and $\alpha = 0.006\frac{N-N_0}{t-t_0}$ & $\beta = 0.05\frac{N-N_0}{t-t_0}$ for the third measurement profile, where N represents the signal point number corresponding to the time t, and N_0 is the signal point number corresponding to the time t_0. Example data profiles after preprocessing are given in Figure 4.

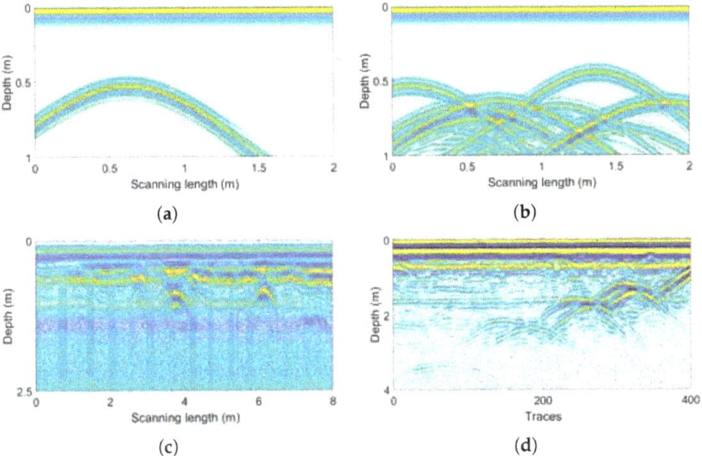

Figure 4. Example data profiles after preprocessing of (**a**) Figure 1c, (**b**) Figure 1d, (**c**) Figure 2a, and (**d**) Figure 2c.

2.3. Wavelet Scattering Network

Wavelet scattering is a null-parameter convolution network originally proposed by Mallat [19] for translation and rotation invariant characterization with specific wavelet approaches. Wavelet transform, which is both a mode recognition and decomposition approach, provides the basic theory for the scattering network to extract either evident or invisible data features. The physical meaning of the wavelet transform is to calculate the joint energy spectrum of signals in the frequency-time domain and thereby to identify both frequency and time information of the distinct modes [32]. The procedure of the wavelet decomposition is complete and reversible, which means no signal features would be lost. That is different from CNN, which is adjustable during training to preserve targeted signal features only. Wavelet scattering networks utilize the wavelet group and scaling functions as convolution kernels to filter signals in preset orientations and bandwidths, while CNN trains undetermined convolution kernels for filtering. This significant factor determines that wavelet scattering networks contain no parameters while performing functionally equivalent to CNN.

A wavelet function group is available by dilating and rotating the mother wavelet:

$$\psi_{2^j r}(t) = 2^{dj}\psi(2^j r^{-1}t) \tag{3}$$

where mother wavelet $\psi \in \mathbf{L}^2(R^d)$, 2^{-j} represents the dilation rate, and r is the rotation coefficient. Wavelet transform decomposes the original signal $x(t)$ through the band-pass filter ψ_λ by convolution calculation.

$$W_\lambda x = x \otimes \psi_\lambda = \int x(\tau)\psi_\lambda(t-\tau)d\tau \tag{4}$$

where $\lambda = 2^j r$ to simplify notation, W_λ is the wavelet transform operator, and $\otimes$ represents convolution calculation. At any transform scale 2^J, signal components with frequency $2^j > 2^{-J}$ are reserved. The low-frequency component not included is decomposed through the scaling functions within the space proportional to 2^J,

$$A_J x = x \otimes \phi_J \tag{5}$$

$$\phi_J(t) = 2^{-dJ}\phi(2^{-J}t) \tag{6}$$

where A_J is the scaling operator, ϕ_J is the scaling function at the scale 2^J, while ϕ is the scaling function at the original scale.

Although wavelet transform can map local signal features, the convolution calculation is translation covariant. It will distinguish similar characteristics at different locations into separate categories, thereby increasing the learning complexity. Translation invariance is significant for classification since the horizontal and vertical movement of the same objects should result in few classification mistakes [33]. Mallat [19] demonstrated that the integration of the modulus $| x \otimes \psi_\lambda |$ is translation invariant and introduced a path-sorted iteration operator on the modulus to create the scattering propagator.

$$U_\lambda x = | W_\lambda x | = | x \otimes \psi_\lambda | \tag{7}$$

$$U[p] = U_{\lambda_m}...U_{\lambda_2}U_{\lambda_1} \tag{8}$$

$$U[p]x = ||| x \otimes \psi_{\lambda_1} | \otimes \psi_{\lambda_2} | ... \otimes \psi_{\lambda_m} | \tag{9}$$

where U_{λ_i} $(i = 1, 2, ..., m)$ is the modulus operator at the ith scattering stages, and $U[p]$ is the scattering propagator. Similar to the wavelet transform, a windowed scattering transform is introduced to extend the frequency scale.

$$S_J[p]x = U[p]x \otimes \phi_J = \int U[p]x(\tau)\phi_J(t-\tau)d\tau \tag{10}$$

$$S_J[p]x = ||| x \otimes \psi_{\lambda_1} | \otimes \psi_{\lambda_2} | ... \otimes \psi_{\lambda_m} | \otimes \phi_J \tag{11}$$

The operation of $U[p]x$ and $S_J[p]x$ is contractive and stable. Therefore, the wavelet scattering network is constructed by continuously calculating the convolution results of Equations (9) and (11), as shown in Figure 5.

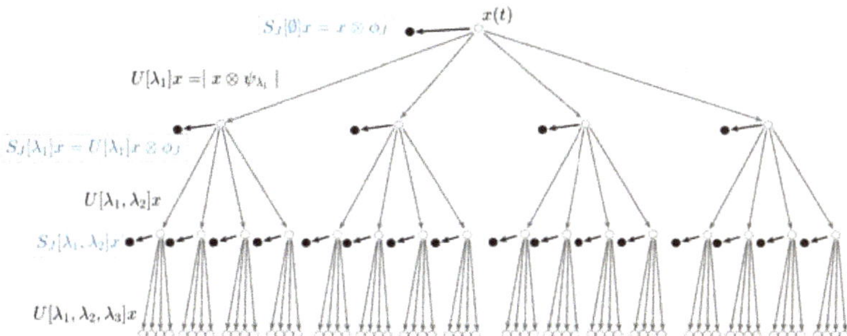

Figure 5. A three layer wavelet scattering network. The operator U_{λ_1} is applied to the original signal x to calculate each $U[\lambda_1]x$ and output $S_J[\varnothing]x$, where $\varnothing$ represents an empty set. Then, the operator U_{λ_2} is applied to each previous layer $U[\lambda_1]x$ to calculate all $U[\lambda_1, \lambda_2]x$ and output $S_J[\lambda_1]x$. This scattering process is operated iteratively to obtain all convolution results.

The wavelet group and scaling functions compose the convolution kernels of the wavelet scattering network, and the modulus operator works as the activation function (seen in the network structure, Figure 5). Morlet wavelet assisted by the Gaussian window is utilized in this research for scattering propagation, as expressed in Equation (12) and shown in Figure 6.

$$\psi(t) = K_{\sigma_t} e^{-\frac{t^2}{2\sigma_t^2}} e^{2\pi i f t} \tag{12}$$

where K_{σ_t} is the normalization constant, σ_t represents the wavelet duration, i is the imaginary unit, and f is proportional to the central frequency.

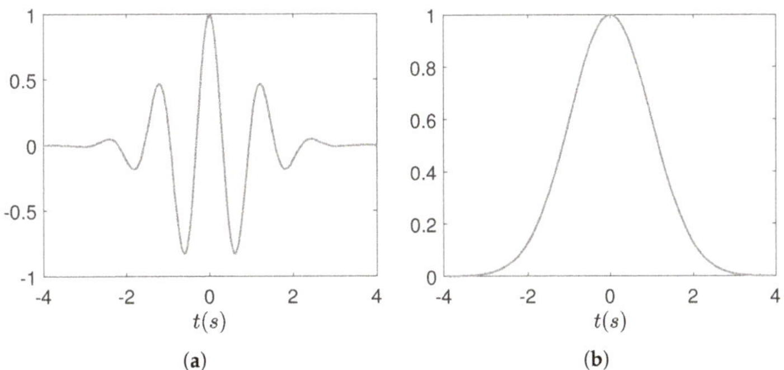

Figure 6. (a) Real component of example Morlet wavelet with parameters $K_{\sigma_t} = 1$, $\sigma_t = 1$, and $2\pi f = 5$. (b) Example Gaussian window as scale function $\phi(t)$.

2.4. Support Vector Machine

Support vector machine (SVM) is a supervised classification approach rooted in the statistical learning theory, mathematically solving the dual optimization problem based on structural risk minimization [34,35]. In linear classification, where the classification model is a linear function of input parameters, data points are divided into different categories by a hyperplane. The optimal classification hyperplane to separate data points is achieved at point-plane distance maximization.

For nonlinear classification, a hypercurved surface is required to separate the data points but difficult in the calculation. Kernel functions are adopted to map the input domain onto the high-dimensional Hilbert space, resulting in problem transformation into linear classification. Therefore, choosing an appropriate kernel function determines the applicability of the SVM classifier. The radial basis function kernel is the most common one with good applications in practical problems and is utilized in our research.

In binary classification, given input training set $T = \{(x_1, y_1), (x_2, y_2), ..., (x_N, y_N)\}$ (where $x_i \in \mathbf{R}^n$, $y_i \in \{-1, 1\}$, $i = 1, 2, ..., N$), SVM classifier solves the following dual optimization problem.

$$\min_{\gamma} \frac{1}{2}\gamma^T Q\gamma - e^T\gamma$$
$$s.t. \quad y^T\gamma = 0 \tag{13}$$
$$0 \leq \gamma_i \leq C, \ i = 1, 2, .., N$$

where γ is the classification hyperplane coefficient vector, Q is a $N \times N$ positive semidefinite matrix with $Q_{i,j} = y_i y_j K(x_i, x_j)$, $K(x_i, x_j) = h(x_i)h(x_j)$ is the kernel function with $h(x_i)$ mapping the input domain onto the high-dimensional Hilbert space, and C is the upper-boundary parameter of γ. Once achieving the optimal solution $\gamma^\star = (\gamma_1^\star, \gamma_2^\star, ..., \gamma_N^\star)$, we can express the constant parameter of the hyperplane using any $\gamma_i^\star$.

$$b^\star = y_j - \sum_{i=1}^{N} \gamma_i^\star y_i K(x_i, x_j) \tag{14}$$

Therefore, the output decision function for any given input x is

$$f(x) = sgn(\sum_{i=1}^{N} \gamma_i^\star y_i K(x_i, x) + b^\star) \tag{15}$$

where sgn represents the sign function $sgn(x) = \frac{x}{|x|}$. In this research, the developed python module "scikit-learn" [36] is utilized for coding the nonlinear SVM classifier.

2.5. Learning Framework

As the general framework that directly classifies 2D GPR profiles overlooks the location and magnitude information [37,38], a specific architecture (seen in Figure 7) is designed for accurately identifying the positions of covered pipelines. The training set is decomposed into two datasets consisting of 1D vertical signals and horizontal signals respectively, followed by separate learning procedures where a multi-layer wavelet scattering network connected with SVM classifier is established. This double-channel framework works to determine the coordinates by recognizing the existence of pipelines on the single signal trace vertically and horizontally. After learning procedures, GPR profiles with visible pipeline locations are reconstructed utilizing data at output coordinates.

A brief scheme of profile reconstruction is provided by Figure 8, where we only illustrate two horizontal trace groups and two vertical trace groups. The blue traces are classified as "negative" traces (non-existence of pipelines) while the red traces are classified as "positive" traces (existence of pipelines) by the two SVM classifiers. The four trace groups form four intersections, and only the intersection D of the two "positive" groups $H2$ & $V2$ is identified as a "positive" section. Therefore, we reconstruct the ultimate profile by the signals inside all "positive" sections.

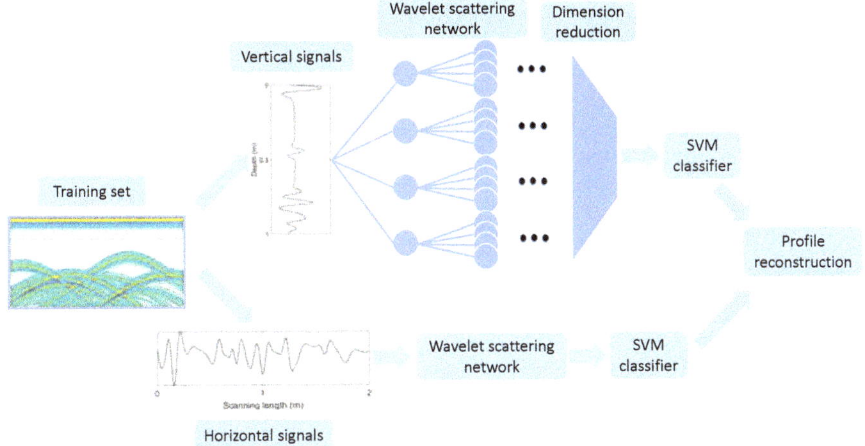

Figure 7. The machine learning architecture for pipeline identification.

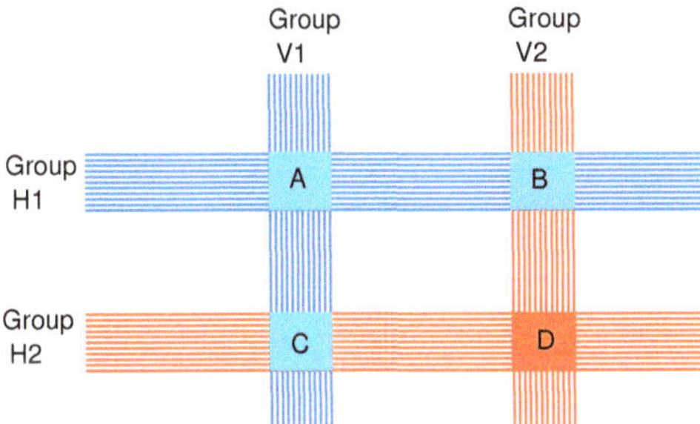

Figure 8. A brief scheme of profile reconstruction.

3. Results

3.1. Simulated Signal Results

First, our proposed learning framework is investigated by the 40 numerically produced GPR B-scan profiles, among which 32 and 8 profiles are randomly separated for the training set and validation set, respectively. Each signal profile is rescaled between 0 and 1 to ensure the learning model suitable for further practical applications. Although the input dataset is small, it indeed contains 40×400 vertical traces and 40×448 horizontal signals for the "wavelet scattering network +SVM" learning procedure, which is sufficient to achieve reasonable results.

A comparison between an independent SVM and our proposed framework at the classification stage is operated to visibly evaluate the learning results. The double-channel classifiers can determine whether pipelines exist on the vertical traces or the horizontal signals, with their learning accuracy shown in Tables 1 and 2. The classifiers containing wavelet scattering networks remarkably outperform the independent SVM, improving validation accuracy to 97.94% and 98.41%. False rates drop below 2.1%, which means that average of 8 vertical traces or horizontal signals per profile are misidentified, and such insignificant errors confuse little of further pipeline recognition. The convincing correct classification rates indicate the feature extraction efficiency of wavelet scattering networks. By contrast, the independent SVM model fails to capture the signal characteristics arising from pipeline existence, as reflected by the unacceptable false rate of over 20%. The false positive rates of classifying both vertical and horizontal signals crumble into an eyesore, as 16.87% & 9.20% of signals not containing pipelines are misidentified. By this comparison, thus the visible results demonstrate the promising eligibility of the proposed framework in identifying pipeline presence.

Table 1. Accuracy and confusion results of vertical traces with or without wavelet scattering networks in the machine learning model.

Models	Training Accuracy	Validation		
		Accuracy	False Positive	False Negative
WaveScat + SVM	99.73%	97.94%	0.91%	1.15%
SVM	78.59%	78.44%	16.87%	4.69%

Table 2. Accuracy and confusion results of horizontal signals with or without wavelet scattering networks in the machine learning model.

Models	Training Accuracy	Validation		
		Accuracy	False Positive	False Negative
WaveScat + SVM	99.70%	98.41%	0.53%	1.06%
SVM	72.10%	72.38%	9.20%	18.42%

The further procedure concentrates on automatically reconstructing GPR profiles to present distinguishable pipeline locations and sizes. Data points repeatedly labeled "positive" (represent pipeline existence) on both the vertical trace and the horizontal signal preserve their values while others are eliminated. The rescaled data less than 0.6 are also eliminated for better locating the upper pipeline interfaces. The results of single-pipeline profiles for validation are presented in Figure 9, where upper panels are input profiles (with pipelines manually marked by dash circles), middle panels are reconstructed by the proposed learning framework ("positive" areas in white and "negative" areas in gray) and lower panels are reconstructed after the single SVM. Although length coordinates of pipelines are convenient to determine from hyperbolic patterns (Figure 9a,b), reliable depth coordinates and diameters are unavailable. In contrast, the pipelines in Figure 9c,d are tangent to the "positive" rectangle with upper interface signals inside, thus determining coordinates by upper interfaces and diameters by minimum rectangle sides. The learning results of central coordinates and diameters are ((0.62 m, 0.636 m), 0.245 m) in Figure 9c and ((0.28 m, 0.669 m), 0.346 m) in Figure 9d, respectively, corresponding brilliantly with the actual sizes (seen in caption of Figure 9, errors within 1 cm). Two insignificant misclassified "positive" areas appear in Figure 9c but no upper interface occurs inside, affecting little on pipeline identification. Therefore, single-pipeline profiles are favorably reconstructed by the proposed learning framework. By comparison, although the upper interface of the pipeline is visible in Figure 9e, the identified coordinates and diameter are ((0.62 m, 0.64 m), 0.14 m) with a 44% diameter error. The pre-buried pipeline is identified as two small pipelines in Figure 9f, which means the independent SVM performs undesirable in reconstructing the profiles. The accuracy of classification determines the accuracy of profile reconstruction and location identification.

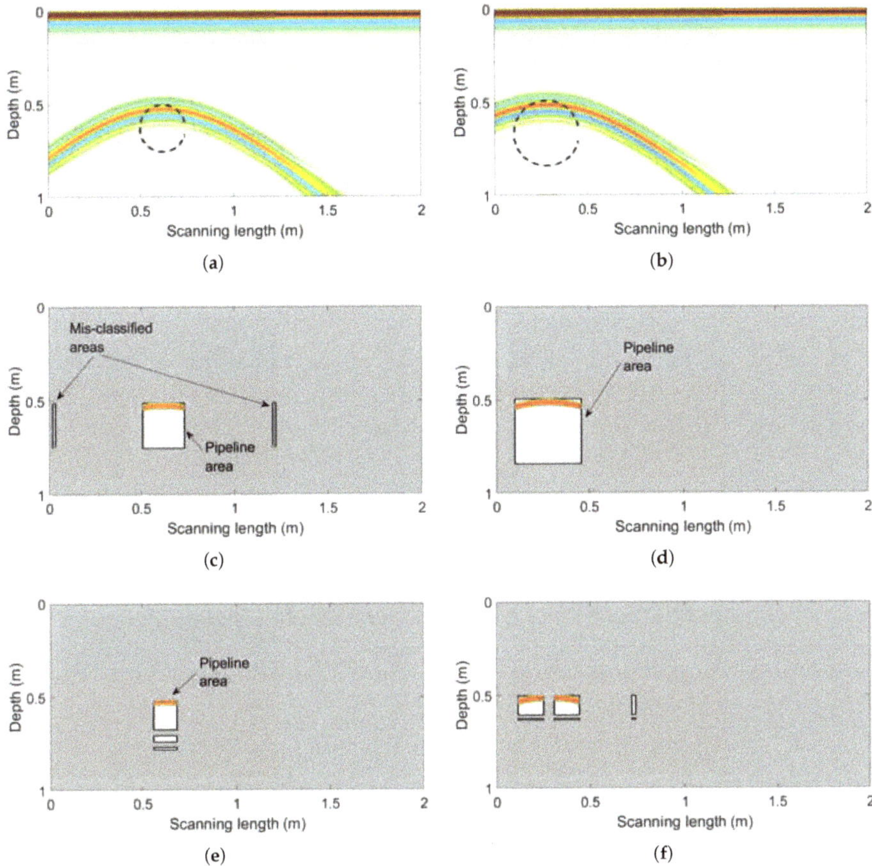

Figure 9. (**a**) The example input profile with a single pipeline (central coordinates (0.615 m, 0.63 m), diameter 0.25 m). (**b**) Another example input profile with a single pipeline (central coordinates (0.275 m, 0.667 m), diameter 0.35 m). (**c**) The reconstructed profile of panel (a) by the proposed learning framework. (**d**) The reconstructed profile of panel (b) by the proposed learning framework. (**e**) The reconstructed profile of panel (a) after the independent SVM for comparison. (**f**) The reconstructed profile of panel (b) after the independent SVM for comparison.

Interpreting reconstructed profiles of multiple pre-buried pipelines becomes complicated since "pseudo-positive" areas appear thus interrupting direct identification, e.g., two "positive" vertical traces and two "positive" horizontal signals can generate four "positive" intersections (seen in Figure 10c). Nonetheless, the intended approach is still to distinguish upper interfaces of pipelines according to their properties that they are approximately parallel to horizontal lines, and then determine the sizes by external "positive" sections. Figure 10a,c shows the learning results of two pipelines manually marked in dashed circles. Three visible hyperbolic patterns occur in the input profile, confusing the location determination of the two pipelines, and diameters are difficult to determine. Contrarily, although four "positive" sections except the two insignificant misclassified areas are reserved in learning results (Figure 10c), the reconstructed profile only contains two distinguishable upper interfaces, further determining coordinates and diameters as aforementioned. The output characterized values of two pipelines are ((1.365 m, 0.263 m), 0.145 m) & ((0.88 m, 0.831 m), 0.395 m), conforming promisingly to the pre-set sizes (seen in caption of Figure 10, errors within 1 cm), which indicates the learning approach is suitable for recognizing two pipelines. However, in the results after the independent SVM classifier (seen in Figure 10e), the upper pipeline is identified but the other pipeline is nearly invisible. The output coordinates and diameter are ((1.365 m, 0.243 m), 0.27 m) with an unacceptable diameter error of 93%. For multiple-pipeline profiles, the pipelines generate a large number of interference signals that obstruct the immediate recognition, as shown in Figure 10b,d. Numerous non-pipeline-related hyperbolic patterns appear in the input profile, hardly providing any magnitude information. In comparison, the six upper interfaces are convenient to identify inside the 'positive' sections since other non-interface signals are shattered and nonparallel to horizontal lines. The output characterized values of left five pipelines are ((0.125 m, 0.56 m), 0.095 m), ((0.51 m, 0.683 m), 0.095 m), ((0.7 m, 0.710 m), 0.195 m), ((1.105 m, 0.676 m), 0.045 m) & ((1.37 m, 0.464 m), 0.045 m), and height and width of the rightest pipeline are (0.377 m, 0.365 m) (the rightest pipeline has crossed two boundaries), which present brilliant correspondence with actual sizes. The horizontal and the diameter errors are within 1 cm while the vertical error is within 2.5 cm. The proposed machine learning approach achieves promising results in multiple pipeline identification. In comparison, all six pipelines are difficult to identify from the reconstructed profile after the independent SVM (seen in Figure 10f), which indicates the classification accuracy determines the further identification accuracy.

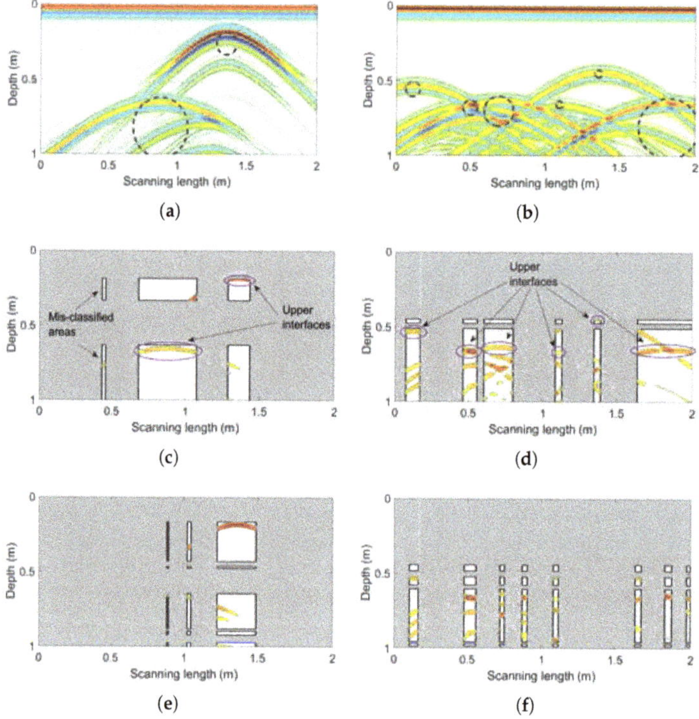

Figure 10. (**a**) The example input profile with two pipelines (central coordinates and diameters ((1.355 m, 0.263 m),0.15 m) and ((0.875 m, 0.823 m), 0.4 m)). (**b**) The example input profile with six pipelines (left five central coordinates and diameters ((0.12 m, 0.56 m), 0.1 m), ((0.505 m, 0.658 m), 0.1 m), ((0.695 m, 0.705 m), 0.2 m), ((1.1 m, 0.66 m), 0.05 m) and ((1.365 m, 0.453 m), 0.05 m), and rightest height and width (0.385 m, 0.37 m)). (**c**) The reconstructed profile of panel (a) by the proposed learning framework. (**d**) The reconstructed profile of panel (b) by the proposed learning framework. (**e**) The reconstructed profile of panel (a) after the independent SVM for comparison. (**f**) The reconstructed profile of panel (b) after the independent SVM for comparison.

3.2. Noise Sensitivity

Second, the noise sensitivity of our learning framework is necessarily evaluated since diverse background noises and shattered subsurface formation affect actual signals from the objects. Random 2D Gaussian white noises with zero mean and standard deviation $\sigma = 0.1$ have supplemented the rescaled GPR profiles as model inputs. A comparison between an independent SVM and our proposed framework at the classification stage is still under consideration to examine learning stability, with the learning accuracy shown in Tables 3 and 4. The classification accuracy of "wavelet scattering network + SVM" has reached 94.81% and 94.98%, noticeably outperforming the independent SVM, and these promising results are adequate for further profile reconstruction. Affected by the complemented noises, the validation accuracy of our framework has decreased around 3%, which indicates noises are worth attention in GPR signal processing. Nonetheless, the accuracy rates maintain around 95%, demonstrating the stability of the learning procedures to the added noises.

Table 3. Accuracy and confusion results of vertical traces (noisy signals) with or without wavelet scattering networks in the machine learning model.

Models	Training Accuracy	Validation		
		Accuracy	False Positive	False Negative
WaveScat + SVM	99.99%	94.81%	2.72%	2.47%
SVM	84.77%	76.25%	14.06%	9.69%

Table 4. Accuracy and confusion results of horizontal signals (noisy signals) with or without wavelet scattering networks in the machine learning model.

Models	Training Accuracy	Validation		
		Accuracy	False Positive	False Negative
WaveScat + SVM	99.76%	94.98%	2.01%	3.01%
SVM	82.42%	67.64%	12.83%	19.53%

The stability of identifying pipeline locations and diameters is further evaluated in reconstructed profiles. Figure 11 illustrates the results of single-pipeline profiles for validation. Although the hyperbolic patterns are still convenient to recognize, they are converted into speckled hyperbolas, increasing difficulty in acquiring reliable depth coordinates and diameters. There are still two insignificant misclassified "positive" areas with no identified upper interfaces, and they occur at different places from Figure 9c. Shattered noise speckles exist inside the pipeline areas, but they hardly affect the determination of pipeline locations and sizes. The learning results of central coordinates and diameters are ((0.6225 m, 0.631 m), 0.24 m) in Figure 11c and ((0.28 m, 0.667 m), 0.345 m) in Figure 11d respectively, corresponding brilliantly (errors within 1 cm) with the actual sizes and the aforementioned non-noise results. Therefore the proposed learning framework is stable in single pipeline identification. In contrast, since the learning accuracy of the independent SVM is below 85%, the corresponding reconstructed profiles fail to present the locations of the pipelines, as shown in Figure 11e,f. The added noises have increased difficulty for the simple learning procedure.

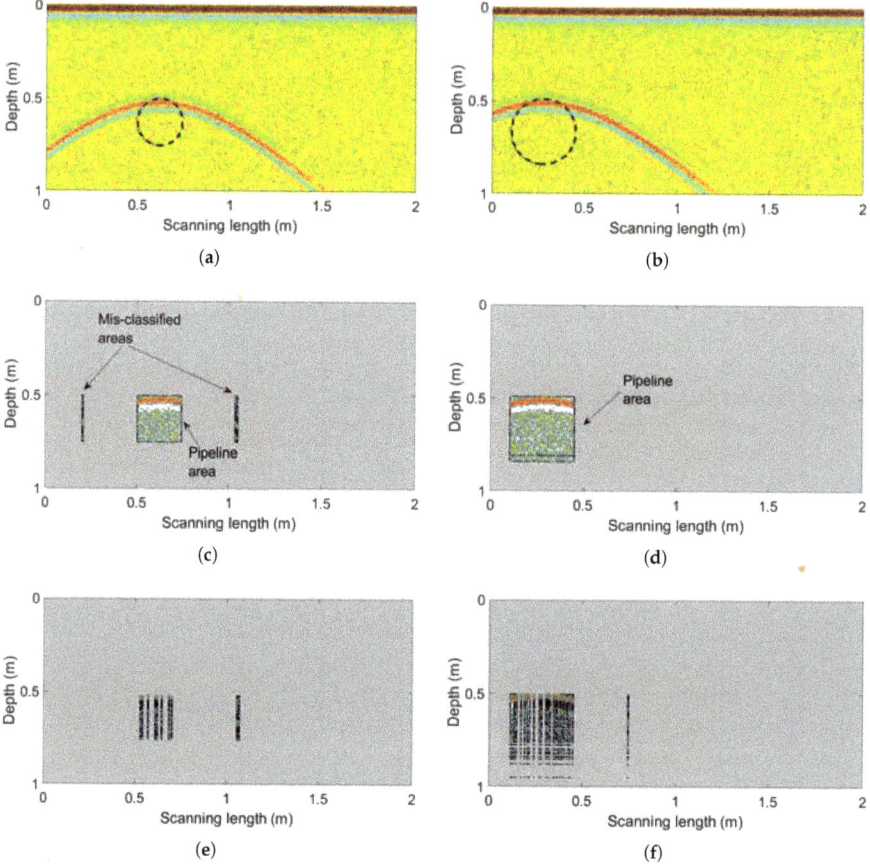

Figure 11. (**a**) The example noisy profile with a single pipeline (seen in Figure 9a). (**b**) Another example noisy profile with a single pipeline (seen in Figure 9b). (**c**) The reconstructed profile of panel (a) by the proposed learning framework. (**d**) The reconstructed profile of panel (b) by the proposed learning framework. (**e**) The reconstructed profile of panel (a) after the independent SVM for comparison. (**f**) The reconstructed profile of panel (b) after the independent SVM for comparison.

Situations of multiple pipeline identification become complicated since noises would ruin the hyperbolic patterns from weak GPR responses. Figure 12c shows the results of the reconstructed double-pipeline profile. Although insignificant, multiple misclassified areas appear and even interrupt the middle-lower "positive" section with "false-negative" identification. Nevertheless, two upper interfaces are distinguishable from the noises, further determining the pipeline characterized values, ((1.36 m, 0.263 m), 0.145 m) & ((0.88 m, 0.839 m), 0.395 m). The horizontal and the diameter errors are within 1 cm while the vertical error is within 2 cm . For multiple-pipeline profiles, the noises interrupt the hyperbolic patterns as shown in Figure 12b,d, which increases difficulty in immediate recognition. Although four upper pipeline interfaces are convenient to identify in Figure 12d, those of the two small pipelines (diameters of 0.05 m) are nearly ruined and hard to determine in the reconstructed profile. Therefore, our machine learning framework is sensitive to noises when overcoming the identification problem with multiple small pipelines (diameters of 0.05 m). Excepting this, the proposed approach presents promising eligibility in GPR profile reconstruction and pipeline identification. In contrast, the comparing results of the independent SVM remain blurring for pipeline recognition since the "positive" traces are shattered-distributed. Only one upper interface in each profile is visible (seen in Figure 12e,f), but the locations and diameters are difficult to determine. The low learning accuracy of the simple learning procedure has affected further profile reconstruction.

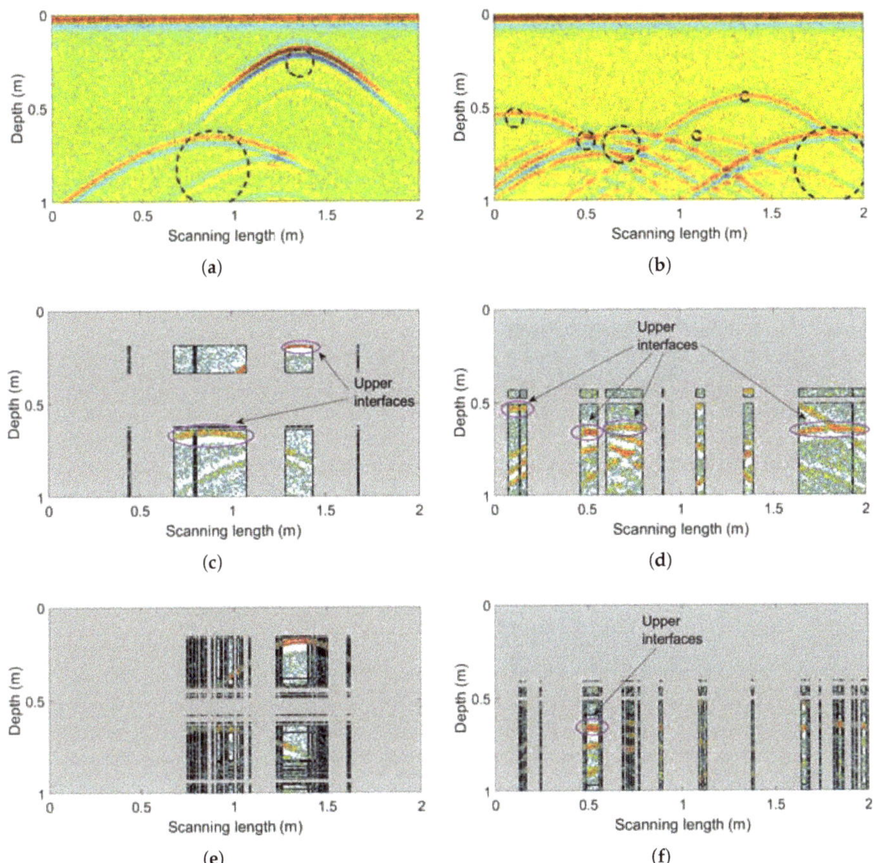

Figure 12. (**a**) The example noisy profile with two pipelines (central coordinates and diameters ((1.355 m, 0.263 m), 0.15 m) & ((0.875 m, 0.823 m), 0.4 m)). (**b**) The example noisy profile with six pipelines (left five central coordinates and diameters ((0.12 m, 0.56 m), 0.1 m), ((0.505 m, 0.658 m), 0.1 m), ((0.695 m, 0.705 m), 0.2 m), ((1.1 m, 0.66 m), 0.05 m) & ((1.365 m, 0.453 m), 0.05 m), and rightest height and width (0.385 m, 0.37 m)). (**c**) The reconstructed profile of panel (a) by the proposed learning framework. (**d**) The reconstructed profile of panel (b) by the proposed learning framework. (**e**) The reconstructed profile of panel (a) after the independent SVM for comparison. (**f**) The reconstructed profile of panel (b) after the independent SVM for comparison.

3.3. Effect of Surface Roughness

Third, the effect of surface roughness is analyzed since the rough surface can result in unstable signals [39,40] affecting further pipeline identification. The surface roughness is complemented to the 40 simulated radargrams by "gprMax", with a maximum amplitude of 5 cm (5% of the depth, 12.5% of the maximum pipeline diameter). A comparison between an independent SVM and our proposed framework is considered to evaluate learning stability, with the learning accuracy shown in Tables 5 and 6. The classification accuracy of "wavelet scattering network +SVM" has reached

96.72% & 97.09%, around 20% larger than that of the independent SVM. Affected by the rough surface, the validation accuracy of our framework has decreased around 1%, which indicates that surface roughness is a significant factor in GPR signal processing. Nonetheless, the high accuracy rates have demonstrated the stability of our learning procedure to surface roughness.

Table 5. Accuracy and confusion results of vertical traces (considering surface roughness) with or without wavelet scattering networks in the machine learning model.

Models	Training Accuracy	Validation		
		Accuracy	False Positive	False Negative
WaveScat + SVM	98.01%	96.72%	2.34%	0.94%
SVM	79.22%	77.81%	6.88%	15.31%

Table 6. Accuracy and confusion results of horizontal signals (considering surface roughness) with or without wavelet scattering networks in the machine learning model.

Models	Training Accuracy	Validation		
		Accuracy	False Positive	False Negative
WaveScat + SVM	98.84%	97.09%	1.98%	0.93%
SVM	78.61%	76.10%	19.00%	4.90%

The effect of surface roughness in identifying pipeline locations and diameters is further investigated in reconstructed profiles. Figure 13 presents the results of single-pipeline profiles for validation. In Figure 13a,b, although the hyperbolic patterns are still convenient to recognize, the signals become fluctuated and noisy due to the rough surface, which increases the difficulty in acquiring pipeline coordinates and diameters. Some insignificant misclassified "positive" areas appear in Figure 13c,d and they will not affect pipeline identification. The pipeline area in Figure 13d is divided into three pieces by the false "negative" areas, but we can identify them as an entire section. The learning results of central coordinates and diameters are ((0.6125 m, 0.6283 m), 0.24 m) in Figure 13c and ((0.2825 m, 0.659 m), 0.355 m) in Figure 13d, respectively, with errors less than 1 cm. The proposed learning framework is applicable in pipeline identification despite the effect of surface roughness. By contrast, the reconstructed profiles after the independent SVM fail to present pipeline areas because of the low learning accuracy, as shown in Figure 13e,f, which indicates that the learning accuracy determines the profile reconstruction accuracy.

The cases of multiple pipeline identification are complicated since the hyperbolic signals have become fluctuated and noisy to ruin each other (shown in Figure 14a,b). In the reconstructed profile (Figure 14c) by our learning framework, the upper surfaces of the two pipelines are distinguishable inside the "positive" sections despite some insignificant misclassified areas. The identified central coordinates and diameters are ((1.36 m, 0.2615 m), 0.145 m) and ((0.88 m, 0.8205 m), 0.4 m), with errors less than 1.5 cm . For multiple-pipeline profiles, the two pipelines with diameters 0.05 m are not distinguishable in the "positive" sections, similar to the results in Section 3.2. This is affected by the surface roughness as the maximum roughness is 0.05 m. Upper surfaces of the residual four pipelines are convenient to recognize, and the horizontal and diameter errors are within 1 cm while the depth errors are within 3 cm. The proposed approach shows promising applicability in multiple pipeline identification. Contrarily, although the upper pipeline in Figure 14e can be identified, the reconstructed profiles after the independent SVM fail to present other pipeline areas, thereby not suitable for pipeline identification. The classification accuracy determines further reconstruction accuracy.

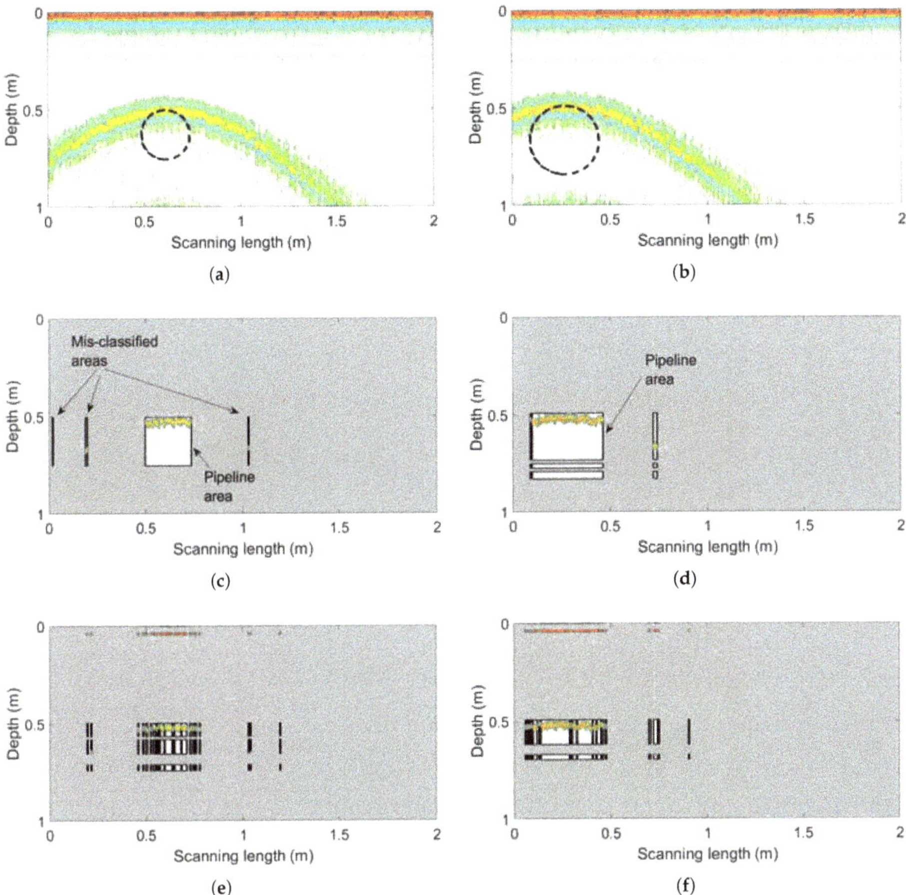

Figure 13. (**a**) The example profile with a single pipeline (seen in Figure 9a) considering surface roughness. (**b**) Another example profile with a single pipeline (seen in Figure 9b) considering surface roughness. (**c**) The reconstructed profile of panel (a) by the proposed learning framework. (**d**) The reconstructed profile of panel (b) by the proposed learning framework. (**e**) The reconstructed profile of panel (a) after the independent SVM for comparison. (**f**) The reconstructed profile of panel (b) after the independent SVM for comparison.

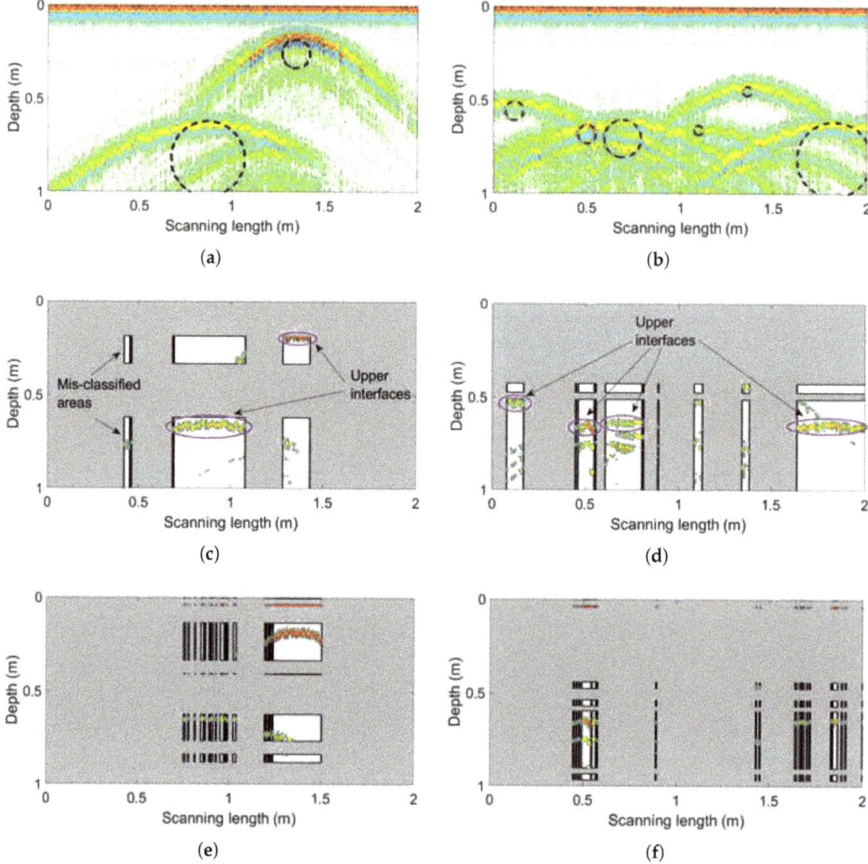

Figure 14. (**a**) The example profile with two pipelines (central coordinates and diameters ((1.355 m, 0.263 m), 0.15 m) & ((0.875 m, 0.823 m), 0.4 m)) considering surface roughness. (**b**) The example profile with six pipelines (left five central coordinates and diameters ((0.12 m, 0.56 m), 0.1 m) , ((0.505 m, 0.658 m), 0.1 m), ((0.695 m, 0.705 m), 0.2 m), ((1.1 m, 0.66 m), 0.05 m) & ((1.365 m, 0.453 m), 0.05 m), and rightest height and width (0.385 m, 0.37 m)) considering surface roughness. (**c**) The reconstructed profile of panel (a) by the proposed learning framework. (**d**) The reconstructed profile of panel (b) by the proposed learning framework. (**e**) The reconstructed profile of panel (a) after the independent SVM for comparison. (**f**) The reconstructed profile of panel (b) after the independent SVM for comparison.

3.4. Applicability in Field Signals

Fourth, three practical radargrams from filed measurements are adopted as the final test set to investigate the applicability in field signals. They are fed into the trained machine learning model from Section 3.2, as the simulated noisy inputs are comparable to practical signals. To convert the three radargrams with the input format, we linearly interpolated in the profiles to 448 × 400 pixels and then rescaled them between 0 to 1 (seen in left panels in Figure 15). After processed by the trained model, the reconstructed profiles are illustrated in the right panels in Figure 15. Although hyperbolic patterns are distinguishable and can provide some location information in input profiles, it is difficult to determine the target sizes. Each pipeline in Figure 15e generates two hyperbolas, which increases the confusion of the locations by immediate identification. Contrarily, both location and diameter information can be acquired from the reconstructed profiles. In Figure 15b, two pipelines are identified inside the 'positive' areas where upper interface signals are visible, and their output characterized values (central coordinates and diameters) are ((3.74 m, 1.02 m), 0.34 m) and ((6.14 m, 0.96 m), 0.34 m), corresponding well with the actual diameters 0.32 m. In Figure 15d, the 5 × 2 cylinders' locations are determined by the "positive" sections, and their identified diameters are 0.1 m or 0.1125 m, similar to the actual diameters 0.1 m. In Figure 15f, only the upper two pipelines are accurately identified with upper interfaces' depth 1.14 m and 1.51 m and diameters 0.54 m and 0.54 m , respectively. The machine learning model failed to vertically identify the deepest pipeline since signals are weak beyond 2 m depth but our training inputs are all within 1 m depth. The depth and the diameter errors have reached 14 cm and 4 cm respectively in the third profile, and this may arise from the significantly different central frequency (250 MHz) of the third detection. Nevertheless, the proposed approach presents promising applicability in field signals.

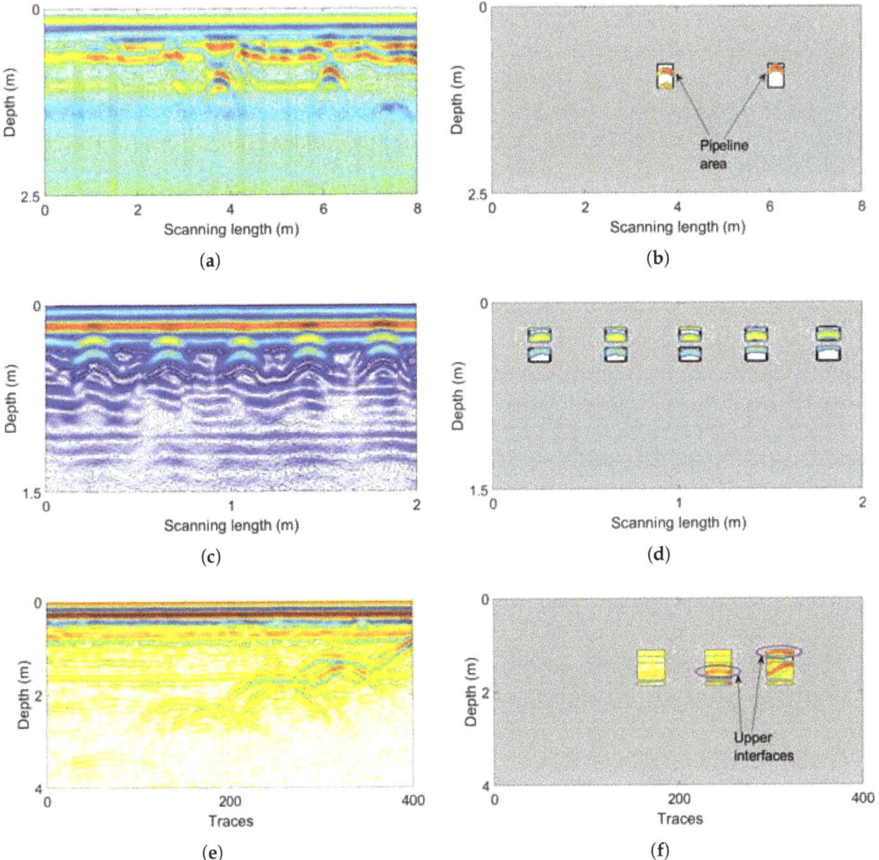

Figure 15. (**a**) Example input profile with two pipelines (diameter 32 cm). (**b**) Reconstructed profile of panel (a) by machine learning. (**c**) Example input profile with 5 × 2 cylinders (diameter 10 cm). (**d**) Reconstructed profile of panel (c) by machine learning. (**e**) Example input profile with three pipelines (diameter 50 cm, depth of upper interface 1 m, 1.5 m and 2 m). (**f**) Reconstructed profile of panel (e) by machine learning.

4. Discussion

Two significant research developments resulting from our novel machine learning approach are demonstrated in the results: outputting size information and training on small datasets.

Object sizes are difficult to determine in GPR detection of subterranean sections. Although the development of radargram interpretation approaches from manual processing to machine learning is taking place, most researches (see, e.g., in [41–43]) on object detection only determined two characteristics: whether the objects were detectable and where the objects were. Some researches have determined the sizes of rebars [44] and small-scale voids [45] inside concrete successfully, but investigations inside the complicated underground sections become difficult. For instance, Pasolli et al. [46] attempted to estimate the buried object size, but they only utilized numerically produced data and the mean error was up to 18.6%; Luo and Lai [47] failed to determine the subsurface void sizes as the identified magnitudes significantly different from the actual sizes. In this paper, both locations and diameters of the subsurface pipelines can be extracted from the reconstructed profiles. The diameter errors are within 1 cm for numerically generated datasets (pipeline diameters: 5–40 cm) and within 4 cm for practical datasets (pipeline diameters: 10–50 cm). These acceptable errors arise from only one or two misclassified vertical or horizontal traces, which indicates a promising performance in determining pipeline sizes.

Dataset requirement is a major issue affecting applications of machine learning in GPR signal processing, as introduced in Section 1. Translation and scaling are reasonable approaches to complement the data shortage, but each profile is usually necessary to generate hundreds of duplicates that bring doubts about training reliability. Besides, producing considerable radargrams by numerical simulation and artificially labeling the large datasets are time-consuming. In this paper, a null-parameter wavelet scattering network, functionally comparable to CNN, is utilized in our machine learning framework. As the convolution kernels are predefined without parameters, we do not train the networks and thereby data requirement decreases. The classification accuracy rates are around 98% and 95% for datasets without and with noises respectively, as well as 97% for considering surface roughness. We also want to mention that the proposed learning framework is only compared with an independent SVM but not with CNN, because the purpose here is to evaluate machine learning on small datasets but the data requirement is large for CNN. Nonetheless, promising performance in both simulated GPR signals and practical radargrams demonstrates the applicability and efficiency of the proposed approach.

The practical applicability is also investigated in this paper but not extended to all aspects. Although cases considering complicated pipeline distribution are analyzed, the subsurface material is simple in both simulated and field-measured profiles, e.g., slightly variant attributes and no underground stratification. Two environmental conditions, the noises and surface roughness, have been analyzed to improve the practical applicability. The added Gaussian noises can help investigate the effects of the environmental noises and the shattered subsurface formations, while surface roughness will vary in different urban locations. Other environmental conditions (e.g., the surface material and the complicated underground formations) and detection settings (e.g., the positioning method of GPR and the bandwidth) are also important in GPR radargram interpretation, and the applicability of our learning framework will be investigated in more complicated cases in the future. Further research is planned in the following four aspects. First, we will extend our exploration depth and consider more complicated subsurface conditions; second, the practical applicability of our learning framework will be further investigated in complicated environmental conditions with different GPR detection settings; third, as the identification of upper interfaces in the reconstructed profile is manual in this paper, we will investigate an automatic approach to overcome this disadvantage; fourth, the research applications will be extended to identify other subterranean objects and even shattered underground structures.

5. Conclusions

In this paper, a wavelet scattering network based machine learning approach intended for non-intensive GPR datasets is investigated for subsurface pipeline identification. Wavelet scattering is a null-parameter convolution network for translation and rotation invariant characterization with specific wavelet approaches. It can extract signal features by predefined convolution kernels without parameters, thereby reducing data requirement. In our learning structure, the double-channel procedures, each containing a multi-layer wavelet scattering network followed by a SVM, work to determine the coordinates by recognizing the existence of pipelines on the single signal trace vertically and horizontally. GPR profiles are then reconstructed by signals inside the "positive" sections (vertical and horizontal traces contain the pipelines) for further determining pipeline locations and sizes.

Promising performance is achieved in both simulated and practical datasets. The classification accuracy rates are around 98% and 95% for datasets without and with noises respectively, as well as 97% for considering surface roughness. Upper interfaces of pipelines are convenient to identify in reconstructed profiles, and further determined locations and diameters conform well with the actual values. Pipelines with diameters of 0.05 m are difficult to identify in noisy profiles, which means their GPR patterns are sensitive to noises. These small pipelines are also unrevealed in profiles considering the rough surface since their diameters are similar to the maximum roughness. Feeding the practical radargrams as the test set into the trained learning model, "positive" sections corresponds promisingly with the object locations. The diameters of the pipelines are accurately determined despite small biases. However, the machine learning model failed to vertically identify the deepest pipeline as signals are weak beyond 2 m depth but our training inputs are all within 1 m. The depth and the diameter errors have reached 14 cm and 4 cm respectively in the third profile, and this may arise from the significantly different central frequency (250 MHz) of the third detection. Reliable detection depth of GPR is determined by the central frequency of the electromagnetic wave and the attribute of the subsurface formation, while detection resolution is limited by the wave frequency and signal bandwidth [48]. For example, high-frequency electromagnetic waves can recognize small objects, but within shallow depth due to sharp energy attenuation. Excepting the noise sensitivity of small pipelines and the failure in recognizing the deep pipeline, the proposed machine learning approach presents promising applicability in both simulated and practical GPR signals.

Author Contributions: Conceptualization, Y.J. and Y.D.; methodology, Y.J.; software, Y.J.; validation, Y.J. and Y.D.; formal analysis, Y.J. and Y.D.; investigation, Y.J. and Y.D.; writing—original draft preparation, Y.J.; writing—review and editing, Y.D. and Y.J.; visualization, Y.J.; supervision, Y.D. All authors have read and agreed to the published version of the manuscript.

Funding: This research received no external funding.

Acknowledgments: Pinning Huang and Jinming Feng in Hydraulic Engineering, Tsinghua University have kindly provided help in this research. The authors thank Feng's work for managing the instruments and experiments. The authors thank Huang for programming support.

Conflicts of Interest: The authors declare no conflicts of interest.

Abbreviations

The following abbreviations are used in this manuscript.

GPR	Ground penetrating radar
CNN	convolutional neural network
SVM	Support vector machine
2D	2-dimension
1D	1-dimension

References

1. Morris, I.; Abdel-Jaber, H.; Glisic, B. Quantitative attribute analyses with ground penetrating radar for infrastructure assessments and structural health monitoring. *Sensors* **2019**, *19*, 1637. [CrossRef] [PubMed]
2. Trinks, I.; Hinterleitner, A. Beyond Amplitudes: Multi-Trace Coherence Analysis for Ground-Penetrating Radar Data Imaging. *Remote Sens.* **2020**, *12*, 1583. [CrossRef]
3. Gu, Z.; Shi, C.; Yang, H.; Yao, H. Analysis of dynamic sedimentary environments in alluvial fans of some tributaries of the upper Yellow River of China based on ground penetrating radar (GPR) and sediment cores. *Quat. Int.* **2019**, *509*, 30–40. [CrossRef]
4. Soldovieri, F.; Brancaccio, A.; Prisco, G.; Leone, G.; Pierri, R. A Kirchhoff-based shape reconstruction algorithm for the multimonostatic configuration: The realistic case of buried pipes. *IEEE Trans. Geosci. Remote Sens.* **2008**, *46*, 3031–3038. [CrossRef]
5. Lombardi, F.; Griffiths, H.D.; Lualdi, M. Sparse Ground Penetrating Radar Acquisition: Implication for Buried Landmine Localization and Reconstruction. *IEEE Geosci. Remote Sens. Lett.* **2018**, *16*, 362–366. [CrossRef]
6. Lai, W.W.; Chang, R.K.; Sham, J.F. A blind test of nondestructive underground void detection by ground penetrating radar (GPR). *J. Appl. Geophy.* **2018**, *149*, 10–17. [CrossRef]
7. Šarlah, N.; Podobnikar, T.; Ambrožič, T.; Mušič, B. Application of Kinematic GPR-TPS Model with High 3D Georeference Accuracy for Underground Utility Infrastructure Mapping: A Case Study from Urban Sites in Celje, Slovenia. *Remote Sens.* **2020**, *12*, 1228. [CrossRef]
8. Travassos, X.L.; Avila, S.L.; Ida, N. Artificial neural networks and machine learning techniques applied to ground penetrating radar: A review. *Appl. Comput. Inform.* **2020**. [CrossRef]
9. Guan, B.; Ihamouten, A.; Dérobert, X.; Guilbert, D.; Lambot, S.; Villain, G. Near-field full-waveform inversion of ground-penetrating radar data to monitor the water front in limestone. *IEEE J. Sel. Top. Appl. Earth Obs. Remote Sens.* **2017**, *10*, 4328–4336. [CrossRef]
10. Klotzsche, A.; Vereecken, H.; van der Kruk, J. Review of crosshole ground-penetrating radar full-waveform inversion of experimental data: Recent developments, challenges, and pitfalls. *Geophysics* **2019**, *86*, H13–H28. [CrossRef]
11. Ayodele, T.O. Types of machine learning algorithms. In *New Advances in Machine Learning*; Zhang, Y., Ed.; Books on Demand: Rijeka, Croatia, 2010.
12. Maas, C.; Schmalzl, J. Using pattern recognition to automatically localize reflection hyperbolas in data from ground penetrating radar. *Comput. Geosci.* **2013**, *58*, 116–125. [CrossRef]
13. Torrione, P.A.; Morton, K.D.; Sakaguchi, R.; Collins, L.M. Histograms of oriented gradients for landmine detection in ground-penetrating radar data. *IEEE Trans. Geosci. Remote Sens.* **2013**, *52*, 1539–1550. [CrossRef]
14. Pham, M.T.; Lefèvre, S. Buried object detection from B-scan ground penetrating radar data using Faster-RCNN. In Proceedings of the IGARSS 2018 IEEE International Geoscience and Remote Sensing Symposium, Valencia, Spain, 22–27 July 2018; pp. 6804–6807.
15. Kang, M.S.; Kim, N.; Lee, J.J.; An, Y.K. Deep learning-based automated underground cavity detection using three-dimensional ground penetrating radar. *Struct. Health Monit.* **2020**, *19*, 173–185. [CrossRef]
16. Liu, H.; Lin, C.; Cui, J.; Fan, L.; Xie, X.; Spencer, B.F. Detection and localization of rebar in concrete by deep learning using ground penetrating radar. *Autom. Constr.* **2020**, *118*, 103279. [CrossRef]
17. Kafedziski, V.; Pecov, S.; Tanevski, D. Detection and classification of land mines from ground penetrating radar data using faster R-CNN. In Proceedings of the 2018 26th Telecommunications Forum (TELFOR), Serbia, Belgrade, 20–21 November 2018; pp. 1–4.
18. Ponti, F.; Barbuto, F.; Di Gregorio, P.P.; Mangini, F.; Simeoni, P.; Troiano, M.; Frezza, F. Deep Learning for Applications to Ground Penetrating Radar and Electromagnetic Diagnostic. In Proceedings of the 2019 PhotonIcs & Electromagnetics Research Symposium-Spring (PIERS-Spring), Rome, Italy, 17–20 June 2019; pp. 547–551.
19. Mallat, S. Group invariant scattering. *Commun. Pure Appl. Math.* **2012**, *65*, 1331–1398. [CrossRef]
20. Sifre, L.; Mallat, S. Rotation, scaling and deformation invariant scattering for texture discrimination. In Proceedings of the IEEE Conference on Computer Vision and Pattern Recognition, Portland, Oregon, USA, 23–28 June 2013; pp. 1233–1240.

21. Bruna, J.; Mallat, S. Invariant scattering convolution networks. *IEEE Trans. Pattern Anal. Mach. Intell.* **2013**, *35*, 1872–1886. [CrossRef]

22. Wang, H.; Li, S.; Zhou, Y.; Chen, S. SAR automatic target recognition using a Roto-translational invariant wavelet-scattering convolution network. *Remote Sens.* **2018**, *10*, 501. [CrossRef]

23. Warren, C.; Giannopoulos, A.; Giannakis, I. gprMax: Open source software to simulate electromagnetic wave propagation for Ground Penetrating Radar. *Comput. Phys. Commun.* **2016**, *209*, 163–170. [CrossRef]

24. Giannopoulos, A. Modelling ground penetrating radar by GprMax. *Constr. Build. Mater.* **2005**, *19*, 755–762. [CrossRef]

25. Soldovieri, F.; Hugenschmidt, J.; Persico, R.; Leone, G. A linear inverse scattering algorithm for realistic GPR applications. *Near Surf. Geophys.* **2007**, *5*, 29–41. [CrossRef]

26. Liu, T.; Su, Y.; Huang, C. Inversion of ground penetrating radar data based on neural networks. *Remote Sens.* **2018**, *10*, 730. [CrossRef]

27. Kang, M.S.; Kim, N.; Im, S.B.; Lee, J.J.; An, Y.K. 3D GPR Image-based UcNet for Enhancing Underground Cavity Detectability. *Remote Sens.* **2019**, *11*, 2545. [CrossRef]

28. Dérobert, X.; Pajewski, L. TU1208 open database of radargrams: The dataset of the IFSTTAR geophysical test site. *Remote Sens.* **2018**, *10*, 530. [CrossRef]

29. Cassidy, N. Ground Penetrating Radar Data Processing, Modelling and Analysis. In *Ground Penetrating Radar: Theory and Applications*; Jol, H.M., Ed.; Elsevier: Oxford, UK, 2008; pp. 141–176.

30. Shamir, O.; Goldshleger, N.; Basson, U.; Reshef, M. Laboratory measurements of subsurface spatial moisture content by ground-penetrating radar (GPR) diffraction and reflection imaging of agricultural soils. *Remote Sens.* **2018**, *10*, 1667. [CrossRef]

31. Jin Y.; Duan Y. A new method for abnormal underground rocks identification using ground penetrating radar. *Measurement* **2020**, *149*, 106988. [CrossRef]

32. Pathak, R.S. The wavelet transform. In *Atlantis Studies in Mathematics for Engineering and Science*; Chui, C.K., Ed.; Springer Science & Business Media: Paris, France, 2009; Volume 4.

33. Haasdonk, B.; Burkhardt, H. Invariant kernel functions for pattern analysis and machine learning. *Mach. Learn.* **2007**, *68*, 35–61. [CrossRef]

34. Cortes, C.; Vapnik, V. Support-vector networks. *Mach. Learn.* **1995**, *20*, 273–297. [CrossRef]

35. Vapnik, V. The Nature of Statistical Learning Theory. In *Information Science and Statistics*; Jordan, M., Lauritzen, S., Lawless, J., Nair, V., Eds.; Springer Science & Business Media: New York, NY, USA, 2013.

36. Pedregosa, F.; Varoquaux, G.; Gramfort, A.; Michel, V.; Thirion, B.; Grisel, O.; Blondel, M.; Prettenhofer, P.; Weiss, R.; Dubourg, V.; et al. Scikit-learn: Machine Learning in Python. *J. Mach. Learn. Res.* **2011**, *12*, 2825–2830.

37. Ozkaya, U.; Seyfi, L. Deep dictionary learning application in GPR B-scan images. *Signal Image Video Process.* **2018**, *12*, 1567–1575. [CrossRef]

38. Ozkaya, U.; Melgani, F.; Bejiga, M.B.; Seyfi, L.; Donelli, M. GPR B Scan Image Analysis with Deep Learning Methods. *Measurement* **2020**, *165*, 107770. [CrossRef]

39. Lambot, S.; Antoine, M.; Vanclooster, M.; Slob, E.C. Effect of soil roughness on the inversion of off-ground monostatic GPR signal for noninvasive quantification of soil properties. *Water Resour. Res.* **2006**, *42*, W03403. [CrossRef]

40. Jonard, F.; Weihermüller, L.; Vereecken, H.; Lambot, S. Accounting for soil surface roughness in the inversion of ultrawideband off-ground GPR signal for soil moisture retrieval. *Geophysics* **2012**, *77*, H1–H7. [CrossRef]

41. Núñez-Nieto, X.; Solla, M.; Gómez-Pérez, P.; Lorenzo, H. GPR signal characterization for automated landmine and UXO detection based on machine learning techniques. *Remote Sens.* **2014**, *6*, 9729–9748. [CrossRef]

42. Hoarau, Q.; Ginolhac, G.; Atto, A.M.; Nicolas, J.M. Robust adaptive detection of buried pipes using GPR. *Signal Process.* **2017**, *132*, 293–305. [CrossRef]

43. Kim, N.; Kim, K.; An, Y.K.; Lee, H.J.; Lee, J.J. Deep learning-based underground object detection for urban road pavement. *Int. J. Pavement Eng.* **2018**, 1–13. [CrossRef]

44. Xiang, Z.; Ou, G.; Rashidi, A. Integrated Approach to Simultaneously Determine 3D Location and Size of Rebar in GPR Data. *J. Perform. Constr. Facil.* **2020**, *34*, 04020097. [CrossRef]

45. Yang, Y.; Lu, J.; Li, R.; Zhao, W.; Yan, D. Small-Scale Void-Size Determination in Reinforced Concrete Using GPR. *Adv. Civ. Eng.* **2020**, *2020*, 2740309. [CrossRef]

Remote Sens. **2020**, *12*, 3655

46. Pasolli, E.; Melgani, F.; Donelli, M. Gaussian process approach to buried object size estimation in GPR images. *IEEE Geosci. Remote. Sens. Lett.* **2009**, *7*, 141–145. [CrossRef]
47. Luo, T.X.; Lai, W.W. GPR pattern recognition of shallow subsurface air voids. *Tunn. Undergr. Space Technol.* **2020**, *99*, 103355. [CrossRef]
48. Annan, A.P. Electromagnetic Principles of Ground Penetrating Radar. In *Ground Penetrating Radar: Theory and Applications*; Jol, H.M., Ed.; Elsevier: Oxford, UK, 2008; pp. 3–40.

Publisher's Note: MDPI stays neutral with regard to jurisdictional claims in published maps and institutional affiliations.

 *remote sensing*

Article

Analysis of the Snow Water Equivalent at the AEMet-Formigal Field Laboratory (Spanish Pyrenees) During the 2019/2020 Winter Season Using a Stepped-Frequency Continuous Wave Radar (SFCW)

Rafael Alonso [1,*], **José María García del Pozo** [1], **Samuel T. Buisán** [2] and **José Adolfo Álvarez** [3]

[1] Department of Applied Physics, University of Zaragoza, 50018 Zaragoza, Spain; chgarcia@unizar.es
[2] Infrastructure Division of the Aragon Regional Office of the Spanish State Meteorological Agency (AEMet), Paseo del Canal, 17, 50007 Zaragoza, Spain; sbuisans@aemet.es
[3] Ebro River Basin Authority (CHE), Paseo de Sagasta, 24-26, 50008 Zaragoza, Spain; jaalvarez@chebro.es
* Correspondence: ralonso@unizar.es; Tel.: +34-976-762665

Abstract: Snow makes a great contribution to the hydrological cycle in cold regions. The parameter to characterize available the water from the snow cover is the well-known snow water equivalent (SWE). This paper presents a near-surface-based radar for determining the SWE from the measured complex spectral reflectance of the snowpack. The method is based in a stepped-frequency continuous wave radar (SFCW), implemented in a coherent software defined radio (SDR), in the range from 150 MHz to 6 GHz. An electromagnetic model to solve the electromagnetic reflectance of a snowpack, including the frequency and wetness dependence of the complex relative dielectric permittivity of snow layers, is shown. Using the previous model, an approximated method to calculate the SWE is proposed. The results are presented and compared with those provided by a cosmic-ray neutron SWE gauge over the 2019–2020 winter in the experimental AEMet Formigal-Sarrios test site. This experimental field is located in the Spanish Pyrenees at an elevation of 1800 m a.s.l. The results suggest the viability of the approximate method. Finally, the feasibility of an auxiliary snow height measurement sensor based on a 120 GHz frequency modulated continuous wave (FMCW) radar sensor, is shown.

Keywords: snow; snow water equivalent (SWE); stepped-frequency continuous wave radar (SFCW); software defined radio (SDR); snowpack multilayer reflectance

Citation: Alonso, R.; Pozo, J.M.G.d.; Buisán, S.T.; Álvarez, J.A. Analysis of the Snow Water Equivalent at the AEMet-Formigal Field Laboratory (Spanish Pyrenees) During the 2019/2020 Winter Season Using a Stepped-Frequency Continuous Wave Radar (SFCW). *Remote Sens.* **2021**, *13*, 616. https://doi.org/10.3390/rs13040616

Academic Editors: Yuri Álvarez López and María García Fernández

Received: 30 December 2020
Accepted: 5 February 2021
Published: 9 February 2021

Publisher's Note: MDPI stays neutral with regard to jurisdictional claims in published maps and institutional affiliations.

1. Introduction

1.1. Snow Water Equivalent Importance

Snow makes a large contribution to the water balance, climate, and economy of many regions. The seasonal accumulation of snow acts as a form of natural regulation of great importance in the hydrological cycle. The parameter to characterize the available water from snow cover is the well-known snow water equivalent (SWE). The SWE is the equivalent depth of water available if the snowpack melted into liquid water. This paper presents a near-surface-based technique for determining the SWE from the measured complex spectral reflectance, in the range from 150 MHz to 6 GHz, and an electromagnetic model for the analysis of the reflectance of the snowpack.

As we can see, physical knowledge of the snow cover properties, and, in particular, the availability of field instrumentation to understand them, is crucial. In addition to the most basic hydrological information [1], these measurements provide data to elaborate mathematical models of the climate [2] and the snowpack evolution [3] as well as for avalanche forecasting [4].

1.2. Instrumentation and Techniques

Among the destructive measuring methods of the snowpack, we can find several standard sampling procedures for stratigraphy that are manually performed [5]. Non-destructive methods—mainly, ground-based remote sensing techniques for characterizing the physical properties of snow—have been exhaustively discussed in the literature [5]. Conventional in situ techniques to non-destructively determine the SWE are based on cosmic-ray neutron attenuation (CRN) [6,7], or acoustic signal delays [8], among others. From them, we can directly extract the SWE, but they give no further information about the structure of the snowpack.

The snowpack is a multilayer structure [9], where each layer has its own physical properties, depending on the crystallization form during snowfall and the weather conditions during and after the layer deposition. From a physics point of view, we can consider each layer as a porous body of mixed air with solid and liquid state water. The liquid water content (LWC) is the key parameter to characterize the presence of liquid water [10–12].

The only physical entity capable of providing information about the internal structure of a snowpack is an electromagnetic wave in the 100 MHz–6 GHz range [13], where the layers are sufficiently transparent to allow penetration depths in the order of meters. The interaction between electromagnetic waves and the snow is determined by the dielectric permittivity of the latter, which is strongly dependent on its air and water content, and the aggregation state of the water [10,14].

The most promising electromagnetic methods are those based on microwave radar analysis. In fact, over the past 40 years, numerous studies on radar applications for snow cover characterization have been conducted. Impulse waveform radars [15,16] and frequency modulated continuous wave (FMCW) radars [17–23] have been used to explore snowpack structure.

Until recently, the construction of an FMCW radar, stepped-frequency continuous wave (SFCW) radar, or pulsed radars required an expensive vector network analyzer (VNA) or developing specific electronical equipment [24]. However, a new technology has been developed, the so-called software defined radio (SDR). An SDR system is a radio in which some or all its physical layer functionalities are software defined. A shared feature of all of them is the possibility to generate and detect signals up to 6 GHz, so they are potentially usable for reconfigurable radar system construction [25–30].

The recent advances in radar technology, propitiated by automotive radars or SDR, have resulted in low cost and small dimension equipment that will allow portable instrumentation with a high vertical resolution to study properties, such as the depth, stratigraphy, and SWE, of snow cover. The aim of this paper is to show the operation of an SDR-SFCW radar, and the application of an electromagnetic model of the multilayer snowpack to analyze the SWE of snow cover.

This paper is organized as follows. In Section 2, we present the theory of SFCW radar. Section 3 shows the electromagnetic model and the procedure to obtain the SWE. Section 4 focuses on the description of the experimental set-up and the measurement method. Section 5 presents the results and a comparison between the experimental and theoretical results. Finally, in Section 6, we summarize the contributions of the present work.

2. Theory of the SFCW Radar

2.1. Measurement Principle

An SFCW [31–33] radar transmits consecutive trains of CW signals of increasing frequency toward a target. The stepped-frequency increment is δf from an initial frequency $f_{0,RF}$. After receiving the signal at a frequency $f_{0,RF}+i\delta f$, with i as an integer, which is delayed in phase as a result of the flight time, a heterodyne process is performed with a local frequency oscillator $f_{0,RF}+i\delta f+f_{BB}$, of unity amplitude. The mixing is performed with the harmonic oscillator in-phase with the harmonic function used in transmission (I) and with the harmonic function offset by one quarter cycle ($\pi/2$ rads) or in-quadrature (Q) as is shown in Figure 1. After low pass filtering, two harmonic functions at the down converted or intermediate frequency (IF), f_{BB}, are obtained for the *i*th frequency

$$U_I(i,t) = \frac{1}{2}A(R)\cos(2\pi f_{BB}t + \phi) = \frac{1}{2}A(R)\cos(2\pi f_{BB}t + 2\pi(f_{0,RF} + i\delta f)\Delta t) \quad (1)$$

$$U_Q(i,t) = \frac{1}{2}A(R)\sin(2\pi f_{BB}t + \phi) = \frac{1}{2}A(R)\sin(2\pi f_{BB}t + 2\pi(f_{0,RF} + i\delta f)\Delta t) \quad (2)$$

where $A(R)$ is the amplitude of the signal received from a target located at an R distance and is proportional to its reflectance, and ϕ represents the phase associated to the time-of-flight. The received IF signals $U_I(i,t)$ and $U_Q(i,t)$ are multiplied, again, by the signal $\cos(2\pi f_{BB}t)$, and low pass filtered (homodyne process) to obtain the complex amplitude, $\hat{S}(i)$, at each *i*th frequency of the scan

$$\hat{S}(i) \propto e^{j2\pi((f_{0,RF} + i\cdot\delta f)\cdot\frac{2R}{c})} = I(i) + jQ(i) \; i = 0,\cdots,N \quad (3)$$

where c is the speed of light in the vacuum, N the number of frequencies of scan, and $I(i)$ and $Q(i)$ the real and imaginary part of the amplitude $\hat{S}(i)$, respectively. Therefore, if we perform the fast Fourier transform (FFT) with respect to the variable i, we will have an important peak in the dimensionless frequency domain given by

$$g_{peak} = \delta f \frac{2R}{c}. \quad (4)$$

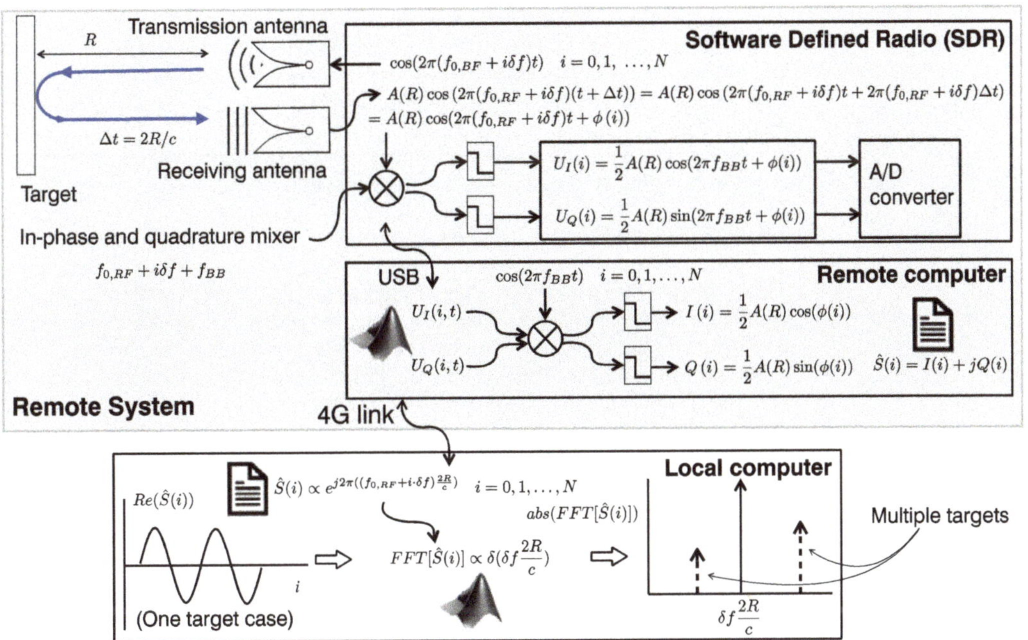

Figure 1. Structure of a stepped-frequency continuous wave (SFCW) radar system. The gray area is the part of the system that is located in the remote location. Software defined radio (SDR) control is performed in C++ from the remote computer and also computes the spectral reflectance and sends it to the local computer. Both computers develop their algorithms using MATLAB®.

The received spectral reflected signal by targets contains the transit time information or the 'optical path' from the target, given by

$$R = \frac{c g_{peak}}{2 \delta f}.$$

(5)

The FFT resolution, when N frequencies are taken, is given by $1/N$. Then, the spatial resolution is given by

$$\delta g_{peak} = \frac{1}{N} = \delta f \frac{2 \delta R}{c} \implies \delta R = \frac{c}{2B},$$

(6)

which is a well-known relationship in the general radar theory, and where $B = N \delta f$ is the bandwidth of the sweep performed by the SFCW radar. The maximum unambiguous range, R_u, can be derived from the periodicity of $\hat{S}(i)$ (3) as

$$2 \pi \delta f \frac{2 R_u}{c} = 2 \pi \implies R_u = \frac{c}{2 \delta f}.$$

(7)

The main disadvantage of an SFCW radar is its long measurement time. In our application, however, the snow cover is a 'static' target, and the measurement time is not relevant. The SFCW radar has very narrow instantaneous bandwidth at each frequency due to the narrow low pass filter width at the end of the DC filtering of the U_I and U_Q signals, resulting in a high signal-to-noise ratio at the receiver. The bandwidth, B, can be very wide, leading, according to (6), to a fine resolution.

Figure 1 presents the different systems of SFCW radar and its operations. At the remote location, the SDR performs the transmission and reception, obtaining a double IF temporal file (U_I and U_Q) at each i-frequency of the reflected electromagnetic wave. The remote computer performs the homodyne detection numerically at each frequency. Finally, a file with the complex amplitudes of snowpack reflectance at each frequency is sent to the local computer.

In the case of multiple targets, we obtain a peak associated with the position of each target with the height proportional to its reflectance (Figure 1).

3. Snowpack Electromagnetic Model

Many authors describe wave propagation in snow layered media with recursive relationships of the transmitted and reflected rays, including radiation diffusion and the thermal emission of slabs in the context of radiative transfer models [14]. However, we can simplify the radiative behavior of the snow layers working at long wavelengths compared to the grain size of the deposited snow. In this way, we can minimize the diffusion effects and assume the snow layers as homogenous and isotropic media with wavelength dependent and complex dielectric permittivities to take into account the attenuation of radiation. We used the 2×2 matrix method that is typically employed in planar multilayer optical structures and extensively described in the literature [34,35].

3.1. Matricial Multilayer Snowpack Electromagnetic Model

The matrix formulation is an extremely useful form of the steady-state solution of Maxwell's equations subjected to the boundary conditions imposed at the interfaces of a multilayer stack with $n + 1$ media where the upper and lower media are semi-infinite, and the number of layers is $n - 1$ (Figure 2). Maxwell's equations reduce to independent sets of equations for the transverse-electric (TE) and transverse-magnetic (TM) polarizations. The TE polarization has the electric-field vector E perpendicular to the plane of incidence, whereas the magnetic-field vector, B, is transverse in the TM case.

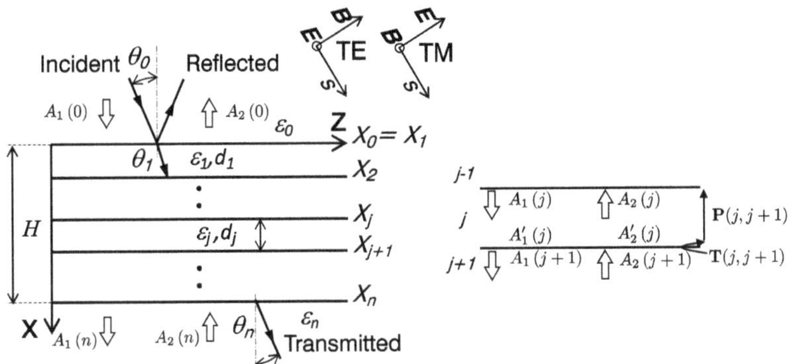

Figure 2. An $n+1$ media structure ($n-1$ layers) of snowpack of height H. The relative dielectric permittivities are, in general, complex magnitudes. The upper and lower media are semi-infinite.

According to Maxwell's equations, the general expression for the *TM* polarization of the x component of electric field in layer j can be written as

$$E_x(j) = \left[A_1(j)e^{ikn_{xj}(x-x_j)} + A_2(j)e^{-ikn_{xj}(x-x_j)}\right]e^{i(kn_z z - \omega t)}, \quad n_j^2 = \epsilon_{rj}, \quad k = \frac{2\pi}{\lambda}, \tag{8}$$

$$n_{xj} = n_j \cos(\theta_j), \quad n_z = n_0 \sin(\theta_0) = \ldots = n_n \sin(\theta_n), \quad n_{xj}^2 = n_j^2 - n_z^2, \tag{9}$$

where A_1 represents the amplitude of a planar wave propagating in the positive x, and z, direction and A_2 represents a plane wave propagating in the negative x direction and positive z direction, k is the wavenumber in vacuum and n_j is the refractive index of the jth slab related to the relative dielectric permittivity.

By replacing solution (8) in the Maxwell's equations, we obtain the general expression of the electric and magnetic fields for *TM* solutions in jth medium

$$E(j) = \left\{ \begin{pmatrix} A_1(j) \\ 0 \\ -\frac{n_{xj}}{n_z}A_1(j) \end{pmatrix} e^{ikn_{xj}(x-x_j)} + \begin{pmatrix} A_2(j) \\ 0 \\ \frac{n_{xj}}{n_z}A_1(j) \end{pmatrix} e^{-ikn_{xj}(x-x_j)} \right\} e^{i(kn_z z - \omega t)} \tag{10}$$

$$B(j) = \frac{n_j^2}{cn_z} \left\{ \begin{pmatrix} 0 \\ A_1(j) \\ 0 \end{pmatrix} e^{ikn_{xj}(x-x_j)} + \begin{pmatrix} 0 \\ A_2(j) \\ 0 \end{pmatrix} e^{-ikn_{xj}(x-x_j)} \right\} e^{i(kn_z z - \omega t)} \tag{11}$$

The relationship between the coefficients $A'_1(j)$, $A'_2(j)$, $A_1(j+1)$, and $A_2(j+1)$ corresponding to the amplitudes on one side and the other of the interface $(j)-(j+1)$, (Figure 2 right), is given by the continuity of the tangential components of the fields. We can summarize this relation in a matrix form

$$\begin{bmatrix} A'_1(j) \\ A'_2(j) \end{bmatrix} = \mathbf{T}_{TM}(j, j+1) \begin{bmatrix} A_1(j+1) \\ A_2(j+1) \end{bmatrix} \tag{12}$$

$$\mathbf{T}_{TM}(j, j+1) = \frac{1}{2} \begin{bmatrix} \frac{n_{j+1}^2}{n_j^2} + \frac{n_{x(j+1)}}{n_{xj}} & \frac{n_{j+1}^2}{n_j^2} - \frac{n_{x(j+1)}}{n_{xj}} \\ \frac{n_{j+1}^2}{n_j^2} - \frac{n_{x(j+1)}}{n_{xj}} & \frac{n_{j+1}^2}{n_j^2} + \frac{n_{x(j+1)}}{n_{xj}} \end{bmatrix} \tag{13}$$

The coefficients $A'_1(j)$ and $A'_2(j)$ on the interface $(j) - (j + 1)$ are related to the coefficients $A_1(j)$ and $A_2(j)$ of the same layer but at the interface $(j - 1) - (j)$ by the propagation matrix $\mathbf{P}(j)$

$$\begin{bmatrix} A_1(j) \\ A_2(j) \end{bmatrix} = \mathbf{P}(j) \begin{bmatrix} A'_1(j) \\ A'_2(j) \end{bmatrix}; \quad \mathbf{P}(j) = \begin{bmatrix} e^{-ikn_{xj}d_j} & 0 \\ 0 & e^{ikn_{xj}d_j} \end{bmatrix} \tag{14}$$

The recursive application of the boundary conditions at the interfaces and the propagation for E and B through the layers, leads to a relationship between 0 and n amplitudes given by

$$\begin{bmatrix} A_1(0) \\ A_2(0) \end{bmatrix} = \mathbf{T}(0,1)\mathbf{P}(1)\mathbf{T}(1,2)\cdots\mathbf{P}(n-1)\mathbf{T}(n-1,n)\begin{bmatrix} A_1(n) \\ A_2(n) \end{bmatrix} = \mathbf{M}_{TM}\begin{bmatrix} A_1(n) \\ A_2(n) \end{bmatrix}, \tag{15}$$

$$\mathbf{M}_{TE} = \begin{bmatrix} M_{11} & M_{12} \\ M_{21} & M_{22} \end{bmatrix} \tag{16}$$

Similarly, we can obtain the transmission matrix for *TE* polarization, but, in our particular case, as the angle of incidence is null, the two polarizations are equivalent. In a general case, the transmission matrix from the $j + 1$ to j layer for *TE* polarization is

$$\mathbf{T}_{TE}(j, j+1) = \frac{1}{2}\begin{bmatrix} 1 + \frac{n_{x(j+1)}}{n_{xj}} & 1 - \frac{n_{x(j+1)}}{n_{xj}} \\ 1 - \frac{n_{x(j+1)}}{n_{xj}} & 1 + \frac{n_{x(j+1)}}{n_{xj}} \end{bmatrix} \tag{17}$$

3.2. Reflectance Calculation for the Snowpack Multilayer

According to the formulation outlined before, the complex reflectance, Γ, of the snowpack, and assuming the condition that there is no upward propagating wave in the medium n, i.e., $A_2(n) = 0$, is given, in the *TE* or *TM* case, by

$$\begin{bmatrix} A_1(0) \\ A_2(0) \end{bmatrix} = \mathbf{M}_{TE,\,TM}\begin{bmatrix} A_1(n) \\ 0 \end{bmatrix} \implies \Gamma = \frac{A_2(0)}{A_1(0)} = \frac{M_{21}}{M_{11}}, \tag{18}$$

therefore, to calculate the reflectance of the snow cover, we calculate only the M_{21} and M_{11} elements of the $\mathbf{M}$ matrix, according to expression 18. This is the amplitude reflectance of a plane wave (*TE* or *TM*) incident from medium 0 with an angle θ_0 and reflected with the same angle.

3.3. Spectral and Spatial Reflectance of Snowpack

According the theory of SFCW radar, in the scan process we obtain the reflectance in N frequencies, i.e.,

$$\Gamma(i) = \frac{M_{21}(f_{0,RF} + i\delta f)}{M_{11}(f_{0,RF} + i\delta f)}, \quad i = 0, \ldots N. \tag{19}$$

The relation between spectral and temporal (spatial) reflectance can be derived by means of FFT.

$$\Gamma_s(t) = \frac{1}{N}\sum_{i=0}^{N}\Gamma(i)e^{j2\pi((f_{0,RF} + i\cdot\delta f)t)}, \quad i = 0, \ldots N, \tag{20}$$

or in function of distance, R,

$$\Gamma_s(R) = \frac{1}{N}\sum_{i=0}^{N}\Gamma(i)e^{j2\pi((f_{0,RF} + i\cdot\delta f)\frac{2R}{c})}, \quad i = 0, \ldots N, \tag{21}$$

with represents the "spatial reflectance" that comes from different distances.

3.4. Complex Dielectric Constant of Dry and Wet Snow

To simulate the reflectance of a multilayer structure, a complex relative dielectric permittivity model for dry and wet snow in the 150 MHz to 6 GHz band is needed. We have used the empirical formula from Tiuri [10]. In the literature, however, there are other models such as the one proposed by Wiesmann [14], but the permittivities calculated by the two models are practically the same.

Figure 3 shows the frequency behavior of dielectric permittivity for dry snow and wet snow with a liquid water content (LWC) of 2% and a relative density of 0.2.

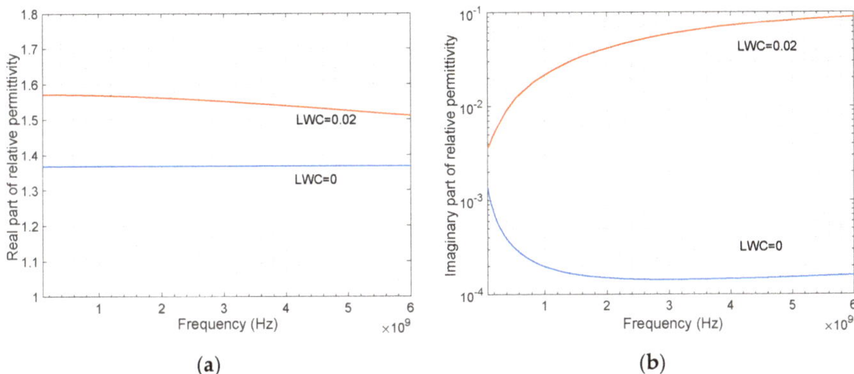

Figure 3. Relative dielectric permittivity of snow from 100 MHz to 6 GHz for dry snow and wet snow with a liquid water content (LWC) of 2% and a relative density of 0.2: (**a**) Real part of the relative permittivity. (**b**) Imaginary part of the relative permittivity.

We can see the remarkable effect of the presence of water in liquid form both on the real part and, particularly, on the imaginary part. The presence of water in liquid form will have an important impact in the penetration depth of radiation into the snowpack.

3.5. Simulations of Snowpack Reflectance

To show the behavior of the reflected spatial signals of the SFCW radar, we analyzed several cases calculated with the matrix method. A sweep from 150 MHz to 6 GHz was used in 15 MHz steps ($N = 390$).

A 1.5×1.5 m piece of sheet metal was incorporated into the soil in the experimental structure to improve the reflectance of the snowpack–soil interface. The relative permittivity considered for this layer (nth) was j50. Matrix calculations to simulate the spectral and spatial reflectances, $\Gamma(i)$ an $\Gamma_s(i)$ respectively were graphically implemented (Figure 4) in interactive software. Experimentally measured traces can be imported into the application to compare with the theoretical traces calculated with the matrix method.

Figure 5a presents the reflectance of the signal produced by the aluminum sheet (orange line) located 2 m away from the origin. This is a reference reflectance in the next graphs. On the same graph, we overlap the radar reflectances in the cases of one-meter snow layer with relative density 0.3 and one-meter snow layer with relative density 0.6. A first reflection on the upper face associated with the air–snow interface can be observed. The position of this first reflection relative to that of metal without snow cover (orange), allows us to know the height of the snow cover, H.

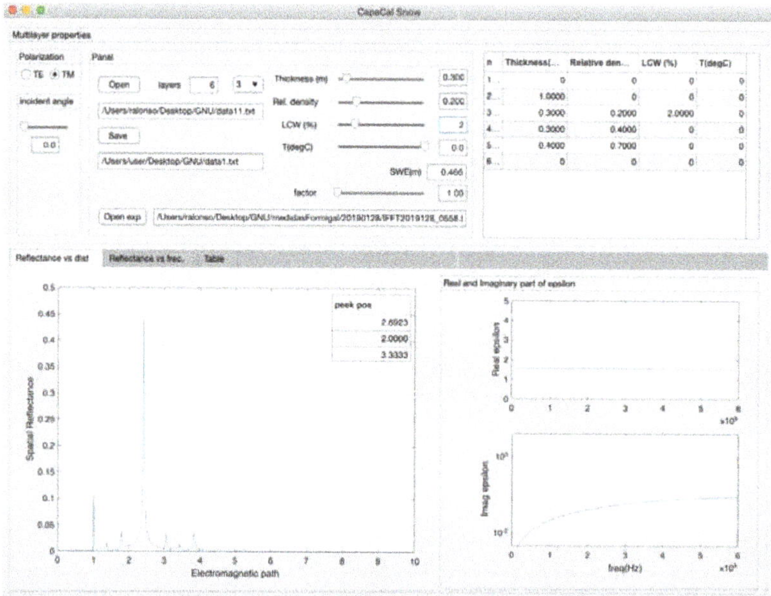

Figure 4. Software developed to calculate the time signal of radar for a snowpack, including the density, wetness, and frequency of all layers. (Adapted with permission from ref. [30]. Copyright 2020 IEEE).

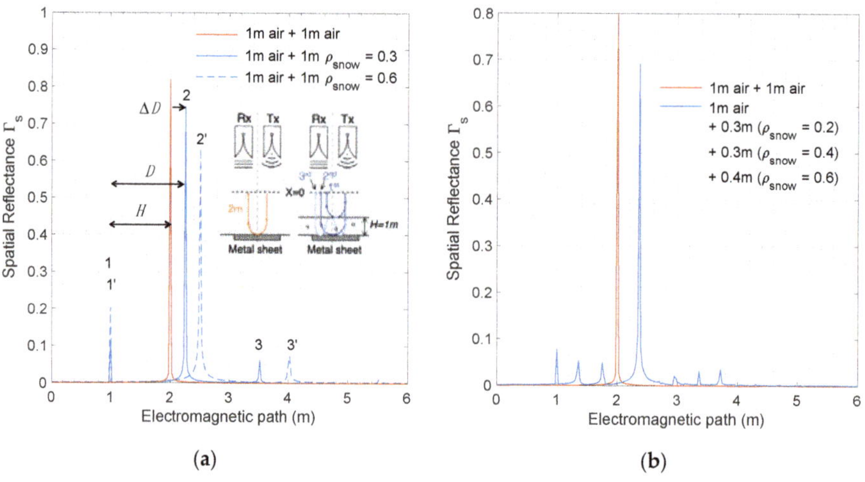

Figure 5. SFCW simulated spatial reflectances with the matrix method: (**a**) Signals for a 1 m of air and 1 m of snow layer with 0 (air), 0.3, and 0.6 relative densities. The orange line represents the reflectance of metal without a snow cover. (**b**) Air + three-layer structure with thicknesses [1,0.3,0.3,0.4] and relative densities [0,0.2,0.4,0.6]. We use an array format to describe the thicknesses and densities of layers.

We also observed how the reflection of the metal layer moves to the right as a result of the flight time inside the snow cover. In the case of a relative density of 0.6, there was a greater displacement. Finally, there are symmetrical peaks (3 in Figure 5a) that are the second reflections in the snow cover as is indicated in the draw included in Figure 5a. More reflections are observed with difficulty; however, it is necessary to note that even the case of dry snow, there is an attenuation (see Figure 3b) that reduces the magnitude of the successive reflections.

Figure 5b shows the superposition of the metal plate signal with a slightly more complex structure than the previous one. Specifically, there are three layers of thicknesses of 0.3, 0.3, and 0.4 m with relative densities of 0.2, 0.4, and 0.6, respectively. We can see how, again the first reflection, positioned at the height of the snow cover. There are some reflections between the different interfaces of the structure. These internal reflections are very small in magnitude compared to the reflectance of the air–snow and snow–metal interfaces.

In Figure 6, we can see the effect of the wetness on the structure of Figure 5b. Increasing the wetness of the layers results in a higher reflectance at the air–snow interface. This effect is due to an increase in the real part of the permittivity as a result of the presence of liquid water. Likewise, for this same reason, there is a slight increase in the optical path as is evident in the rightward displacement of reflection on the sheet metal as the content in liquid water increases. Finally, the effect of the imaginary part is seen in the drastic reduction of reflectance in the sheet metal and all other secondary peaks. Note that we reduced the vertical scale to see them.

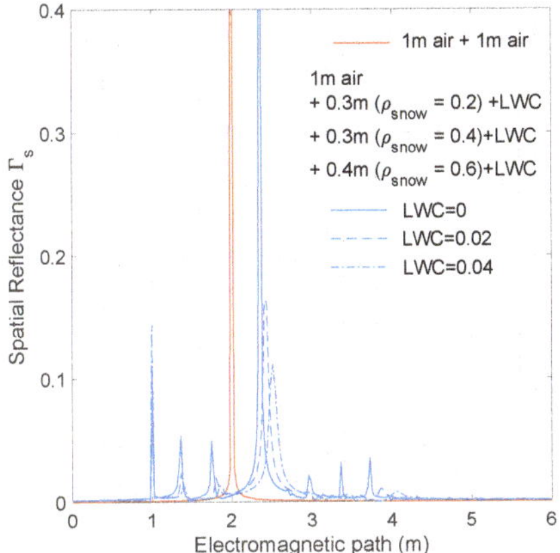

Figure 6. SFCW radar simulated spatial reflectances with the matrix method. Effect of wetness (LWC) in the reflectance (blue lines). Thicknesses (m) = [1,0.3,0.3,0.4]. Relative densities of the structure = [0,0.2,0.4,0.6], LWC = 0 (solid line), 0.02 (dashed line), and 0.04 (dot-dashed line).

3.6. Estimation Procedure of SWE

Figure 7 shows the effect of the layer density on the position on the reflectance of the snow–metal interface of 1-m-thick snow of relative densities of 0, 0.2, 0.4, 0.6, and 0.8. We can observe a linear displacement of the peaks, proportional to the relative density of the layer (ρ).

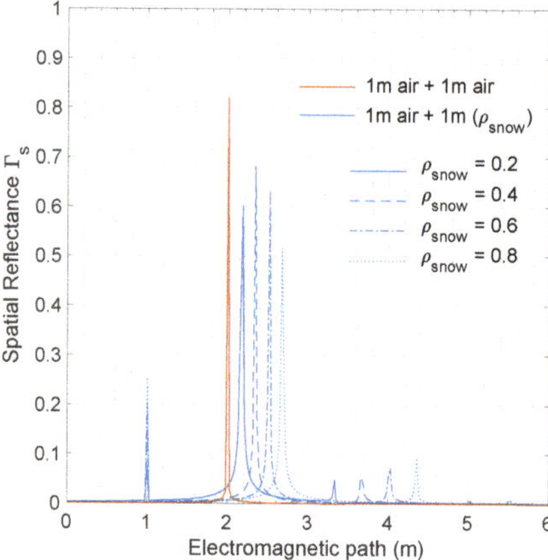

Figure 7. SFCW radar simulated spatial reflection of a 1 m snowpack with the relative densities 0, 0.2, 0.4, 0.6, and 0.8.

In view of this effect, and according to the definition of the snow water equivalent for a multilayer (n + 1 media) particularized to one-layer snowpack

$$\text{SWE}(m) = \sum_{i=1}^{n-1} \rho_i d_i(m) = \rho d(m),\tag{22}$$

we can conclude that there is a linear relationship between the SWE and the position of reflection on the sheet metal when it is covered by a layer of dry snow. This relationship, a priori, is not obvious. Indeed, the position of the snow–metal peak is determined by the optical or electromagnetic path, which depends linearly on the refractive index of the layer, while the SWE is density related. However, we have just seen the proportionality between them. We can check this fact in the case of dry snow. In fact, using the empirical expression proposed in [10,14], we can observe that the real part of the square root of the relative dielectric permittivity is approximately linear with the snow density as shown in Figure 8—i.e., the real part of the refraction index of dry snow is proportional to the density [17]. Then, we can assume a relationship between the real part of the refractive index, n', and the relative density given by

$$n' = \text{Re}(n) = \text{Re}\left(\sqrt{\varepsilon}\right) \simeq 1 + a\rho \tag{23}$$

where Re denote the real part of a complex number. Factor a in the above equation is obtained from a linear fit of Figure 8 ($a = 0.844$).

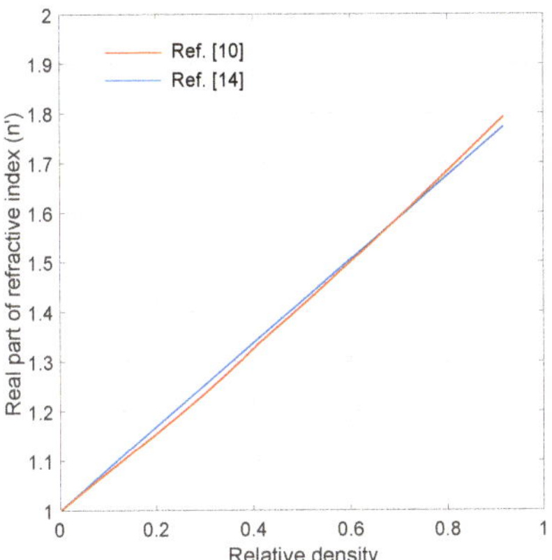

Figure 8. Real part of refractive index vs. the snow density at a frequency of 1 GHz and 0% LWC. Results using [10,14] are represented by blue and orange lines, respectively. The linear fit gives $n'(\rho_{\text{relative}}) = 1 + 0.8439\rho_{\text{relative}}$.

Then, the 'flight time' or 'optical/electromagnetic path', D, into the snow structure, which is the distance between the air–snow peak and the snow–metal one, can be written as

$$D = \sum_{i=1}^{n-1} n'_i d_i = \sum_{i=1}^{n-1} (1 + a\rho_i)d_i = \sum_{i=1}^{n-1} d_i + a \sum_{i=1}^{n-1} \rho_i d_i = H + a\,\text{SWE} , \tag{24}$$

where n'_i is the real part of refractive index of the ith slab, of thickness d_i, and H is the snow depth. Then, we have confirmed the relationship between the optical path and the SWE, which justifies those discussed in relation to Figure 7. The displacement position (see Figure 5a) of the snow–metal sheet reflection peak with respect to the position of the sheet in the air, ΔD, is given by

$$\Delta D(m) = D - H = a\,\text{SWE}(m) = 0.8439\,\text{SWE}(m) \tag{25}$$

$$\text{SWE(m)} = 1.1850\,\Delta D(m) \tag{26}$$

Figure 9 shows the comparison between the SWE vs. ΔD calculated with the previous expression (solid line) and the one provided by the displacement of the metal reflection by means of the matrix model (circles), for a layer 1 m thick and densities from 0.2 to 0.8 (SWE from 0.2 to 0.8 m). Accordingly, we can conclude that the SWE of a snowpack is intrinsically related to its electromagnetic thickness. The measurement of the SWE by this approximate procedure is independent of the measurement of the height of the snow cover—we should only need to know the displacement of the snow–metal reflection relative to the position without snow cover, to obtain the SWE.

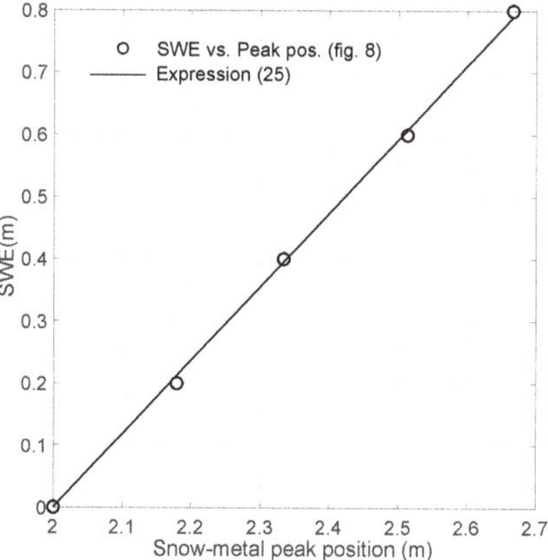

Figure 9. Comparison between the SWE and the displacement of the metal reflectance for the structure of Figure 7 by matrix simulation (circles) and the analytical expression 25 (solid line) assuming linearity of the real part of refractive index vs. the density.

4. Materials and Methods

4.1. Test Site Description

The experimental validation of our SFCW radar was carried out at the Formigal–Sarrios test site located in the Spanish Pyrenees (42°45′40.6″N 0°23′31.8″W) at an elevation of 1800 m a.s.l. (Figure 10). The site was the Spanish location established by the Spanish State Meteorological Agency (Agencia Estatal de Meteorología (AEMet)) in the World Meteorological Organization Solid Precipitation Intercomparison Experiment (WMO-SPICE) [36]. Currently, the site is equipped with sensors to continuously record the meteorological and snow properties. This installation, managed by the territorial delegation of AEMet in Aragon, is an exceptional laboratory for cryosphere studies since it has underground electrical infrastructure, broadband communications, real-time video feeds of the different experiments, a small warehouse with work tools as well as has easy access in the winter, as it is located inside the ski resort Aramón-Formigal. It is a closed area with restricted access and is maintained by the ski resort staff.

Figure 10. The AEMet Formigal–Sarrios test site located in the Spanish Pyrenees (42°45′40.6″ N 0°23′31.8″ W) at an elevation of 1800 m a.s.l. The cosmic ray neutron gauge (CRN) N014 from the Automatic Hydrologic Information System of the Ebro River Basin (SAIH-CHE) is buried by the snow. The SFCW radar is located near the CRN.

It is located in a safe area where a stable snowpack is available from mid-November to May, reaching up to 2 m of snow thickness. Due to its flatness and low exposure to the winds, it has a very homogeneous snow cover through the area of experimentation.

We compared our results with the SWE cosmic ray neutron attenuation gauge (CRN) #N014 [37] from Automatic Hydrologic Information System of the Ebro River Basin (Sistema Automático de Información Hidrológica de la Confederación Hidrográfica del Ebro (SAIH-CHE)), located 10 m away from the SFCW radar measurement area (Figure 11b).

Figure 11. (**a**) Diagram of the SFCW radar measurement process. (**b**) The SAIH-CHE N014 CRN gauge in soil. (**c**) Prototype of the SFCW radar and metal sheet.

4.2. Field Measurement Radar Setup

The SFCW radar was implemented using a SDR platform. Currently, SDR is one of the emerging areas of research in low cost and accurate radars. The radio frequency oscillator, the receiver ultrawide band amplifier, the demodulator, and the analog-to-digital converters were programmed in low level software in a high-performance SDR USRP X300 with a daughter coherent board UBX 10–6000 MHz from Ettus Research. The final system was a coherent and full-duplex wideband transceiver that covered frequencies from 10 to 6 GHz (see Figure 11a). SDR control was performed in C++ from the remote computer located in the experimental field, which also processes the IF signals at each frequency of scan to, finally, obtain the complex amplitude of the reflected signal vs. frequency.

The data are requested by a local computer from this remote system via a 4G link. The FFT of the signals and the visualization is performed on the local computer. Both computers develop their algorithms using MATLAB®. A detailed picture of the processes being carried out is shown in Figure 1 in Section 2.1. Finally, the system used aluminum Vivaldi transmitter and receiver antennas (UWB3 from RFSpace) placed next to each other and protected with two radomes. Both antennas also included a heater system (Figure 12a) to avoid undesirable freezing problems. It has been verified that the radiant parameters (S_{11}) of the antennas were not modified, from 150 MHz to 6 GHz, by incorporating the heaters.

(a) (b) (c)

Figure 12. (**a**) Detail of the emission and reception antennas. At the bottom, we can see the heating resistors. (**b**) Overview of the complete system showing the flatness of snowpack in the work area. (**c**) Virtual "0" calibration plate (CP) placed at 254 cm of the metal sheet.

Finally, an aluminum sheet of 1.5×1.5 m (see Figure 11c) was placed on the ground in an orthogonal direction to the antenna's axis and was the n-medium of the stack with an intrinsically imaginary behavior of relative permittivity. This sheet was buried underneath the snowpack during the entire snow-covered period. Table 1 summarizes the main parameters of the radar.

Table 1. SFCW radar specifications.

Parameter	Value	Parameter	Value
Start frequency	150 MHz	Stop frequency	6 GHz
Number of steps (N)	390	Stepped-frequency, δf	15 MHz
Chirp period	$\approx$80 s	Unambiguous range, R_u [2]	10 m
Resolution (in vacuum), δR [1]	$\approx$2.5 cm	Intermediate frequency (IF)	31.250 kHz
Sampling rate	3.125 MS/s	Antennas wide band	675 MHz–12 GHz
Directivity of antennas	11dBi@2.4GHz	FOV [3] of Tx and Rx antennas	<15°

[1] Expression 7, [2] Expression 8, [3] Field of View.

A small weather station, CR300 unit, and a temperature gauge from Campbell Scientific, was included in the system to systematically record the temperatures and activate antenna heaters when the outside temperature dropped below 4 °C.

For calibration purposes, a temporal metallic calibration plate (CP) was placed at a position between the soil and antennas and represents a virtual position that we call '0' and a reference signal to obtain the spatial reflectance of snow cover. In our system, this was located 254 cm from the metal platform on the soil. This CP was also used to optimize the receiver gain in each frequency of the chirp. The measured file generated, $\hat{S}_{CP}(f_{0,RF} + i\delta f)$, represents a complex amplitude array reference for the succeeding measurements as we will see later. We have selected this position for the definition of the origin to be able to perform calibrations also in the winter period if necessary. In addition, the choice of this origin simplifies the interpretation of spatial reflectance signals by being all of the same sign.

4.3. Measurement Process

The remote computer, controlled by the local computer, runs a C++ program in which the SDR transmitted and received 390 frequencies between 150 MHz and 6 GHz. The duration of this sweep was approximately 80 s. The IF used (f_{BB}) was 31.25 kHz and was sampled at a rate of 3.125 MHz. The received signal at each frequency $U_I(i,t)$ and $U_Q(i,t)$, was multiplied numerically by the signal at f_{BB} frequency, and low pass filtered to obtain the magnitude and phase of the amplitude received at each frequency. This file, $\hat{S}(f_{0,RF} + i\delta f)$, was sent via the 4G link to the local computer.

The measured spectral amplitude is divided by that measured with the calibration plate, $\hat{S}_{CP}(i)$. The spectral reflectance is, then, calculated as

$$\Gamma(i) = \frac{\hat{S}(f_{0,RF} + i\delta f)}{\hat{S}_{CP}(f_{0,RF} + i\delta f)}, \quad i = 0, \ldots, N. \tag{27}$$

Subsequently, the FFT of $\Gamma(i)$ was performed to obtain the spatial reflectance, $\Gamma_s(R)$, the file with time and date identification was archived, and a real-time graphical representation was made. The definition of reflectance given in expression 18 requires the use of plane waves according to the theoretical model. In the experimental situation, this is not possible due to the proximity of the antennas to the snow cover, so the experimental reflectance given by (26) is not exactly that given in (18), but in normal incidence will be proportional. For this reason, we have not used a different symbol for both cases. In the graphs, Γ_s, vs. electromagnetic path the most important information is the spatial position of the peaks and their relative magnitudes.

On 25 January 2019, before the first important snowfall of 2019, the SFCW radar was operative. A previous series of measurements of the distance of metal sheet on the soil (2.54 m) were made with the radar to later obtain the height of the snow cover.

In Figure 13a, we can see the comparison between the measurement of the position of the sheet without snow cover, and the reflectance with a thickness of 61 cm of snow. The reflectance position on the snow–metal interface moved to the right as a result of the increased flight time due to the refractive index of the snow cover. Reflection on the air–snow interface was also observed symmetrically, from which it is possible to deduce the geometric height of the snow cover. According to expression 25, we can, thus, obtain the SWE. The position of the first peak associated with the air–snow interface was 1.923 m, and considering that the reference position of the sheet metal was 2.538 m, we obtain a depth of 61 cm. On the other hand, the displacement of the snow–metal interface was 2.667 m − 2.538 m = 0.129 m. Expression 25 allows us to obtain the SWE as 0.129 × 1.185 = 0.153 m.

The reflectance simulation by means of the matrix model was superimposed. To fit the theoretical reflectance to the experimental one, it was necessary to adjust the density of the layer of height 0.61 m, to 0.26, and then the SWE using expression 21 is 0.26 × 0.61 = 0.158 m, which is similar to the previous result and confirms the validity of the proposed model.

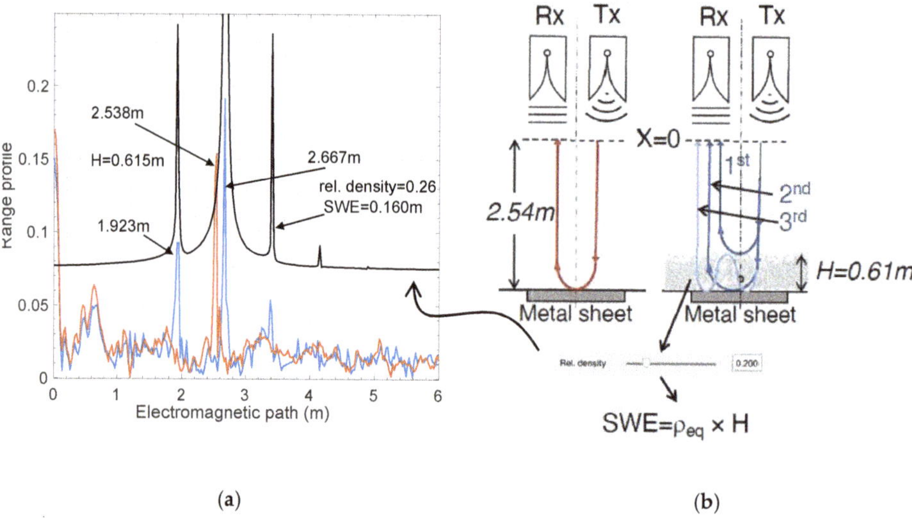

Figure 13. (**a**) Comparison of the experimental spatial reflectance obtained for a snow layer 61 cm deep and the one simulated by the matrix model. (**b**) Outline of the model adjustment procedure with the experimental curve using the density slider of the visual program.

4.4. Milimeter Wave Radar Depth Sensor

In January 2020, to reinforce the measure of the snow depth, we incorporated a millimeter wave radar sensor (mmWave). The rapid evolution of radars has generated, in the last years, the so-called mmWave sensors, characterized by their small dimensions, low cost, and high processing performances, which were unthinkable a few years ago. In many of them, the transmitter, the receiver, the antennas, the low noise amplifier, the mixer, and the pre-processing are included in a single chip.

Currently, we use the sensor MMIC TRX_120_001 of Silicon Radar [38]. This integrated circuit operates in the band for industrial, scientific, and medical (ISM) purposes of 120 GHz (λ = 2.5 mm). These components, halfway between photonics and microwaves, are practically handled as photonic devices and components based on plastic materials—such as lenses, guides, etc.—are typically used in this technology. Due to its rapid development, the application fields are yet to be determined. An example of this fact is the application proposed.

Snow height measurement is typically performed using ultrasonic sensors by measuring the flight times of a pulse emitted by the transducer and reflected by the snow surface. However, the low acoustic reflectance of the first layer, in particular, in the case of fresh snow with a high air content, is very small, generating noisy signals. In addition, achieving good accuracy requires compensation of the propagation speed with the temperature, which greatly complicates achieving high accuracy.

The use of an FMCW radar at 120 GHz and a bandwidth of 6 GHz, based on Silicon Radar MMIC TRX_120_001 IC, at a normal incidence toward the snow allowed us to obtain sufficient accuracy to verify the height measurement of the SFCW radar. As we have already seen in the examples calculated with the matrix method, the reflectance of the first snow layer in the 1–6 GHz band was very small. This is a consequence of the slight change in the refractive index between the air and the first surface of the snow cover, especially for low-density, fresh snow. However, at frequencies of 120 GHz, there is a low penetration of waves in the snowpack, as the relative permittivity models foresee; nevertheless, there is a strong surface reflectance dominated by diffusion rather than optical reflection by the refractive index change. This effect is a consequence of the use of wavelengths comparable to the surface grain of the snowpack. In Figure 14, we can see the implementation of this auxiliary system.

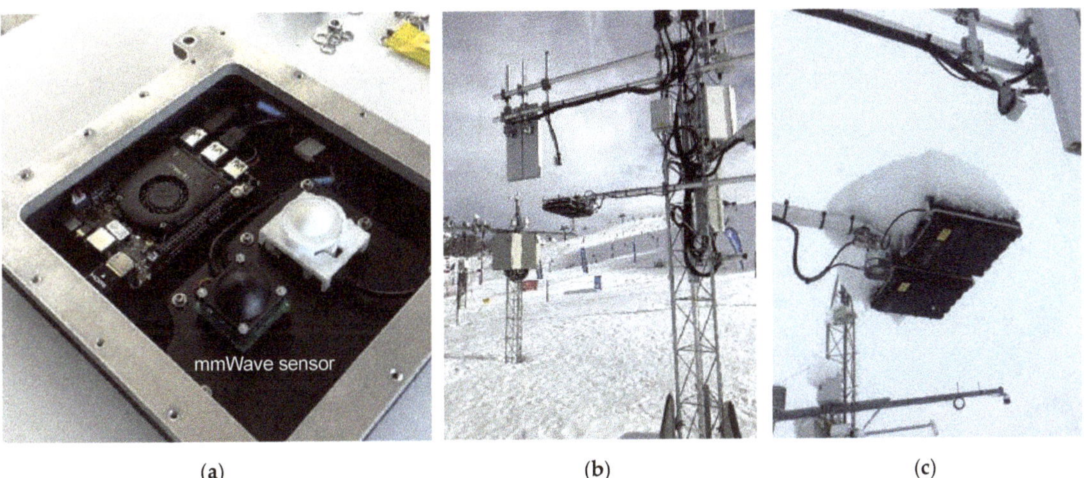

(a) (b) (c)

Figure 14. (**a**) 120 GHz FMCW mmWave radar sensor with a collimation lens manufactured in plastic. (**b**) Position of the height sensor near the SFCW radar antennas and outside its field of view. (**c**) Bottom view of the system.

5. Results

In this section, we describe the main results obtained during the winter period from 1 November 2019 to the 1 May 2020.

Figure 15 shows two examples of range profiles made on the 1 and 20 of March 2020. The orange curve represents the reflectance of the metal sheet registered before the winter period. This is the reference reflectance for the measurement of the geometric height, H, as well as the SWE of the snowpack. Figure 15a clearly shows the profile of a signal as predicted by the matrix model of the snowpack: an important reflectance of the snow–metal interface and a smaller one corresponding to the air–snow interface, as well as the secondary reflection close to 3.5 m of the electromagnetic path. This measure corresponds to a period with low night temperatures, which implies that at the LWC of the snow was small, which justifies the appearance of the secondary reflection.

We observe a lower height of the reflectance of first peak compared to the snow–metal one, as predicted by the matrix model (see Figure 5a). Figure 15b shows the measure made in the early morning of the 20 March, after a period of high temperatures even at night. In this case, we see a more intense reflectance peak than in the previous case associated with air–snow reflection, the disappearance of the snow–metal interface reflection, and, of course, the secondary reflection. This situation corresponds to the one shown in Figure 6a, but with a LWC greater than 4%. In this situation, it is not possible to measure the SWE; however, it is possible to measure the geometric height.

In these figures and the following ones, it is not possible to see the transitions between layers as seen in Figure 5b of the model. These internal reflections are very small and, to see them, we would need to improve the signal-to-noise ratio of the radar. However, it is also possible that the strong transformations produced in the snowpack in the Pyrenees, with wide thermal fluctuations and strong solar irradiations can produce smoothing of these interfaces and transform the multilayer structure to a single-layer structure, from an electromagnetic point of view, in a short time.

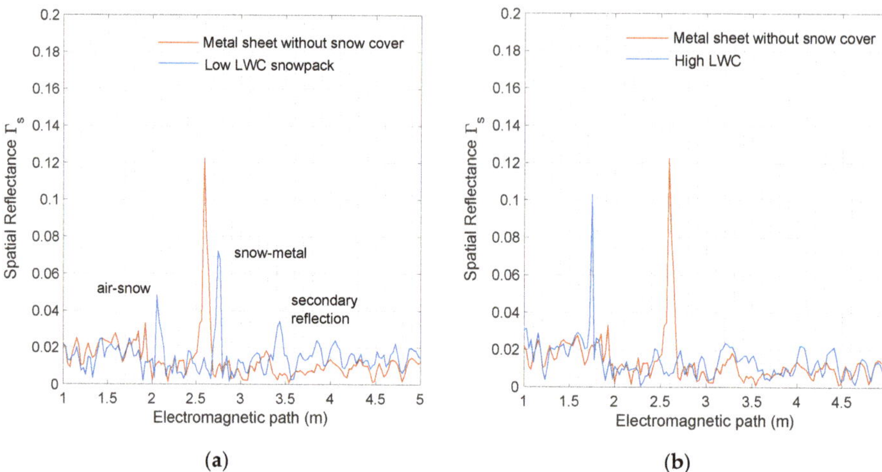

(a)

(b)

Figure 15. Spatial reflectances of two measurements compared with the metal sheet without snow cover case (orange): (**a**) 1 March 2020, 1:18 AM; (**b**) 20 March 2020, 3:24 AM.

In Figure 16a, we depict the measured range profile from 8 March, after a significant snowfall, and an elevation of temperatures with an increase of the LWC, also produced a very wet surface layer magnifying the air–snow reflectance even more (see Figure 6a). However, in Figure 16b, just a few hours later, in the early morning of 9 March, we observed the refreezing of the structure, with a decrease in the air–snow reflectance, due to the reduction of the LWC, and the appearance of the displaced snow–metal reflection, 2.92 m − 2.54 m with respect to the reference reflectance, indicating about 0.44 m of SWE according to expression 25.

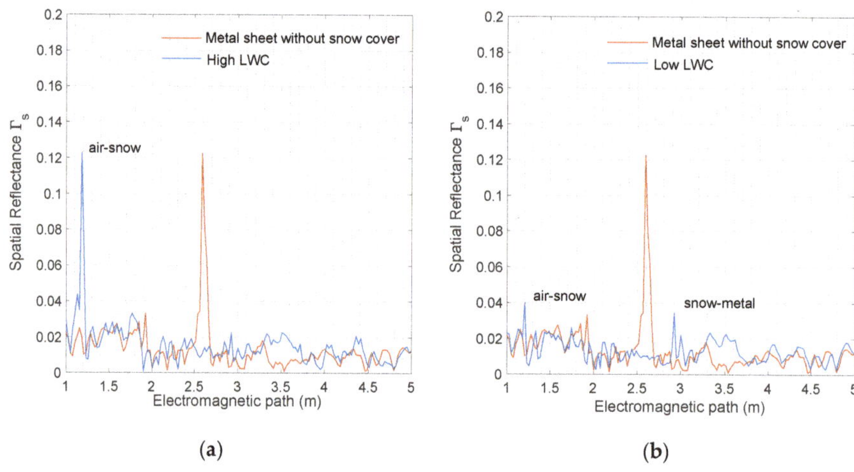

(a)

(b)

Figure 16. Spatial reflectances of two measurements compared with the metal sheet without snow cover case (orange): (**a**) 8 March 2020, 4:46 PM; (**b**) 9 March 2020, 1:45 AM.

In Figure 17, we can compare the profiles of reflections of a cold period (20 January) that was somewhat sunny, where dry and little-transformed snow predominates, with a warmer period (1 April) with a high content of liquid water. In the first case (Figure 17a), the lower air–snow reflection was lower than that of the snow–metal, which moved to the right. In Figure 17b, we only see a reflection on the upper face of the snow cover.

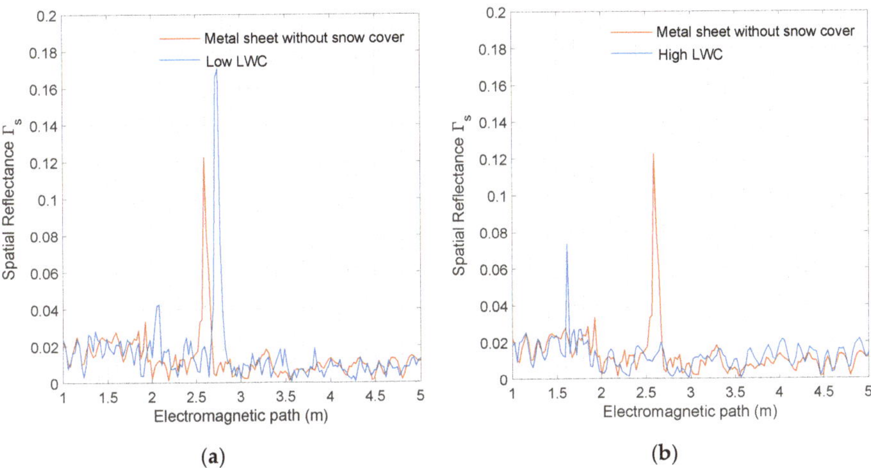

Figure 17. Spatial reflectance of two measurements compared with the metal sheet without snow cover case (orange): (**a**) 20 January 2020, 7:02 AM; (**b**) 1 April 2020, 3:53 PM.

Figure 18 shows a summary of the conditions under which the experiment was developed. The upper graph shows the temperatures recorded throughout the measurement period. This information is important since the snowpack is exposed to periods with temperatures higher than 0 °C, even at night.

The central chart represents an image of the space-time profiles recorded over the measurement period. In this image, it is possible to appreciate the time intervals in which the reflections of the air–snow interface or the snow–metal interface predominate.

In the lower image, the measured depths of snow cover and the SWE are represented. Depths measured with an ultrasonic sensor located in the position of the CRN probe, are compared with the measurements carried out with the mmWave sensor described in Section 4.4. Finally, the comparison between the SWE measured by the CRN probe and that obtained from the spatial profiles by measuring the displacement of snow–metal reflection in accordance with the expression 25 is shown.

We observed in the central image that, toward the end of the winter period, the reflection on the metal sheet was weaker due to the greater LWC compared with in the early part of winter. However, it was possible to find areas where the reflection was recovered, and it was, therefore, still possible to obtain the SWE. In the central chart, it is not possible to see them; however, a detailed analysis of the profiles allows one to locate those reflections. We have represented them by points in the graph below in Figure 18. This effect focuses on the periods located in the second half of April and the first fortnight of May, in which only some useful points can be recovered.

In the center chart, a reference line was plotted at the 2.54 m position, and the magnitudes of interest provided by this chart are indicated. First is the geometric height, *H*, which is the distance between the reference line and the upper reflection (air–snow). The maximum height reached in early March is also noted. Secondly is the electromagnetic path, *D*. Finally, the displacement of the reflection of the snow–metal interface, ΔD, represents the variation of the electromagnetic path from that in the air. This distance directly contains information about the SWE of the snowpack, regardless of the height of it.

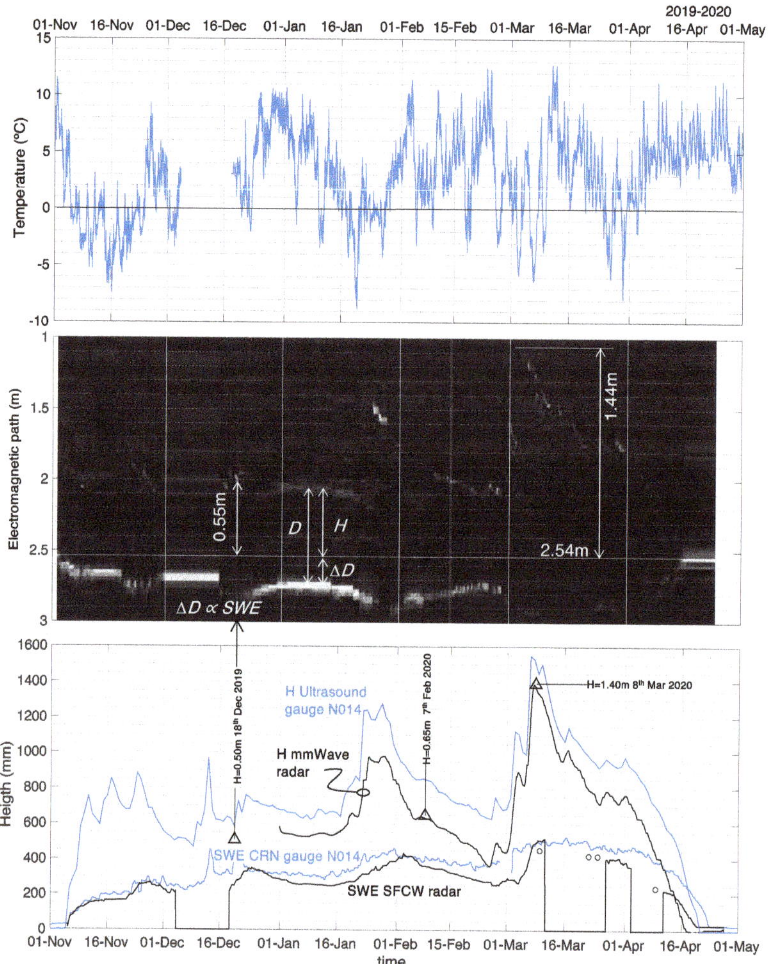

Figure 18. Evolution of certain magnitudes of interest from 1 November 2019 to 1 May 2020. In the figure above: Temperatures recorded in the position of the radar SFCW. Central figure: Graphic chart of the evolution of reflectance profiles over time. Figure below: Height comparison between the millimeter wave radar sensor and the CHE ultrasonic probe (N014) and snow water equivalent (SWE) measured by the SFCW radar and the CHE CRN N014.

The lower graph depicts the depth of the snow cover measured by the ultrasonic sensor located in the position of the CHE tower N014 (blue) and the one measured by the mmWave sensor (black), located about 10 m from the previous one, and which became operational in January 2020. We can see differences of about 20 cm that were proven to correspond to the irregularity of the terrain.

However, several manual depth measurements were also made in the SFCW measurement area, and the correct height measurement was checked. Manual depth measurements are represented with triangles. The first measurement, with a portable depth rod, took place on 8 December 2019, when the mmWave sensor was not yet installed; however, the information can be checked with that provided by the SFCW, being coincident with the manual testing. This comparison is possible for other points, being correct as well.

The two lower curves represent the SWE provided by the CRN gauge (blue) and the SWE obtained by the SFCW radar (black), which was slightly lower than the previous one. Again, the terrain irregularities between the two measurement points would justify this difference. The presence of the metal sheet can modify the drainage properties of the natural soil, retaining more liquid water. This effect would explain that, in some periods, there was the same accumulated water with less snow height. This situation must be corrected in the future, replacing the opaque sheet metal with a semi-buried wired metallic mesh that modifies the properties of the natural soil as little as possible. As the minimum wavelength in the vacuum used by the radar SFCW is 5 cm, a wired mesh with centimeter-order openings would remain effective as a reflector and would not significantly modify the natural soil drainage.

In the second half of April, there was a mismatch between the height and SWE measurements for both the radar SFCW and SAIH-CHE's N014 system. This is because the height sensors are closer to the towers than the measurement position of the SWE. This fact is not significant for thicknesses greater than 20 cm, but for lower thicknesses, the tower carrying all the elements acts as a hot spot making the snow layer disappear more quickly in its vicinity than in the measurement area of the SWE.

On the black curve, there is a period between 4 and 17 December in which the SFCW system was not operational.

6. Conclusions

This work shows the validation of the proposed SFCW radar technique, based in a RDS system, and an electromagnetic model of snowpack, to obtain its SWE and depth, in real time, for hydrological purposes. The RDS technology has been shown to be adequate for this application.

We conclude that the electromagnetic matrix model proposed, well-known in optics, was suitable for this application and reproduce the fundamental features of the reflectances measured by the SFCW, allowing an adequate interpretation of the reflection coefficients based, mainly, on the thickness, the density, and the LWC of layers of the snow cover.

The application of the proposed electromagnetic model of the snowpack has allowed us to derive a simple expression, under the assumption of dry snow, for SWE determination. The SWE can be obtained from the displacement of the position associated with snow–metal reflection, with respect to the position of metal reflection without snow cover. This measurement is independent of the snow height and indicates that the SWE of the snow-pack is linearly related to that displacement. In other words, the increase of "time-of-flight" into snowpack, with respect to vacuum, is directly proportional to SWE. The application of the SWE model to the data captured by SFCW radar during winter 2019–2020 has shown that the agreement with the SWE CRN gauge from Automatic Hydrologic Information System of the Ebro River Basin (SAIH-CHE) was reasonable.

The height of the snowpack can be measured as the first air-snow reflection position with respect to the measurement of the position of the metal sheet in the soil. Therefore, the SFCW technique can be, also, used for measurement of the snow height without an auxiliary sensor.

The new snow depth measurement probe based on a 120 GHz FMCW radar sensor was operational even in freshly fallen snow situations with low density.

The hardware, software, and communications infrastructure are operational and reliable in a harsh environment, providing an operative platform for future experiments.

It was not possible to observe the internal structure of the snowpack. This possibility would require an improvement in the measurement system. Future work must continue with a precise estimation of the error by comparison with gravimetric measurements of SWE. Further improvements of the SFCW hardware and software are necessary to increase the signal-to-noise ratio, improve the 'ghost' peak detection algorithms, and increase the resolution to measure the snowpack stratigraphy. The main improvement in the hardware system will be the design and construction of new ultra-wide band antennas, with linear polarization and good directionality to minimize direct coupling between the transmitting and detecting antennas. The starting design will be a dual ridge horn antenna (DRHA).

Improving the drainage properties of the metal plate, as outlined in Section 5, will also be necessary.

The improvement of the system sensitivity could even allow for removal of the metal plate, using the reflectance of the natural soil as a reference with the advantages that this could have both from a totally non-perturbative analysis of the snowpack and the possibility of making a portable instrument without the need for infrastructure under the snow cover.

Author Contributions: Conceived and designed the experiments, R.A., S.T.B., J.A.Á., and J.M.G.d.P.; Performed the experiments, R.A. and J.M.G.d.P.; Analyzed the data, R.A. and J.A.Á.; Contributed reagents/materials/analysis tools, R.A., S.T.B., and J.A.Á.; Wrote the paper, R.A. and J.M.G.d.P. All authors have read and agreed to the published version of the manuscript.

Funding: This work was supported by DGA-FSE under grant T20_17R to the Photonics Technologies Group of University of Zaragoza.

Acknowledgments: We would like to thank everyone who helped us in this project, especially to those at the Infrastructure Division of the Aragon Regional Office of AEMet and at the Automatic Hydrologic Information System of the Ebro River Basin Authority (SAIH-CHE), for allowing free use of their facilities, for their support and the good advice given during the installation of our radar system. Finally, to Aramon-Formigal Ski Resort for the facilities offered.

Conflicts of Interest: The authors declare no conflict of interest.

References

1. Dutra, E.; Viterbo, P.; Miranda, P.M.A.; Balsamo, G. Complexity of snow schemes in a climate model and its impact on Surface energy and hydrology. *J. Hydrometeorol.* **2012**, *13*, 521–538. [CrossRef]
2. Pomeroy, J.W.; Gray, D.M.; Brown, T.; Hedstrom, N.R.; Quinton, W.L.; Granger, R.J.; Carey, S.K. The cold regions hydrological model: A platform for basing process representation and model structure on physical evidence. *Hydrol. Processes* **2007**, *21*, 2650–2667. [CrossRef]
3. Marks, D.; Dozier, J. Climate and energy exchange at the snow surface in the alpine region of the Sierra Nevada: 2. Snow cover energy balance. *Water Resour. Res.* **1992**, *28*, 3043–3054. [CrossRef]
4. Lehning, M.; Bartelt, P.; Brown, B.; Russi, T.; Stökli, U.; Zimmerli, M. SNOWPACK model calculations for avalanche warning based upon a new network of weather and snow stations. *Cold Reg. Sci. Technol.* **1999**, *30*, 145–157. [CrossRef]
5. Kinar, N.J.; Pomeroy, J.W. Measurement of the physical properties of the snowpack. *Rev. Geophys.* **2015**, *53*, 481–544. [CrossRef]
6. Paquet, E.; Laval, M.; Basalaev, L.M.; Belov, A.; Eroshenko, E.; Kartyshov, V.; Struminsky, A.; Yanke, V.L. An Application of Cosmic-Ray Neutron Measurements to the Determination of the Snow Water Equivalent. In Proceedings of the 30th International Cosmic Ray Conference, Merida, Mexico, 3–11 July 2007; Universidad Nacional Autónoma de México: Mexico City, Mexico, 2008; Volume 1, pp. 761–764.
7. Kodama, M.; Nakai, K.; Kawasaki, S.; Wada, M. An application of cosmic-ray neutron measurements to the determination of the snow-water equivalent. *J. Hydrol.* **1979**, *41*, 85–92. [CrossRef]
8. Kinar, N.J.; Pomeroy, J.W. Automated Determination of Snow Water Equivalent by Acoustic Reflectometry. *IEEE Trans. Geosci. Remote Sens.* **2009**, *47*, 3161–3167. [CrossRef]
9. Pielmeier, C.; Schneebeli, M. Developments in the stratigraphy of snow. *Surv. Geophys.* **2003**, *24.5-6*, 389–416. [CrossRef]
10. Tiuri, M.; Sihvola, A.; Nyfors, E.; Hallikaiken, M. The complex dielectric constant of snow at microwave frequencies. *IEEE J. Ocean. Eng.* **1984**, *9*, 377–382. [CrossRef]
11. Sihvola, A.; Tiuri, M. Snow Fork for Field Determination of the Density and Wetness Profiles of a Snow Pack. *IEEE Trans. Geosci. Remote Sens.* **1986**, *GE-24.5*, 717–721. [CrossRef]
12. Denoth, A.; Foglar, A.; Weiland, P.; Mätzler, C.; Aebischer, H.; Tiuri, M.; Sihvola, A. A comparative study of instruments for measuring the liquid water content of snow. *J. Phys. D* **1984**, *56*, 2154–2160. [CrossRef]
13. Mätzler, C. Applications of the interaction of microwaves with the natural snow cover. *Remote Sens. Rev.* **1987**, *2*, 259–387. [CrossRef]
14. Wiesmann, A.; Mätzler, C. Microwave Emission Model of Layered Snowpacks. *Remote Sens. Environ.* **1999**, *70*, 307–316. [CrossRef]
15. Lundberg, A.; Thunehed, H.; Bergström, J. Impulse Radar Snow Surveys -Influence of Snow Density. *Nord. Hydrol.* **2000**, *31*, 1–14. [CrossRef]
16. Lundberg, A.; Thunehed, H. Snow Wetge Influence on Impulse Radar Snow Surveys. Theoretical and Laboratory study. *Nord. Hydrol.* **2000**, *31*, 89–106. [CrossRef]
17. Gubler, H.; Hiller, M. The use of Microwave FMCW Radar in Snow and Avalanche Research. *Cold Reg. Sci. Technol.* **1984**, *9*, 109–119. [CrossRef]

18. Holmgren, J.; Sturm, M.; Yankielun, N.E.; Koh, G. Extensive measurements of snow depth using FMCW radar. *Cold Reg. Sci. Technol.* **1998**, *27.1*, 17–30. [CrossRef]
19. Marshall, H.P.; Koh, G. FMCW radars for snow research. *Cold Reg. Sci. Technol.* **2008**, *52*, 118–131. [CrossRef]
20. Okorn, R.; Brunnhofer, G.; Platzer, T.; Heilig, A.; Schmid, L.; Mitterer, C.; Schweizer, J.; Eisen, O. Upward-looking L-band FMCW radar for snow cover monitoring. *Cold Reg. Sci. Technol.* **2014**, *103*, 31–40. [CrossRef] [PubMed]
21. Yankieluna, N.; Rosenthalb, W.; Davis, R.E. Alpine snow depth measurements from aerial FMCW radar. *Cold Reg. Sci. Technol.* **2004**, *40*, 123–134. [CrossRef]
22. Pasian, M.; Barbolini, M.; Dell'Acqua, F.; Espín-López, P.F.; Silvestri, L. Snowpack Monitoring Using a Dual-Receiver Radar Architecture. *IEEE Trans. Geosci. Remote Sens.* **2019**, *57*, 1195–1204. [CrossRef]
23. Marshall, H.P.; Schneebeli, M.; Koh, G. Snow stratigraphy measurements with high-frequency FMCW radar: Comparison with snow micro-penetrometer. *Cold Reg. Sci. Technol.* **2007**, *47*, 108–117. [CrossRef]
24. Panzer, B.; Gomez-Garcia, D.; Leuschen, C.; Paden, J.; Rodriguez-Morales, F.; Patel, A.; Markus, T.; Holt, B.; Gogineni, P. An ultra-wideband, microwave radar for measuring snow thickness on sea ice and mapping near-surface internal layers in polar firn. *J. Glaciol.* **2013**, *59*, 244–254. [CrossRef]
25. Available online: https://www.ettus.com (accessed on 7 February 2021).
26. Costanzo, S.; Spadafora, F.; Di Massa, G.; Borgia, A.; Costanzo, A.; Aloi, G.; Pace, P.; Loscrì, V.; Moreno, H.O. Potentialities of usrp-based software defined radar systems. *Prog. Electromagn. Res. B* **2013**, *53*, 417435. [CrossRef]
27. Carey, S.C.; Scott, W.R. Software defined radio for stepped frequency, Ground Penetrating Radar. In Proceedings of the IEEE International Geoscience and Remote Sensing Symposium (IGARSS), Fort Worth, TX, USA, 23–28 July 2017; pp. 4825–4828. [CrossRef]
28. Marimuthu, J.; Bialkowski, K.S.; Abbosh, A.M. Software-Defined Radar for Medical Imaging. *IEEE Trans. Microw. Theory Technol.* **2016**, *64*, 643–652. [CrossRef]
29. Prabaswara, A.; Munir, A.; Suksmono, A.B. GNU Radio based software-defined FMCW radar for weather surveillance application. In Proceedings of the 6th International Conference on Telecommunication Systems, Services, and Applications (TSSA), Bali, Indonesia, 20–21 October 2011; pp. 227–230. [CrossRef]
30. Alonso, R.; Pozo, J.M.G.d.; Peruga, I.; Buisán, S.; Álvarez, J.A. Analysis Of Snow Water Equivalent (Swe) Of Snowpack By An Ultra Wide Band Step Frequency Continuous Wave Radar (Sfcw). In Proceedings of the 14th European Conference on Antennas and Propagation (EuCAP), Copenhagen, Denmark, 15–20 March 2020; pp. 1–5. [CrossRef]
31. Nguyen, C.; Park, J. *Stepped-Frequency Radar Sensors. Theory, Analysis and Design*; Springer International Publishing: Berlin/Heidelberg, Germany, 2016; ISBN 978-3-319-12271-7.
32. Øyan, M.J.; Hamran, S.-E.; Hanssen, L.; Berger, T.; Plettemeier, D. Ultrawideband gated step frequency ground-penetrating radar. *IEEE Trans. Geosci. Remote Sens.* **2012**, *50*, 212–220. [CrossRef]
33. Taylor, J.D. *Ultrawideband Radar Applications and Design*; CRC Press: Boca Raton, FL, USA, 2012; ISBN 9781420089868.
34. Kurosawa, K.; Pierce, R.M.; Ushioda, S.; Hemminger, J.C. Raman-scattering and attenuated-total-reflection studies of surface-plasmon polaritons. *Phys. Rev. B* **1986**, *33*, 789–798. [CrossRef]
35. Born, M.; Wolf, E. *Principles of Optics: Electromagnetic Theory of Propagation, Interference and Diffraction of Light*, 7th ed.; Cambridge University Press: Cambridge, UK, 1975; ISBN 978-0-521-64222-4.
36. Buisán, S.T.; Earle, M.E.; Collado, J.L.; Kochendorfer, J.; Alastrué, J.; Wolff, M.; Smith, C.D.; López-Moreno, J.I. Assessment of snowfall accumulation underestimation by tipping bucket gauges in the Spanish operational network. *Atmos. Meas. Tech.* **2017**, *10*, 1079–1091. [CrossRef]
37. Available online: http://www.saihebro.com/saihebro/index.php?url=/datos/ficha/estacion:N014 (accessed on 7 February 2021).
38. Available online: https://siliconradar.com/products/single-product/120-ghz-radar-transceiver/ (accessed on 7 February 2021).

 *remote sensing*

Article

Analysis and Validation of a Hybrid Forward-Looking Down-Looking Ground Penetrating Radar Architecture

María García-Fernández , Guillermo Álvarez-Narciandi , Yuri Álvarez López * and Fernando Las-Heras Andrés

Area of Signal Theory and Communications, University of Oviedo, 33003 Oviedo, Spain; garciafmaria@uniovi.es (M.G.-F.); alvareznguillermo@uniovi.es (G.Á.-N.); flasheras@uniovi.es (F.L.-H.A.)
* Correspondence: alvarezyuri@uniovi.es; Tel.: +34-985-182281

Abstract: Ground Penetrating Radar (GPR) has proved to be a successful technique for the detection of landmines and Improvised Explosive Devices (IEDs) buried in the ground. In the last years, novel architectures for safe and fast detection, such as those based on GPR systems onboard Unmanned Aerial Vehicles (UAVs), have been proposed. Furthermore, improvements in GPR hardware and signal processing techniques have resulted in a more efficient detection. This contribution presents an experimental validation of a hybrid Forward-Looking–Down-Looking GPR architecture. The main goal of this architecture is to combine advantages of both GPR architectures: reduction of clutter coming from the ground surface in the case of Forward-Looking GPR (FLGPR), and greater dynamic range in the case of Down-Looking GPR (DLGPR). Compact radar modules working in the lower SHF frequency band have been used for the validation of the hybrid architecture, which involved realistic targets.

Keywords: Ground Penetrating Radar; Synthetic Aperture Radar; imaging; landmine; Improvised Explosive Device; radar

 check for
updates

Citation: García-Fernández, M.; Álvarez-Narciandi, G.; Álvarez López, Y.; Las-Heras Andrés, F. Analysis and Validation of a Hybrid Forward-Looking Down-Looking Ground Penetrating Radar Architecture. *Remote Sens.* **2021**, *13*, 1206. https://doi.org/10.3390/rs13061206

Academic Editors: Deodato Tapete and Qi Wang

Received: 30 December 2020
Accepted: 18 March 2021
Published: 22 March 2021

Publisher's Note: MDPI stays neutral with regard to jurisdictional claims in published maps and institutional affiliations.

1. Introduction

1.1. Background

Non-invasive detection of buried objects can be conducted using different techniques, such as ultrasound, magnetic induction (metal detectors), and Ground Penetrating Radar (GPR). The latter is suitable to detect either metallic or non-metallic targets [1], making it appropriate for infrastructure inspection [2,3], archaeological survey [4], material characterization [5] or landmine and Improvised Explosive Devices (IEDs) detection [6,7].

GPR systems can be classified according to different parameters and criteria. Most GPR systems operate in a monostatic or quasi-monostatic configuration (in which the receiving and transmitting antennas are close to each other in terms of wavelengths). These systems can be further grouped in Forward-Looking GPR (FLGPR) and Down-Looking GPR (DLGPR). In the former the GPR antennas are looking ahead from the platform or vehicle performing the GPR scanning [8], whereas in the latter the antennas are pointing normal to the ground [9,10]. An alternative configuration consists of bistatic GPR systems, in which the transmitting and receiving antennas are located far from each other (in terms of wavelengths) [11].

In the case of DLGPR, the distance between the GPR antennas and the scanned region is smaller than in FLGPR, thus enabling greater dynamic range as well as better resolution. However, the GPR signal reflection at the air–ground interface is also greater in DLGPR, thus jeopardizing the detection of shallow targets (as is the case of most landmines and IEDs).

In contrast, FLGPR architectures avoid the normal reflection on the air–ground interface, thus minimizing the clutter coming from the air–ground interface, but at the expense of worse dynamic range due to greater propagation losses. FLGPR systems are thus

137

suitable for detecting shallow targets. In general, FLGPR systems are mounted onboard terrestrial vehicles, as the oblique incidence allows keeping a safe distance between the vehicle and the scanning area where landmines, IEDs and Unexploded Ordnances (UXOs) can be buried [12]. However, there is a clear risk for the vehicle (and its operator/driver) if a buried explosive target is not detected. To improve the detection probability, several signal processing techniques have been proposed in the field of FLGPR systems [13].

Recently, GPR systems onboard Unmanned Aerial Vehicles (UAVs) have been developed [14–18]. In fact, both academia and industry show great interest in these systems due to their advantages in terms of safety, speed and capability to inspect difficult-to-access areas. It is worth noting that, unlike terrestrial vehicles, contact with the ground, and thus the risk of accidental detonation of buried explosives, is avoided. Most UAV-based GPR systems are based on DLGPR architectures [14–17] to maximize range resolution and dynamic range (at the expense of stronger clutter contributions from the air–ground interface), although side-looking GPR has also been assessed in [18].

The development of a GPR system combining the advantages of FLGPR and DLGPR architectures would result in better detection capabilities, which is of special interest for the detection of buried explosives (landmines, IEDs, UXOs). A feasibility study has been presented in [19], where the proposed hybrid FLGPR-DLGPR was evaluated using ray-tracing and Finite-Difference Frequency-Domain (FDFD) simulation methods, comparing the achieved performance against conventional FLGPR and DLGPR.

1.2. Aim and Scope

This contribution presents the implementation and validation of the hybrid FLGPR-DLGPR proposed in [19]. From the technical point-of-view, one of the major challenges to be addressed is related to the synchronization between the transmitter and the receiver, which must be physically decoupled for operational purposes. In particular, the use of radiofrequency cables connecting two radar modules placed several meters away would not be feasible in the case of UAV-based GPR architectures. The main goal is to prove that non-metallic targets buried only a few centimeters under the ground can be detected, as the air-ground reflection is mitigated. The portable setup presented in [20] has been fitted to accommodate the hybrid FLGPR-DLGPR architecture for its validation.

In this sense, it is important to remark that the proposed hybrid FLGPR-DLGPR system is intended to complement DLGPR architectures, which have been successfully used for detecting deeper targets.

2. Methodology

2.1. Overview of the Hybrid FLGPR-DLGPR Architecture

In order to introduce and compare the proposed hybrid FLGPR-DLGPR architecture [19], the most common GPR architectures (described in Section 1) and the proposed one are depicted in Figure 1. Figure 1a corresponds to a DLGPR system where the GPR antennas are oriented towards the ground with normal incidence. Figure 1b shows an FLGPR architecture, being the GPR antennas several meters away from the imaging domain and oriented at grazing incidence. The proposed hybrid FLGPR-DLGPR architecture is depicted in Figure 1c,d, where the transmitter antenna (at grazing incidence) is placed at several meters from the receiver (pointing normal to the ground). Figure 1c corresponds to the case where the transmitter is static and the receiver moves over the scanning area, whereas the transmitter moves synchronously with the receiver in Figure 1d.

These four configurations were compared considering a realistic scenario in Section 4 ("Analysis") of [19], where a 10 cm $\times$ 4 cm size plastic landmine is buried in a sandy soil with a rough surface. Simulations using FDFD were conducted in the 3.5–5.5 GHz frequency band and they were post-processed using a Synthetic Aperture Radar (SAR) algorithm. The resulting GPR-SAR images can be found in Figure 2. The buried landmine cannot be detected with DLGPR (Figure 2a) nor with FLGPR (Figure 2b) configurations: in the case of the DLGPR most of the incident power is reflected at the air–ground interface, whereas the FLGPR architecture exhibits poor resolution. The hybrid FLGPR-DLGPR architecture successfully combines the advantages of FLGPR and DLGPR as it can be noticed in Figure 2c,d: the air–ground reflection is mitigated with respect to DLGPR, while keeping a good resolution in the GPR-SAR image. As a result, reflections coming from the buried target can be identified. It can be noticed that there are no significant differences between considering a static transmitter (Figure 2c) or moving it synchronously with the receiver (Figure 2d). Thus, for the sake of simplicity, the hybrid FLGPR-DLGPR architecture in which the transmitter is static has been considered to perform the experimental validation.

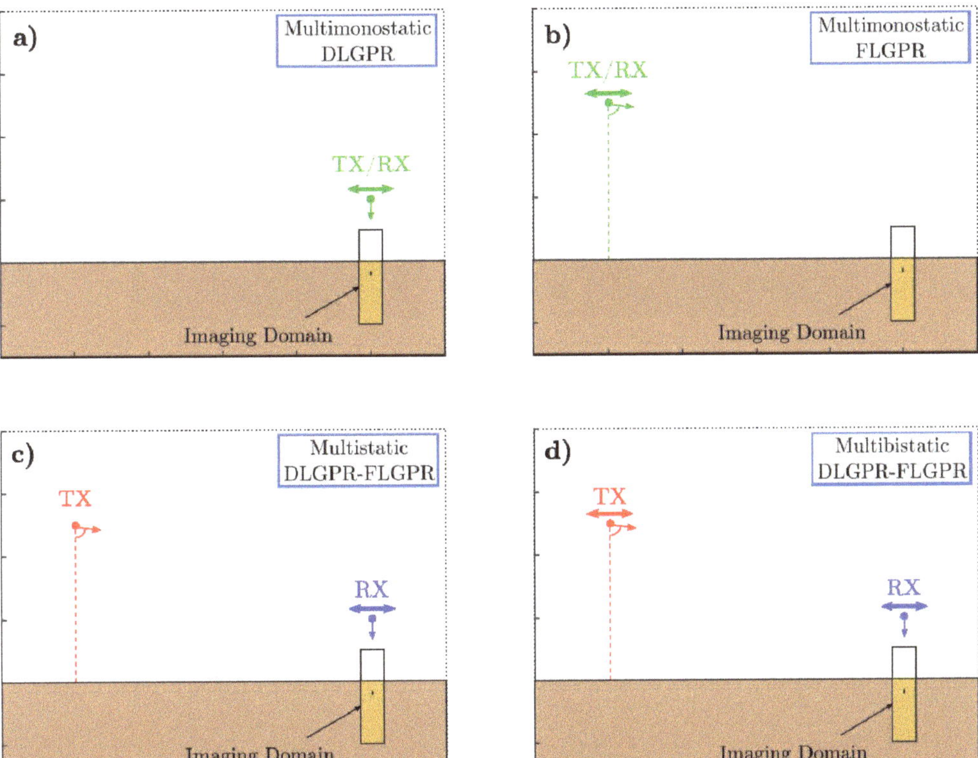

Figure 1. Schemes of different Ground Penetrating Radar (GPR) architectures [21]: conventional Down-Looking GPR (DLGPR) (**a**) and Forward-Looking GPR (FLGPR) (**b**) configuration; proposed hybrid FLGPR-DLGPR architectures in which the transmitter is static (**c**) and the transmitter moves synchronously with the receiver (**d**).

A scheme of the hybrid FLGPR-DLGPR architecture to be implemented and validated is shown in Figure 3. The transmitter will be kept static whereas the receiver will be moved over the scenario under test (that is, the GPR imaging domain). Transmitter and receiver modules must be synchronized to enable the coherent processing of the measurements, not only in range, but also in cross-range (i.e., along the axes orthogonal to the vertical range axis). Assuming ideal conditions, the transmitting antenna will be pointed towards the center of the imaging domain. Part of the incident field (E^{inc}) will be reflected (E^{refl}) in the ground surface, but little or none reflected field will be captured by the receiving antenna, as the specular reflections from the ground surface will not reach it. Another part of the incident field will be transmitted into the ground (E^{trans}) and, in case of a buried target is present, it will be scattered back to the ground surface (E^{scatt}). Finally, a portion of this scattered field might eventually reach the receiving antenna.

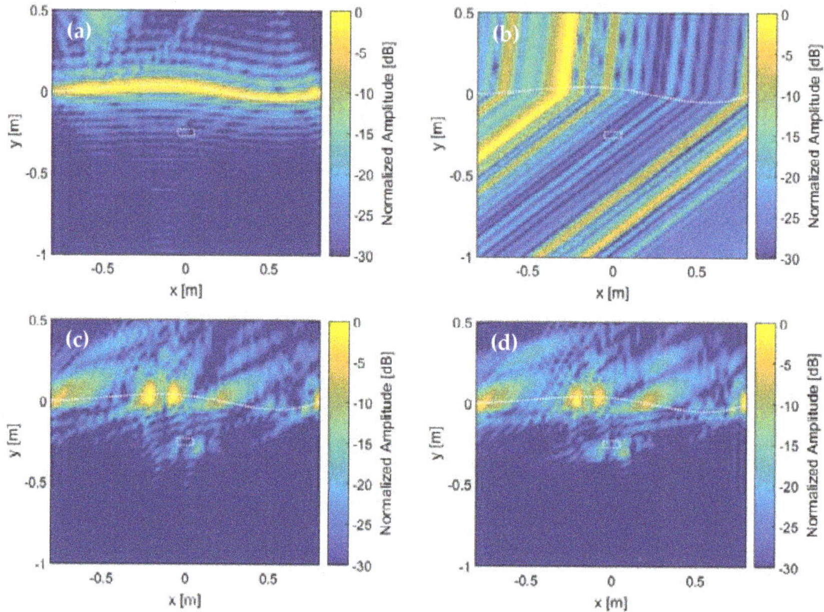

Figure 2. Synthetic Aperture Radar (SAR) image from Finite-Difference Frequency-Domain (FDFD) simulations with a flat-top dielectric target buried under a rough surface (dashed line indicates the soil surface) extracted from [19]: conventional DLGPR (**a**) and FLGPR (**b**) configuration; proposed hybrid FLGPR-DLGPR architectures in which the transmitter is static (**c**) and the transmitter moves synchronously with the receiver (**d**).

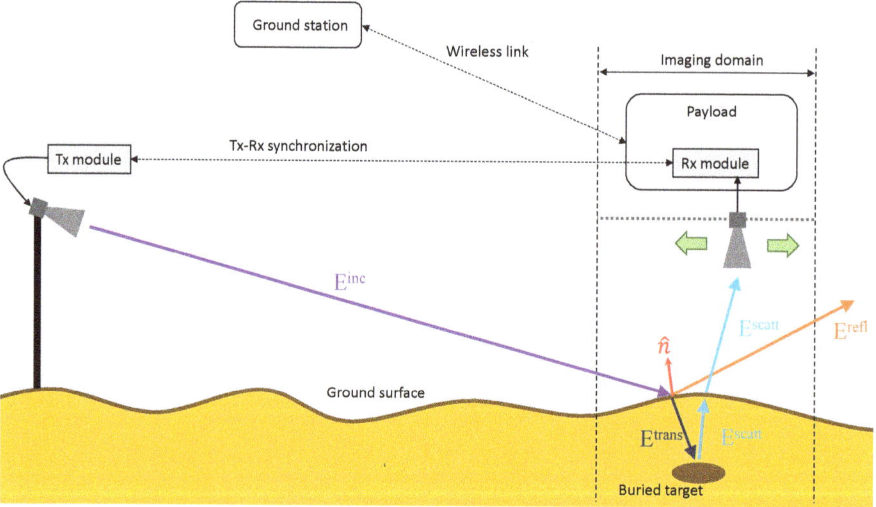

Figure 3. Scheme of the proposed hybrid FLGPR-DLGPR.

Radar measurements are collected at the receiver during the data acquisition and sent to a ground station via a wireless link for further processing. In particular, SAR processing is applied to achieve greater cross-range resolution.

2.2. Radar System Implementation and Processing

As indicated at the end of Section 1, the main issue concerning the implementation of the hybrid FLGPR-DLGPR is the wireless synchronization of the transmitting and receiving radar modules. A set of two PulsOn P440 radar modules [22], working in the 3–5 GHz frequency band, has been selected as these modules can be synchronized without requiring a physical connection between them. Therefore, they are suitable for bistatic and multistatic radar architectures, such as the proposed one. To perform the synchronization the receiving module looks for the first strong arriving signal to determine the time-of-flight. In the hybrid FLGPR-DLGPR architecture, the direct ray between the transmitter and the receiver might be very weak, as the receiving antenna is pointing towards the ground. This caused synchronization issues in the first experiments with these radar modules, as reflections on the ground were interpreted by the radar as the first strong arriving signal.

To overcome these problems, the hardware architecture shown in Figure 4 has been implemented. An additional receiving antenna, denoted as DR Rx antenna, pointing towards the transmitting antenna, has been included in the receiving front-end of the system. Both receiving antennas are connected to a power combiner, the output of which is connected to the receiving radar module. In order to better distinguish between the contributions captured by this DR Rx antenna and the ones captured by the DL-GPR Rx antenna, a delay line, consisting of a 2 m coaxial cable, has also been added between the latter and the power combiner (see Figure 4).

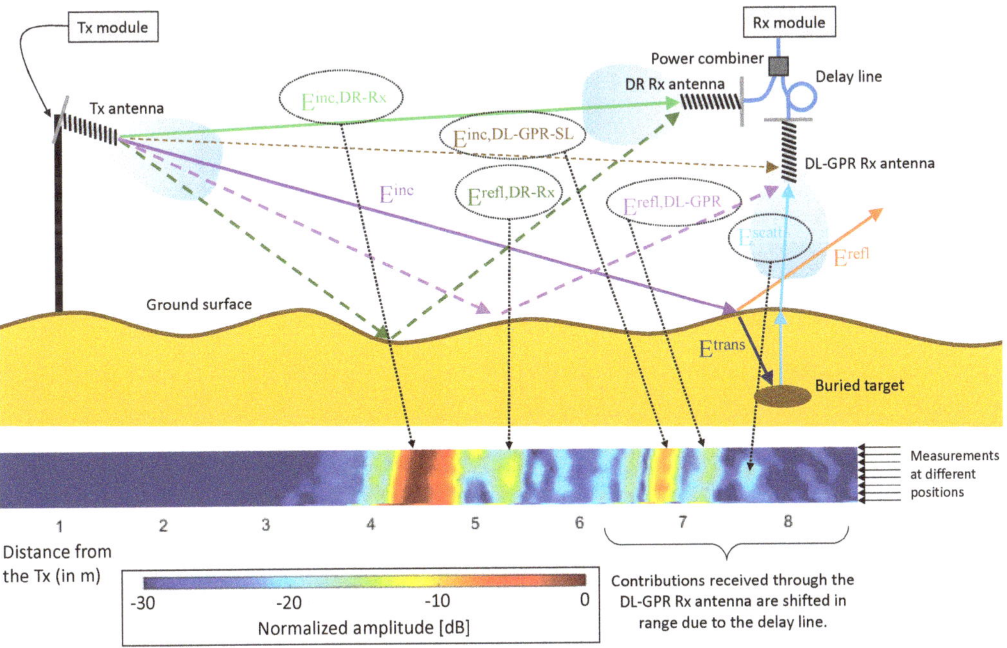

Figure 4. Scheme of the implemented hybrid FL-DL GPR system. Identification of the different received contributions in the retrieved reflectivity plotted as a function of the distance from the transmitter for several measurements.

As shown in Figure 4, the main contributions captured by the DR Rx antenna will be the direct Tx-Rx contributions, $E^{inc,DR-Rx}$. Specular reflections in the ground, $E^{refl,DR-Rx}$, will also be captured, but delayed from $E^{inc,DR-Rx}$ and exhibiting less amplitude, as shown in the reflectivity chart plotted in Figure 4. Concerning the signals captured by the DL-GPR Rx antenna, the first to be received would be the direct Tx-Rx contribution, $E^{inc-DL-GPR-SL}$, captured through a sidelobe of the DL-GPR Rx antenna. Depending on the sidelobe level and the antenna polarization, this contribution may exhibit an amplitude comparable to $E^{inc,DR\text{-}Rx}$ (as it is the case in the example shown in Figure 4). The next contribution that can be identified is the ray reflected on the ground, $E^{refl,DL\text{-}GPR}$. Finally, contributions corresponding to the field scattered by the targets located within the scanning area, E^{scatt}, are received.

Thanks to the use of the delay line, contributions received through the DL-GPR Rx antenna can be better distinguished from contributions received through the DR Rx antenna, just by means of spatial filtering.

Concerning the radar signal processing, the following procedure is applied for each measurement:

(1) The peak of the received time-domain radar signal, which will correspond to $E^{inc,DR\text{-}Rx}$, is identified, so that the time-of-flight is extracted.

(2) A spatial filter is applied to get rid of the contributions captured by the DR Rx antenna. Then, the phase shift introduced by the 2 m-length delay line is compensated.

(3) Contributions $E^{inc\text{-}DL\text{-}GPR\text{-}SL}$ and $E^{refl,DL\text{-}GPR}$ are filtered out by means of another spatial filter, so that the remaining signal peaks would correspond to E^{scatt}.

Once all the measurements have been collected and processed, standard GPR-SAR is applied. First, for each position of the receiver (x, y, z), radar measurements $E_{scatt}(t, x, y, z)$ are transformed into the frequency domain, $E_{scatt}(f, x, y, z)$. Then, the reflectivity at each point of the imaging domain, $\rho(x', y', z')$, is calculated (1):

$$\rho(x',y',z') = \sum_{N_f} \sum_{N_{Rx,meas}} E_{scatt}(f,x,y,z)e^{+jk_0(f)(R_{Tx}+R_{Rx})}, \tag{1}$$

being N_f the number of discrete frequencies in which the working frequency band of the radar modules is discretized, and $N_{Rx,meas}$ the number of measurement points. $k_0(f)$ is the wavenumber in free-space. R_{Tx} and R_{Rx} are the distance between the transmitting antenna and the point of the imaging domain (x', y', z') where the reflectivity is computed, and the distance between this point of the imaging domain and the receiving antenna (DL-GPR Rx antenna), respectively (2), (3):

$$R_{Tx} = \left((x_{Tx} - x')^2 + (y_{Tx} - y')^2 + (z_{Tx} - z')^2 \right)^{1/2}, \tag{2}$$

$$R_{Rx} = \left((x - x')^2 + (y - y')^2 + (z - z')^2 \right)^{1/2}. \tag{3}$$

For the sake of clarity, results presented in this contribution are obtained considering free-space propagation (i.e., the relative permittivity of the entire imaging domain is set to 1).

2.3. Description of the Hardware of the Measurement Setup

The measurement setup described in [20] has been adapted to perform multistatic measurements, resulting in the hardware implementation shown in Figure 5. The transmitting radar module connected to the transmitting antenna has been mounted on a wooden frame located about 5 m away from the center of the scanned area. The height of the transmitting antenna is $z_{Tx} = 105$ cm. The DR Rx and DL-GPR Rx antennas are mounted on the plastic box shown in Figure 6, where the delay line consisting of a 2 m coaxial cable can be observed. The receiving radar module is connected to a micro-computer that also collects positioning and geo-referring information. The geo-referred radar measurements are sent to a ground station (a laptop) using an ad-hoc wireless link [20].

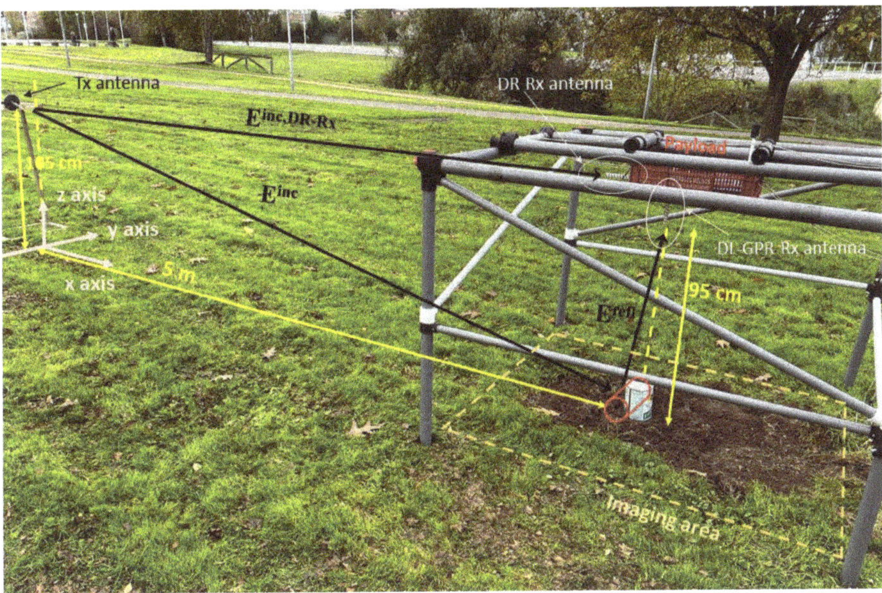

Figure 5. Picture of the implemented FLGPR-DLGPR system. Tested target: metallic can (the profile of the metallic can when laid down is plotted in red color).

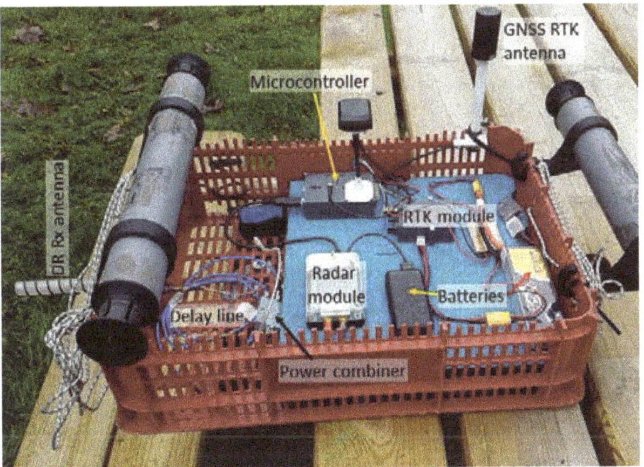

Figure 6. Picture of the devices mounted on the plastic box of the portable scanner.

The plastic box shown in Figure 6 is mounted onto a sliding arm that can be manually displaced along a distance of approximately 150 cm in the x-axis, as described in [20]. The sliding arm can also be moved the same distance in the y-axis. A linear sweep is performed by displacing the plastic box along the sliding arm on top of the frame of the setup. After performing a round-trip sweep along the x-axis, the sliding arm is manually moved along the y-axis to another position. Due to the size of the plastic box, the dimensions of the scanning area are restricted to 100 cm $\times$ 100 cm, approximately. The DL-GPR RX antenna is around 85 cm above-ground.

As mentioned before, a set of two radar modules working in the 3–5 GHz frequency band has been selected. Furthermore, helix antennas have been used for the following reasons: (i) they exhibit good performance in terms of impedance matching and radiation pattern directivity in the 3 to 5 GHz frequency band, which is the working frequency band of the radar modules, and (ii) they have good polarization purity, thus enabling better polarization discrimination of the received signals. In this sense, the DR Rx antenna has the same polarization as the Tx antenna, whereas the DL-GPR Rx antenna has orthogonal polarization with respect to the Tx antenna because of the polarization rotation introduced by the reflection coefficient of the ground.

Positioning and geo-referring information is provided by a dual-band Global Navigation Satellite System (GNSS)–Real Time Kinematics (RTK) system, consisting of a GNSS-RTK antenna and a GNSS-RTK module [23]. RTK can provide positioning information within 1–2 cm accuracy. Finally, the hardware shown in Figure 6 is powered using a LiPo battery. Transmitting and receiving radar modules are equipped with their own battery modules.

3. Results

3.1. Testing Using a Metallic Target Above Ground

First, the hybrid FLGPR-DLGPR system was tested using a calibration target placed on the ground. The calibration target, a 15 cm-diameter by 17 cm-height cylindrical metallic can, was placed in the position shown in Figure 5. Two tests were conducted, one with the metallic can as shown in Figure 5 (axis of the cylindrical can parallel to the z axis, i.e., orthogonal to the ground surface), and another with the metallic can laid down so that the axis of the cylindrical can was parallel to the y axis (i.e., the axis was parallel to the ground surface). It was found that the field scattered by the metallic can was greater in this second scenario, making the detection of the metallic can easier. In consequence, this configuration was chosen for validation purposes. Measurements were conducted in a continuous sweep along the x-axis, as shown in Figure 7. For each position along the y-axis, two sweeps along the x-axis were performed (a round-trip sweep).

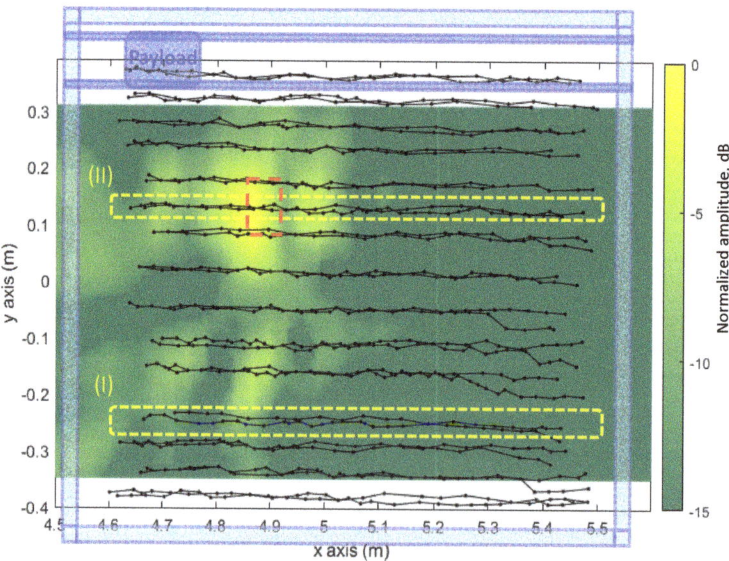

Figure 7. Measurement positions in the XY plane superimposed on the GPR-SAR image retrieved on the $z = 7$ cm XY plane. Black dots (•) indicate individual acquisition positions. Measurement positions corresponding to the same sweep along the x-axis are connected with a solid black line. The placement of the metallic can, being the axis of the cylindric can aligned with the y axis, is highlighted with a dashed red line. For reference purposes, a sketch of the layout of the portable setup [20] is depicted in blue color.

To better illustrate the processing technique, two individual sweeps denoted as (I) and (II) (highlighted with dashed yellow rectangles in Figure 7) have been selected. It should be noted that the individual sweep denoted as (II) intersects the position of the metallic can laid down on the ground. Each individual measurement is processed following the methodology explained in Section 2.2. Results after the second step (i.e., after removing the contributions of the DR Rx antenna and compensating the delay line) for the (I) and (II) sweeps are depicted in Figure 8a,b, respectively. The contribution corresponding to $E^{inc\text{-}DL\text{-}GPR\text{-}SL}$ is the one exhibiting the maximum amplitude. It can be noticed that its location in range increases as the receiver moves away from the transmitter. In the case of Figure 8b, reflections coming from the metallic can (i.e., the scattered field, E^{scatt}) are observed in the range distance of about 6 m.

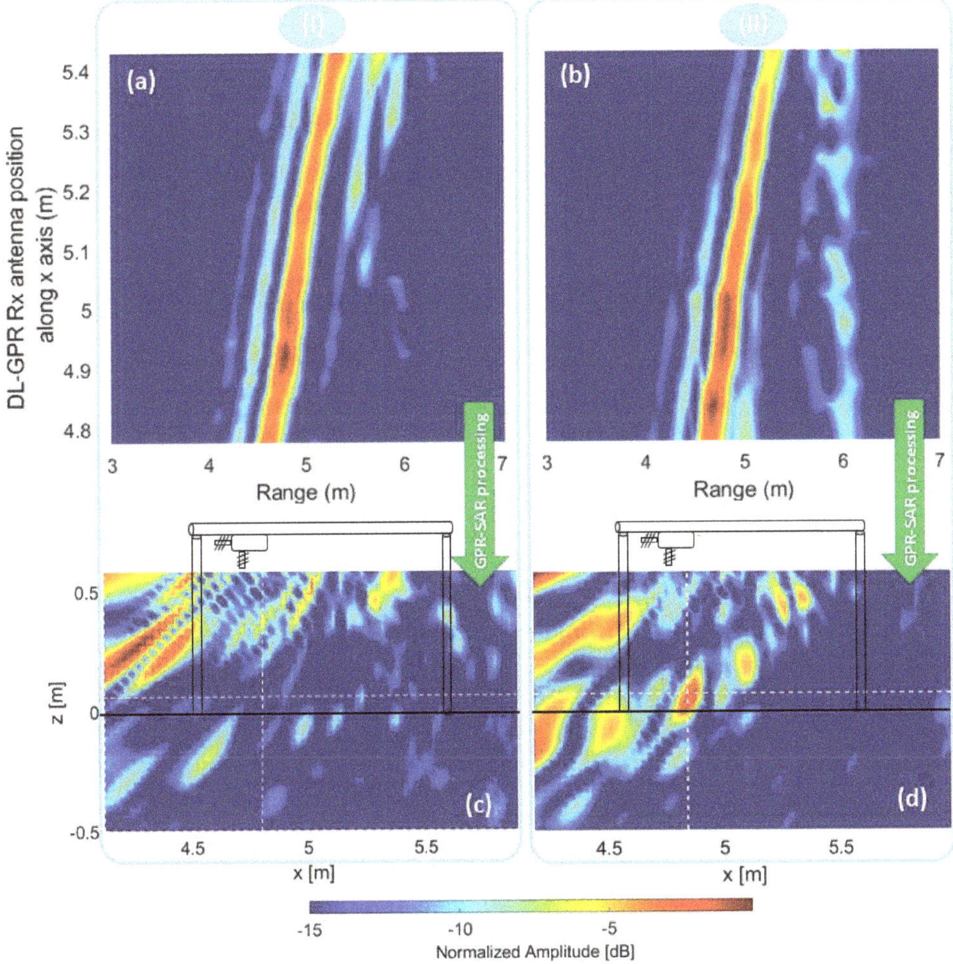

Figure 8. Individual processing of the cuts corresponding to the *x*-axis sweeps highlighted with dashed yellow rectangles in Figure 7, not intersecting (I) and intersecting (II) the position of the metallic can. (**a**) and (**b**) correspond to the range measurements for each acquisition position along the *x*-axis, and (**c**) and (**d**) are the reflectivity images in the XZ plane after applying GPR-SAR processing. For reference purposes, the air–ground interface (*z* = 0 m) and a sketch of the layout of the portable setup are depicted in black color in (**c**) and (**d**).

Next, reflectivity images from each of the selected individual sweeps, denoted as (I) and (II), are retrieved applying GPR-SAR processing. It should be noted that in this case, as only single sweeps are considered, a two-dimensional (2D) GPR-SAR processing is conducted in the vertical plane parallel to the movement direction during each sweep, i.e., an XZ plane. In addition, although during each sweep along the x-axis radar acquisitions were limited to $x \in [4.6, 5.5]$ m, the imaging domain was extended to $x \in [4.0, 6.5]$ m to analyze if potential contributions due to air–ground reflections appear in the GPR-SAR image. Results for the sweeps (I) and (II) are plotted in Figure 8c,d, respectively.

As observed in Figure 8c,d, clutter appears mainly outside the scanning area (in particular between $x = 4$ m and $x = 4.5$ m). No air–ground contributions are observed within the scanned section, i.e., for $x \in [4.6, 5.5]$ m, thus confirming that the proposed hybrid FLGPR-DLGPR minimizes the air–ground reflection contribution effectively. Finally, in the case of Figure 8d that corresponds to sweep (II), which intersects the position of the metallic can, a reflectivity peak can be observed at the position where the metallic can was placed ($x = 4.8$ m and $z = 0.1$ m approximately).

Three-dimensional (3D) GPR-SAR imaging results considering the entire acquisition domain shown in Figure 7 are depicted in Figure 9. Reflectivity cuts in the main planes are centered at the position of the metallic can. The echo due to the reflection on the metallic target can be clearly noticed, being approximately 10–15 dB above clutter level. Concerning air–ground reflections, no significant contributions can be observed, in agreement with the theoretical analysis shown in Figure 2 for the hybrid FLGPR-DLGPR architecture. As noticed in Figure 8c,d, there are some echoes at $x = 4$ m in the XZ plane, which can be due to air–ground reflections captured by a sidelobe of the DL-GPR Rx antenna. Nevertheless, as indicated before, the region between $x = 4$ m and $x = 4.5$ m falls outside the area scanned with the portable setup.

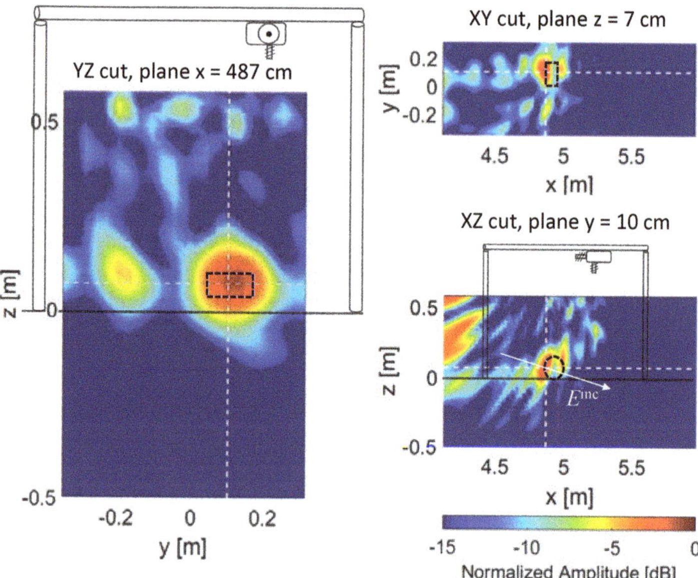

Figure 9. Imaging results for the metallic can placed on the ground. GPR-SAR cuts are centered at the position of the target, the profile of which is highlighted with a black dashed line. The white arrow depicted in the XZ cut represents the direction of the incident field ($-12°$ with respect to the horizontal plane). For reference purposes, the air–ground interface ($z = 0$ m) and a sketch of the layout of the portable setup are depicted in black color in the YZ and XZ cuts.

If Figure 8d and the XZ cut of Figure 9 are compared, it can be noticed that less clutter within the scanning area is observed in the latter. The reason is that 3D GPR-SAR processing is applied in Figure 9 (using all radar measurements), whereas GPR-SAR imaging results for only a single scan along the x-axis were presented in Figure 8d. Thus, clutter and noise present in the different scans along the x-axis are partially cancelled, whereas the echo due to the reflection on the metallic can is reinforced (added coherently).

As observed in Figures 7 and 9, the imaging domain has been limited from $y = -0.35$ m to $y = 0.32$ m due to the fact that in this first validation test, sweeps along the y-axis were spaced 5 cm on average, which is greater than $\lambda/2$ at the highest working frequency (5 GHz) so aliasing appeared when considering x-axis sweeps outside the range from $y = -0.35$ m to $y = 0.32$ m. Concerning sweeps along the x-axis, samples were taken every 3.2 cm on average.

The range resolution provided by the hybrid FLGPR-DLGPR can be extracted from the GPR-SAR results depicted in Figure 9. The reflectivity -3 dB beamwidth along the incident field direction (denoted with a white arrow in Figure 9, XZ cut) is 7 cm, which is in agreement with the range resolution of the 2 GHz-bandwidth radar: $\Delta Range = c/(2\ BW) = 7.5$ cm, being c the speed of light, and BW the radar frequency bandwidth.

3.2. Testing Using a Buried Anti Tank Plastic Landmine

Next, the hybrid FLGPR-DLGPR system to detect shallow targets was tested. For this purpose, a model of non-metallic (plastic) anti-tank landmine was buried at 2–3 cm depth, as shown in Figure 10, resembling a realistic scenario. In this case, sweeps along the y-axis were spaced 3 cm to avoid aliasing. Only GNSS-RTK positioning information provided with the highest accuracy (*fix* operation mode) was considered, thus resulting in an acquisition domain ranging from $y = -0.35$ m to $y = +0.18$ m (RTK was not in *fix* operation mode for x-axis sweeps beyond $y > +0.18$ m due to the poor satellite coverage in the field where the experiments were performed).

Figure 10. Picture of the measurement setup with an anti-tank plastic landmine buried on the ground. A picture of the anti-tank landmine before burying it can be observed in the right picture.

GPR-SAR imaging results are shown in Figure 11. The echo due to the reflection on the buried plastic landmine is observed at $z = -5$ cm, slightly deeper than its true position since the soil permittivity is not taken into account ($\varepsilon_r = 1$ is used for GPR-SAR processing). Signal-to-clutter ratio can be estimated in 5 dB for both cases.

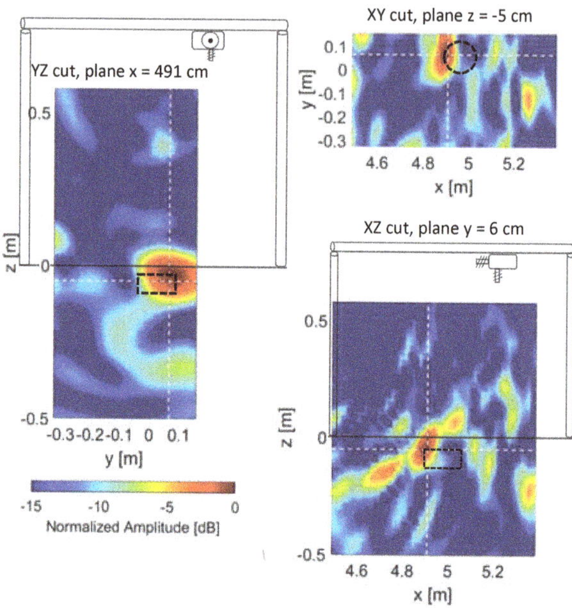

Figure 11. Imaging results for the buried plastic landmine. GPR-SAR cuts are centered at the position of the target, the profile of which is highlighted with a black dashed line. For reference purposes, the air–ground interface ($z = 0$ m) and a sketch of the layout of the portable setup are depicted in black color in the YZ and XZ cuts.

For verification purposes, a second measurement of the plastic landmine buried at 2–3 cm depth has been conducted. This measurement consisted of a single linear sweep along the x-axis performed by manually displacing the receiving devices in 2-cm steps from $x = 5$ m to $x = 6.30$ m. The measurement setup is depicted in Figure 12: instead of using the portable setup [20], a simpler supporting structure was implemented to simplify the manual displacement of the platform containing the receiving devices. In this case, 2D GPR-SAR imaging is conducted (XZ plane, $y = 0$ cm).

Figure 12. Picture of a 2D-measurement setup (acquisition along x axis). A picture of the anti-tank plastic landmine can be seen in the right picture.

Imaging results are plotted in Figure 13, where the reflectivity peak corresponding to the buried landmine can be identified at about $x = 5.7$ m. As in the previous examples shown in this contribution, air–ground reflections are negligible. Inside the imaging domain, a high level of clutter can be noticed only around $x = 5$ m and $x = 5.2$ m (which might be due to reflections in the supporting wooden structure). Furthermore, the reflection happening in the rear wooden supporting structure can be noticed in Figure 13.

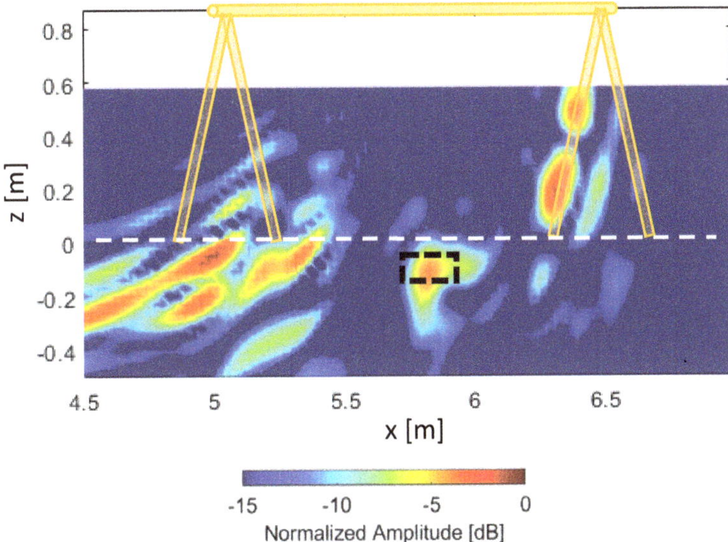

Figure 13. Imaging results for the buried plastic landmine. The position of the buried plastic landmine is highlighted with a black dashed line, and the average position in the *z* axis of the surface of the ground is depicted with a white dashed line. The profile of the 2D-measurement setup has been superimposed.

4. Discussion

4.1. Quantitative Analysis

To complement the qualitative validation of the detection capabilities of the hybrid FLGPR-DLGPR system, a quantitative analysis based on a simplified propagation model is presented next. This quantitative analysis provides an estimation of the signal-to-clutter ratio at the receiver. In particular, the power reflected on the buried target (solid arrows, Figure 14) will be considered as the signal, and the power reflected on the ground due to specular reflection (dashed arrows, Figure 14) as the clutter. For the sake of simplicity, it will be considered that the main contribution of the field backscattered on the buried target will be normal to the soil–air interface, as depicted in Figure 14. The analysis has been performed considering both TE (Transverse Electric) and TM (Transverse Magnetic) polarizations.

For this quantitative analysis, a metallic target will be considered, and the calculations will be conducted at the center frequency ($f = 4$ GHz) of the working frequency band.

The power transmitted by the radar module [22] is approximately -20 dBm, the gain of the transmitting and receiving helix antenna is G = 11 dB at 4 GHz and its directivity has been taken into account in the model. The soil constitutive parameters are: $\varepsilon_r = 7$ and $\sigma = 0.02$ S/m.

Three configurations will be analyzed. The first one (Table 1, second column) corresponds to a monostatic case where both the Tx and Rx antennas are located at $x = 5$ m. In the second case (Table 1, third column), the Tx antenna is located at $x = 3$ m. In the third case (Table 1, fourth column), the Tx antenna is located at $x = 0$ m, as in the examples presented in this contribution with the portable setup. In all the cases, the Tx and Rx antennas are placed at $z = 1$ m above ground (similarly to the examples presented in this contribution), and the receiver is located at $x = 5$ m. The metallic plate is considered to be buried 5 cm deep.

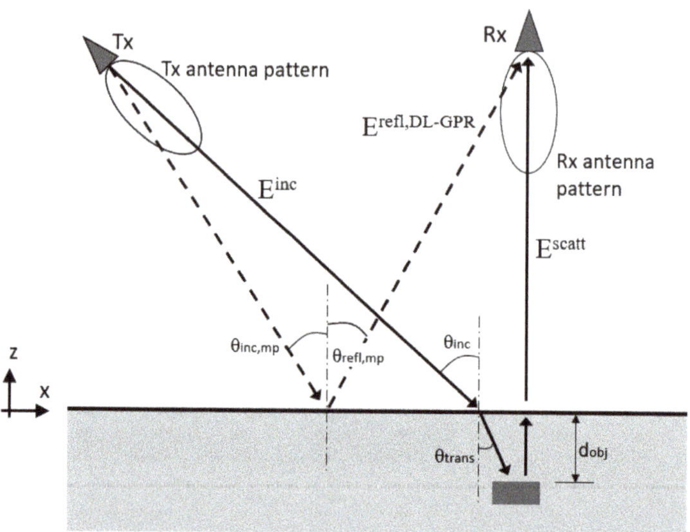

Figure 14. Scheme of the propagation model analyzed in Section 4.1. $z = 0$ corresponds to the air–ground interface. The Rx antenna is located at $x = 5$ m, $z = 1$ m.

Table 1. Quantitative analysis of monostatic DLGPR and hybrid FLGPR-DLGPR configurations. The definition of the terms listed in the first (left) column of the table is provided in the Appendix A.

	Monostatic DLGPR	Hybrid FLGPR-DLGPR		Hybrid FLGPR-DLGPR, This Contribution	
	TE/TM	TM	TE	TM	TE
	Tx at $x = 5$ m	Tx at $x = 3$ m		Tx at $x = 0$ m	
	Rx at $x = 5$ m	Rx at $x = 5$ m		Rx at $x = 5$ m	
θ_{inc} [°]	0	63.4		78.7	
θ_{trans} [°]	0	19.7		21.7	
$\lvert \rho_{air\text{-}ground} \rvert$	0.45	0.12	0.69	0.28	0.85
$\lvert \rho_{ground\text{-}target} \rvert$	1	1		1	
$\lvert \rho_{ground\text{-}air} \rvert$	0.45	0.45		0.45	
$P_{rx,ant,Escatt}$ [dBm]	-52.80	-55.66	-58.74	-60.12	-66.40
$\theta_{inc,mp}$ [°]	0	45		68.2	
$\lvert \rho_{air\text{-}ground,mp} \rvert$	0.45	0.32	0.57	0.02	0.74
$P_{rx,ant,Erefl,DL\text{-}GPR}$ [dBm]	-55.41	-72.41	-67.46	-110.47	-80.73
SCR [dB]	2.61	16.75	8.72	49.32	14.33

From the quantitative analysis presented in Table 1 it can be concluded that:

(1)   The power backscattered by the buried plate is greater in the case of the DLGPR with respect to the hybrid FLGPR-DLGPR. This is consistent with the fact that propagation

losses are smaller in the DLGPR system. Thus, targets buried deeper could be missed by the hybrid FLGPR-DLGPR as soil propagation losses would be larger.

(2) The reflection at the air–ground interface is weaker in the case of the hybrid FLGPR-DLGPR, so that the contribution due to the reflection of the metallic plate will not be masked by the air–ground reflection. Note that, in the case of the DLGPR, the level of the contributions from the air–ground interface and from the buried metallic plate is similar, whereas in the proposed hybrid FLGPR-DLGPR the reflection from the buried plate is notably higher than the contribution due to specular reflection on the ground. This difference is also greater when considering TM polarization due to a low reflection coefficient at angles close to the Brewster's angle.

4.2. Justification of the Applicability of the Hybrid FLGPR-DLGPR Architecture

As indicated in Section 1.2, the proposed hybrid FLGPR-DLGPR is devoted to detect shallow targets that could be masked by the air–ground reflection if DLGPR systems were used. In this sense, it is important to remark that anti-personnel landmines are shallowly buried (less than 5 cm), some being buried up to 15–20 cm [24,25]. Anti-tank mines are also buried not deeper than 20 cm [26,27]. Only in a few cases anti-tank mines and IEDS can be found at depths greater than 20 cm, resulting in a detection challenge even for DLGPR systems in case of wet soils (e.g., Table 2.3 of [28] shows that wet loamy soils can exhibit up to 600 dB/m losses at 1 GHz).

The issue of shallow targets masked by the air–ground reflection has been broadly discussed in the literature. In fact, as stated in [29], the air–ground reflection is the largest single source of clutter (including volumetric inhomogeneities such as rocks or roots, and surface vegetation). Despite this can be mitigated for deeply buried targets using time gating, it remains a challenge for shallowly buried ones, especially in the case of a rough ground surface. In this regard, although the usage of GPR systems for landmines and IEDs detection has several advantages, the antennas must be elevated above the ground in order to keep a safe distance, which results in a strong air–ground reflection [30]. Therefore, in these cases, especially when the buried targets are also small and present low dielectric contrast, it can be helpful to resort to alternative architectures, such as the hybrid FLGPR-DLGPR approach proposed in this contribution, to enhance the detection capabilities.

5. Conclusions

A hybrid FLGPR-DLGPR architecture has been implemented and validated successfully combining the advantages of FLGPR (minimization of the clutter due to air–ground reflections) and DLGPR (greater dynamic range and resolution).

Results presented in Section 3 confirm the capability of the implemented hybrid FLGPR-DLGPR architecture to detect shallowly buried objects, being supported by the discussion presented in Section 4 based on a quantitative analysis. In all the presented results clutter due to air-ground reflection within the scanning area is weaker than the reflection on the buried target, as the incident field on the ground within the scanning area is mainly reflected away from the DL-GPR Rx antenna.

The fact of using the 3 to 5 GHz frequency band makes the system more sensitive to geo-referring and positioning errors if compared to previous works where a DL-GPR radar and a sub-3 GHz frequency band were considered [17,20]. The reasons of using a higher frequency band were (i) the limited availability of multistatic wirelessly-synchronized radar modules, and (ii) the fact that this architecture was conceived to detect shallowly buried targets, so penetration depth capabilities were not as critical as in [17,20]. In these works, targets were buried deeper, so that the air–ground reflection characteristic of DL-GPR configurations did not mask the buried targets significantly. From the GPR-SAR imaging results depicted in [17,20], it can be noticed that the air–ground reflection resulted in a 6–8 cm thick strip, and the targets were buried more than 10 cm deep.

One of the key issues for the successful implementation of the system is the use of two radar modules that are synchronized without requiring physical connection between them.

Remote Sens. **2021**, *13*, 1206

An additional receiving antenna pointing towards the transmitting antenna enabled perfect reception of the direct ray used by the receiving module for synchronization purposes. Signals captured by each receiving antenna could be easily filtered thanks to the addition of a delay line connected at the end of the receiving antenna pointing downwards.

A non-metallic plastic target (an anti-tank landmine) was selected for validation purposes, testing the performance of the system with a 2D setup (where the receiver is manually positioned) and with a 3D acquisition system (in which geo-referring information is provided by a GNSS-RTK module). In both cases, the buried target could be identified.

6. Patents

The work presented in this contribution is related to the patent "Airborne Systems and Detection Methods Localisation and Production of Images of Buried Objects and Characterisation of the Composition of the Subsurface". Publication No. WO/2017/125627. International Application No. PCT/ES2017/000006. Priority date: 21 January 2016. International filing date: 18 January 2017. University of Oviedo, University of Vigo. European Patent EP3407007B1 granted on 30 September 2020. Available online: https://patentscope.wipo.int/search/en/detail.jsf?docId=WO2017125627 (accessed on 5 December 2020).

Author Contributions: Conceptualization, M.G.-F., Y.Á.L., and F.L.-H.A.; methodology, M.G.-F., Y.Á.L., and G.Á.-N.; software, M.G.-F, G.Á.-N. and Y.Á.L.; validation, all the authors; resources, all the authors; data curation, M.G.-F., G.Á.-N., and Y.Á.L.; writing—original draft preparation, Y.Á.L.; writing—review and editing, all the authors; supervision, F.L.-H.A. and Y.Á.L.; funding acquisition, F.L.-H.A. and Y.Á.L. All authors have read and agreed to the published version of the manuscript.

Funding: This research was funded by the Ministerio de Defensa—Gobierno de España and the University of Oviedo under Contract 2019/SP03390102/00000204 / CN-19-002 ("SAFEDRONE"); by the Xunta de Galicia–Axencia Galega de Innovación (GAIN) under project 2018-IN855A 2018/10 ("RadioUAV: drones para aplicaciones más allá de lo visible"); by the Government of the Principality of Asturias (PCTI) and European Union (FEDER) under Grant IDI/2018/000191; and by the Instituto Universitario de Tecnología Industrial de Asturias (IUTA) under Project SV-19-GIJON-1-17 ("RadioUAV").

Acknowledgments: The authors would like to acknowledge the CIED-COE (Col. José Luis Mingote Abad and Br. Ramón Javier Pacheco) for their advice concerning the placement of the IEDs and landmines, and Cap. Santiago García Ramos for the supervision of the Contract 2019/SP03390102/00000204/CN-19-002 ("SAFEDRONE"), and also Carey M. Rappaport and Ann Morgenthaler for their previous contributions and advice in the theoretical study and analysis of this architecture, presented in publication [19].

Conflicts of Interest: The authors declare no conflict of interest.

Appendix A

Definition of the terms listed in the first (left) column of Table 1:

$|\rho_{\text{air-ground}}|$: module of the reflection coefficient at the air-ground interface in the path between the Tx antenna and the target.

$|\rho_{\text{ground-target}}|$: module of the reflection coefficient at the ground-target interface.

$|\rho_{\text{ground-air}}|$: module of the reflection coefficient at the ground-air interface (normal incidence is considered).

$P_{\text{rx,ant,Escatt}}$: power received at the DL-GPR Rx antenna coming from the target, considering the Rx antenna gain.

$|\rho_{\text{air-ground,mp}}|$: module of the reflection coefficient at the air-ground interface at the specular reflection point.

$P_{\text{rx,ant,Erefl,DL-GPR}}$: power received at the DL-GPR Rx antenna due to specular reflection on the ground, and considering the directive gain function of the Rx antenna.

SCR: signal-to-clutter ratio at the Rx antenna, defined as the ratio between $P_{rx,ant,Escatt}$ and $P_{rx,ant,Erefl,DL\text{-}GPR}$. If positive, the power reflected in the buried plate is greater than the power contribution due to specular reflection at the air-ground interface.

References

1. Jol, H.M. *Ground Penetrating Radar Theory and Applications*; Elsevier Science: Amsterdam, The Netherlands, 2009.
2. Santos-Assunçao, S.; Perez-Gracia, V.; Caselles, O.; Clapes, J.; Salinas, V. Assessment of Complex Masonry Structures with GPR Compared to Other Non-Destructive Testing Studies. *Remote Sens.* **2014**, *6*, 8220–8237. [CrossRef]
3. Morris, I.; Abdel-Jaber, H.; Glisic, B. Quantitative Attribute Analyses with Ground Penetrating Radar for Infrastructure Assessments and Structural Health Monitoring. *Sensors* **2019**, *19*, 1637. [CrossRef] [PubMed]
4. Conyers, L.B. Ground-Penetrating Radar Mapping Using Multiple Processing and Interpretation Methods. *Remote Sens.* **2016**, *8*, 562. [CrossRef]
5. De Chiara, F.; Fontul, S.; Fortunato, E. GPR Laboratory Tests for Railways Materials Dielectric Properties Assessment. *Remote Sens.* **2014**, *6*, 9712–9728. [CrossRef]
6. Przemyslaw, K.; Godziuk, A.; Kapruziak, M.; Olech, B. Fast Analysis of C-Scans From Ground Penetrating Radar via 3-D Haar-Like Features With Application to Landmine Detection. *IEEE Trans. Geosci. Remote Sens.* **2015**, *53*, 3996–4009.
7. Lombardi, F.; Lualdi, M.; Picetti, F.; Bestagini, P.; Janszen, G.; Di Landro, L.A. Ballistic Ground Penetrating Radar Equipment for Blast-Exposed Security Applications. *Remote Sens.* **2020**, *12*, 717. [CrossRef]
8. Comite, D.; Ahmad, F.; Dogaru, T.; Amin, M.G. Adaptive Detection of Low-Signature Targets in Forward-Looking GPR Imagery. *IEEE Geosci. Remote Sens. Lett.* **2018**, *15*, 1520–1524. [CrossRef]
9. Rosen, E.M.; Ayers, E. Assessment of Down-Looking GPR Sensors for Landmine Detection. In Proceedings of the SPIE, Orlando, FL, USA, 10 June 2005.
10. Comite, D.; Galli, A.; Catapano, I.; Soldovieri, F. Advanced imaging for down-looking contactless GPR systems. In Proceedings of the International Applied Computational Electromagnetics Society Symposium—Italy (ACES), Florence, Italy, 26–30 March 2017.
11. Zhang, Y.; Orfeo, D.; Burns, D.; Miller, J.; Huston, D.; Tian, X. Buried nonmetallic object detection using bistatic ground penetrating radar with variable antenna elevation angle and height. In *Nondestructive Characterization and Monitoring of Advanced Materials, Aerospace, and Civil Infrastructure*; SPIE Proceedings: Portland, OR, USA, 2017; Volume 10169. [CrossRef]
12. Kimata, T.; Shigematsu, K.; Nakajima, H.; Morita, J. Detection of the buried landmine/projectile using LS-band FLGPR vehicle. In Proceedings of the SPIE Defense + Commercial Sensing, Baltimore, ML, USA, 14–18 April 2019.
13. Comite, D.; Ahmad, F.; Dogaru, T.; Amin, M. Coherence-Factor-Based Rough Surface Clutter Suppression for Forward-Looking GPR Imaging. *Remote Sens.* **2020**, *12*, 857. [CrossRef]
14. Colorado, J.; Perez, M.; Mondragon, I.; Mendez, D.; Parra, C.; Devia, C.; Martinez-Moritz, J.; Neira, L. An integrated aerial system for landmine detection: SDR-based Ground Penetrating Radar onboard an autonomous drone. *Adv. Robot.* **2017**, *31*, 791–808. [CrossRef]
15. Schreiber, E.; Heinzel, A.; Peichl, M.; Engel, M.; Wiesbeck, W. Advanced Buried Object Detection by Multichannel, UAV/Drone Carried Synthetic Aperture Radar. In Proceedings of the 2019 13th European Conference on Antennas and Propagation (EuCAP), Krakow, Poland, 31 March–5 April 2019; pp. 1–5.
16. Šipoš, D.; Gleich, D. A Lightweight and Low-Power UAV-Borne Ground Penetrating Radar Design for Landmine Detection. *Sensors* **2020**, *20*, 2234. [CrossRef] [PubMed]
17. García-Fernández, M.; López, Y.Á.; Andrés, F.L.-H. Airborne Multi-Channel Ground Penetrating Radar for Improvised Explosive Devices and Landmine Detection. *IEEE Access* **2020**, *8*, 165927–165943. [CrossRef]
18. Schartel, M.; Burr, R.; Bähnemann, R.; Mayer, W.; Waldschmidt, C. An experimental study on airborne landmine detection using a circular synthetic aperture radar. *arXiv* **2020**, arXiv:2005.02600. Available online: http://arxiv.org/abs/2005.02600 (accessed on 7 December 2020).
19. Garcia-Fernandez, M.; Morgenthaler, A.; Alvarez-Lopez, Y.; Las Heras, F.; Rappaport, C. Bistatic Landmine and IED Detection Combining Vehicle and Drone Mounted GPR Sensors. *Remote Sens.* **2019**, *11*, 2299. [CrossRef]
20. García-Fernández, M.; Álvarez López, Y.; De Mitri, A.; Castrillo Martínez, D.; Álvarez-Narciandi, G.; Las-Heras Andrés, F. Portable and Easily-Deployable Air-Launched GPR Scanner. *Remote Sens.* **2020**, *12*, 1833. [CrossRef]
21. Garcia-Fernandez, M. Novel Measurement System Based on Electromagnetic Sensors on Board Unmanned Aerial Vehicles for Subsurface Imaging and Antenna Measurement Applications. Ph.D. Thesis, University of Oviedo, Gijón, Spain, 2019.
22. P-440 Ultra Wideband (UWB) Radio Transceiver. Available online: https://usermanual.wiki/Humatics/P440A/pdf (accessed on 1 December 2020).
23. Topcon B111 Receiver. Available online: https://www.topconpositioning.com/oem-components-technology/gnss-components/b111 (accessed on 1 December 2020).
24. Keeley, R. Understanding Landmines and Mine Action. September 2003. Available online: http://web.mit.edu/demining/assignments/understanding-landmines.pdf (accessed on 23 January 2021).
25. Marble, J.A.; Yagle, A.E. Measuring Landmine Size and Burial Depth with Ground Penetrating Radar. In Proceedings of the SPIE Defense and Security, Orlando, FL, USA, 13–15 April 2004.

26. Walls, R.; Brown, T.; Clodfelter, F.; Coors, J.; Laudato, S.; Lauziere, S.; Patrikar, A.; Poole, M.; Price, M. Ground penetrating radar field evaluation in Angola. In Proceedings of the SPIE Defense and Security Symposium, Orlando, FL, USA, 17–21 April 2006.
27. Geneva International Centre for Humanitarian Demining (GICHD) and Stockholm International Peace Research Institute (SIPRI). The Humanitarian and Developmental Impact of Anti-Vehicle Mines. Geneva, September 2014. Available online: https://www.gichd.org/en/resources/publications/detail/publication/the-humanitarian-and-developmental-impact-of-anti-vehicle-mines/ (accessed on 23 January 2021).
28. Daniels, D.J. *Ground Penetrating Radar*, 2nd ed.; The Institution of Electrical Engineers: London, UK, 2004; ISBN 9780863413605. [CrossRef]
29. Rappaport, C.M.; El-Shenawee, M.; Zhan, H. Suppressing GPR Clutter from Randomly Rough Ground Surfaces to Enhance Nonmetallic Mine Detection. *Subsurf. Sens. Technol. Appl.* **2003**, *4*, 311–326. [CrossRef]
30. Feng, X.; Sato, M.; Zhang, Y.; Liu, C.; Shi, F.; Zhao, Y. CMP Antenna Array GPR and Signal-to-Clutter Ratio Improvement. *IEEE Geosci. Remote Sens. Lett.* **2009**, *6*, 23–27. [CrossRef]

remote sensing

Article

Deep Convolutional Denoising Autoencoders with Network Structure Optimization for the High-Fidelity Attenuation of Random GPR Noise

Deshan Feng [1,2], Xiangyu Wang [1,2,*], Xun Wang [1,2], Siyuan Ding [1,2] and Hua Zhang [1,2]

1 School of Geosciences and Info-Physics, Central South University, Changsha 410083, China; fengdeshan@csu.edu.cn (D.F.); wangxun@csu.edu.cn (X.W.); syding@csu.edu.com (S.D.); zhanghua@csu.edu.cn (H.Z.)

2 The Key Laboratory of Metallogenic Prediction of Nonferrous Metals, Ministry of Education, Changsha 410083, China

* Correspondence: nephilim.csu@gmail.com or 195001029@csu.edu.cn; Tel.: +86-189-7485-3132

Abstract: The high-fidelity attenuation of random ground penetrating radar (GPR) noise is important for enhancing the signal-noise ratio (SNR). In this paper, a novel network structure for convolutional denoising autoencoders (CDAEs) was proposed to effectively resolve various problems in the noise attenuation process, including overfitting, the size of the local receptive field, and representational bottlenecks and vanishing gradients in deep learning; this approach also significantly improves the noise attenuation performance. We described the noise attenuation process of conventional CDAEs, and then presented the output feature map of each convolutional layer to analyze the role of convolutional layers and their impacts on GPR data. Furthermore, we focused on the problems of overfitting, the local receptive field size, and the occurrence of representational bottlenecks and vanishing gradients in deep learning. Subsequently, a network structure optimization strategy, including a dropout regularization layer, an atrous convolution layer, and a residual-connection structure, was proposed, namely convolutional denoising autoencoders with network structure optimization (CDAEsNSO), comprising an intermediate version, called atrous-dropout CDAEs (AD-CDAEs), and a final version, called residual-connection CDAEs (ResCDAEs), all of which effectively improve the performance of conventional CDAEs. Finally, CDAEsNSO was applied to attenuate noise for the H-beam model, tunnel lining model, and field pipeline data, confirming that the algorithm adapts well to both synthetic and field data. The experiments verified that CDAEsNSO not only effectively attenuates strong Gaussian noise, Gaussian spike impulse noise, and mixed noise, but it also causes less damage to the original waveform data and maintains high-fidelity information.

Keywords: GPR; noise attenuation; Gaussian spike impulse noise; deep convolutional denoising autoencoders (CDAEs); deep convolutional denoising autoencoders with network structure optimization (CDAEsNSO)

Citation: Feng, D.; Wang, X.; Wang, X.; Ding, S.; Zhang, H. Deep Convolutional Denoising Autoencoders with Network Structure Optimization for the High-Fidelity Attenuation of Random GPR Noise. *Remote Sens.* **2021**, *13*, 1761. https://doi.org/10.3390/rs13091761

Academic Editors: Yuri Alvarez and Maria Garcia

Received: 24 March 2021
Accepted: 29 April 2021
Published: 1 May 2021

Publisher's Note: MDPI stays neutral with regard to jurisdictional claims in published maps and institutional affiliations.

1. Introduction

Ground penetrating radar (GPR) is a surface geophysical method that utilizes high-frequency broadband electromagnetic waves (1 MHz–10 GHz) to detect and locate structures or objects in the shallow subsurface [1,2]. GPR has numerous characteristics, including a high resolution, strong anti-interference ability, and high efficiency, and this technique is nondestructive; consequently, GPR has been extensively used in many fields, such as geological exploration, water conservancy engineering, and urban construction [3]. There has been an increasing tendency to use nondestructive testing techniques that do not alter the reinforcement elements of vulnerable structures, such as the combined methodology, which uses GPR and infrared thermography (IRT) techniques for the detection and evaluation of corrosion. In [4], cracked cement concrete layers that are located below the asphalt layer in the case of rigid pavements were similarly investigated. Therefore, detection is a difficult task, and nondestructive surveys are, in many cases, applied to detect these types of damage. However, the difficulty in data interpretation limits their use [5]. Furthermore, the GPR profiles are affected by various factors, such as complex and varying detection environments, the instrument system, and the data acquisition mode, which results in various forms of clutter and noise that reduce the quality of the radar signal. Therefore, it is particularly important to research fast and effective noise attenuation algorithms to obtain GPR data with a high signal-to-noise ratio (SNR) [6,7].

In recent years, many scholars worldwide have conducted a substantial amount of research on methods of attenuating GPR noise. Common noise attenuation algorithms include the curvelet transform, empirical mode decomposition (EMD), and the wavelet transform. To improve the clarity of GPR data in the process of underground pipeline positioning, a new method based on the curvelet transform that reduces clutter and profile artifacts to highlight significant waves was proposed, as reducing noise and removing undesirable items, such as clutter and artifacts, are important for highlighting these echoes; the experiments show that the qualitative and quantitative results of this method are satisfactory [8]. However, because of strong linear interference, the conventional curvelet transform is ineffective for noise removal in this case, because it cannot adaptively remove noise according to the signal characteristics; hence, a method, called the empirical curve wave transform, which can suppress interference signals, was proposed and compared with the conventional curvelet transform. The results confirmed the effectiveness of the method [9]. In addition, to remove noise from GPR echo signals, a denoising method that was based on ensemble EMD (EEMD) and the wavelet transform was presented in [10]; as compared with other common methods, the EEMD-wavelet method improves the SNR. Ref. [11] first used a complete EEMD (CEEMD) method to perform time-frequency analysis of data for processing GPR signal data. The CEEMD method was proven to solve the mode mixing problem in EMD and significantly improve the resolution for EEMD processing when the GPR signal to be processed has a low SNR, thereby effectively avoiding the disadvantages of both EMD and EEMD. The results show that, in a comparison with EMD and EEMD methods, CEEMD obtains higher spectral and spatial resolution, and it also proves that CEEMD has better characteristics. To further reduce random GPR noise that is based on denoising using EMD, an EMD technique in combination with basis pursuit denoising (BPD) was developed and provided satisfactory outputs [12]. Ref. [13] extended $f - x$ EMD to form a semiadaptive dip filter for GPR data to adaptively separate reflections at different dips. Ref. [14] used the two-dimensional Gabor wavelet transform to process signals and proposed a new denoising method to be solved when extracting the reflected signals of buried objects. In a comparison with the $f - k$ filter, the effectiveness of this method was proven. Another alternative is the drumbeat-beamlet (dreamlet) transform. Because the dreamlet foundation automatically satisfies the wave equation, it can provide an effective way to represent the physical wave field. Ref. [15] theoretically deduced the representation of the damped dreamlet and reported its geometric explanation and analysis. Furthermore, a GPR denoising approach that was based on the empirical wavelet transform (EWT) in combination with semisoft thresholding was proposed, and a spectrum segmentation strategy was designed that accounted for different frequency characteristics of different signals; this method achieved better performance than CEEMD and the synchrosqueezed wavelet transform (SWT) [16].

Nevertheless, all of the above algorithms are based on domain transformation. A significant number of scholars have researched strategies involving the sparse representation (SR) of signals or signal processing combined with morphology in order to further improve the noise attenuation performance and increase the data fidelity. According to the correlation of a signal, the eigenvalues and corresponding eigenvectors were obtained by decomposing the covariance matrix of GPR data, and a linear transformation was applied to the GPR data to obtain the principal components (PCs), where the lower-order PCs represent the strongly correlated target signals of the raw data and the higher-order PCs represent the uncorrelated noise; thus, the target signal was extracted, and uncorrelated noise was effectively filtered out by principal component analysis (PCA) [17]. Implementing the SR of a signal is an effective method that can use the sparsity and compressibility of noisy data to estimate the signal from noisy data; in this method, signal estimation can be achieved by relinquishing some unimportant bases and eliminating random noise. Ref. [18] derived a damped SR (DSR) of a signal; a damping operator is employed in the DSR to obtain greater accuracy in signal estimation. Additionally, based on the physical wavelet, a seismic denoising method that was based on sparse Bayesian learning (SBL) was developed in [19]. In the SBL algorithm, the physical wavelet can be estimated based on various seismic and even logging data and correctly describe the different characteristics of these different seismic data. Moreover, the physical wavelet can adaptively estimate the trade-off regularization parameter that is used to determine the quality of noise reduction according to the updated data mismatch and sparsity. In the iterative process. Through comprehensive and real seismic data examples, the effectiveness of the SBL method has been proven.

Another conventional technique, namely, time-domain singular value decomposition (SVD), introduces pseudosignals that did not previously exist when eliminating the direct waves and poorly suppresses the random noise surrounding the nonhorizontal phase axes. To resolve these inadequacies, an SVD method in the local frequency domain of GPR data based on the Hank matrix was proposed, and a comparison showed that this method could improve the suppression of random noise in proximity to nonhorizontal phase reflections [20]. In addition, a new dictionary learning method, namely structured graph dictionary learning (SGDL), was recently proposed by adding the local and non-local similarities of the data via a structured graph, thereby enabling the dictionary to contain more atoms with which to represent seismic data; the SGDL method was shown to effectively remove strong noise and retain weak seismic events [21]. In [22], the authors addressed the denoising of high-resolution radar image series in a nonparametric Bayesian framework; this method imposes a Gaussian process (GP) model on the corresponding time series of each pixel and effectively denoises the image series by implementing GP regression. Their method exhibited improved flexibility in describing the data and superior performance in preserving the structure while denoising, especially in scenarios with a low SNR. Furthermore, the authors of [23] proposed a modified morphological component analysis (MCA) algorithm and applied their technique to the denoising of GPR signals. The core of their MCA algorithm is its selection of an appropriate dictionary by combining the undecimated discrete wavelet transform (UDWT) dictionary with the curvelet transform dictionary (CURVELET). The modified MCA algorithm was compared with SVD and PCA to confirm the superior performance of the algorithm. The authors first put forward the expression of the basic principles and the methods of mathematical morphology. Subsequently, they combined the Ricker wavelet and low-frequency noise to form a synthetic dataset example for testing the MCA method in order to verify the feasibility and performance of the MCA method. According to the results of the synthesis example, the proposed method can effectively suppress the large-scale low-frequency noise in the original data and, at the same time, it can slightly suppress the small signals that exist in the original data. Finally, the proposed method was applied to field microseismic data, and the results are encouraging in [24]. The authors of [25] developed a novel algorithm based on the difference in seismic wave shapes and introduced mathematical morphological filtering (MMF) into the attenuation of coherent noise. The morphological operation is calculated in the trajectory direction of the rotating coordinate system, and the rotating coordinate system is established along the coherent noise trajectory to distribute the energy of the coherent noise in the horizontal direction. When compared with other existing technologies, this MMF method is more effective in rejecting outliers and reduces artifacts. A new method was proposed for enhancing the GPR signal. It is based on a subspace method and a clustering technique. The proposed method makes it possible to improve the estimation accuracy in a noisy context. It is used with a compressive sensing method to estimate the time delay of layered media backscattered echoes coming from the GPR signal [26].

Most of the above noise attenuation algorithms are based on the SR strategy of signals and they adopt domain transformation to process the data. Nevertheless, these approaches are all based on a fixed transformation basis and cannot self-adjust according to the characteristics of various signals. Hence, these methods cannot accurately represent the signal when encountering complex GPR signal data. Thus, it is necessary to develop an adaptive transform basis denoising method that is based on the characteristics of GPR data. A deep convolutional denoising autoencoder (CDAE) is one possible solution, which is a new method of random noise attenuation based on a deep learning architecture that is a type of unsupervised neural network learning algorithm. Deep CDAEs are mainly composed of two types of networks: encoders and decoders. In the context of this research, the encoder encodes noisy GPR profile data into multiple levels of abstraction to extract the 1D latent vectors containing important features, while the decoder decodes the 1D latent vectors containing the feature information to reconstruct the noise-free signal and, thus, eliminate random noise. Such algorithms are often used in the fields of noise attenuation and image generation.

Models that are based on deep learning show great promise in terms of noise attenuation. However, the disadvantages of these methods are that a large number of training samples are required and the computational costs are very high. Refs. [27,28] showed that denoising autoencoders constructed using convolutional layers with a small sample size can be used to effectively denoise medical images and they can combine heterogeneous images to increase the sample size, thereby improving denoising performance. In [29], the authors proposed using deep fully convolutional denoising autoencoders (FCDAEs) instead of deep feedforward neural networks (FNNs), and their experimental results showed that deep FCDAEs perform better than deep FNNs, despite having fewer parameters. In addition, a very novel data preprocessing method is proposed. This method uses data points between adjacent samples to obtain a set of training data. To obtain a better SR, they constructed standard penalty points that are based on the combination of the standard penalty points of L_1 and L_2, and a comparison with normal denoising autoencoders verified the superiority of this method [30]. Ref. [31] proposed the deep evolving denoising autoencoder (DEVDAN); it has an open structure in the generation phase and differentiation phase, which can automatically and adaptively add and discard hidden units; in the generation phase, they use the dataset (unlabeled) to improve the prediction performance of the discriminative model, optimize and modify the discriminant model from the data in the generation phase, and, finally, achieve a dynamic balance and improve the accuracy of the overall model prediction. Ref. [32] developed a new denoise/decomposition method that is based on deep neural networks, called DeepDenoiser. The DeepDenoiser network uses a mask strategy. First, the input signal is decomposed into signals of interest and uninteresting signals. These uninteresting signals are defined as noise. The composition of this noise includes not only the usual Gaussian noise, but also various nonseismic signals. Subsequently, nonlinear functions are used to map the representation into the mask, and these nonlinear functions are finally used to learn and train the data SR in the time-frequency domain. The DeepDenoiser network that is obtained through training can suppress noise according to the minimum change of the required waveform when the noise level is very high, thereby greatly improving the SNR. DeepDenoiser has clear applications in seismic imaging, microseismic monitoring and environmental noise data preprocessing. More recently, Ref. [33] proposed a new method that is based on the deep denoising autoencoder (DDAE) to attenuate random seismic noise.

In summary, the conventional noise attenuation methods can be roughly divided into four categories: 1. methods based on a fixed transformation basis, 2. methods based on a sparse representation, 3. methods based on morphological component analysis, and 4. methods based on deep learning. Table 1 summarizes these strategies.

Table 1. Conventional noise attenuation methods.

Methods	Typical Case
fixed transformation basis	curvelet transform, EMD, wavelet transform
sparse representation	PCA, SVD, SBL
morphological component analysis	MCA, MMF
deep learning	FNNs, FCDAEs, DECDAN, DDAE

GPR signals attenuate more rapidly than seismic signals, and GPR waveforms are more complicated due to the different methods of observation. If a deep CDAE is directly applied to attenuate the noise of theoretical synthetic GPR data and field data, the GPR profile will be distorted and the attenuation of noise will be incomplete due to overfitting, an incorrect size of the local receptive field, and the representational bottlenecks and vanishing gradients that are encountered in deep learning. To solve these problems, the authors have modified the structure of deep CDAEs and optimized the network structure consisting of a dropout regularization layer, an atrous convolutional layer, and a residual-connection structure. Furthermore, a modified deep CDAE strategy that is based on network structure optimization is proposed, namely, convolutional denoising autoencoders with network structure optimization (CDAEsNSO), which consists of atrous-dropout CDAEs (AD-CDAEs) and residual-connection CDAEs (ResCDAEs), all of which effectively improve the performance of conventional CDAEs. CDAEsNSO exhibits a strong noise attenuation capability and good adaptability to different data and various types of noise and it does less damage to the information in the original profile, thereby maintaining a high level of fidelity.

2. Materials and Methods

2.1. Deep Convolutional Denoising Autoencoders

In a denoising autoencoder, the neural network tries to find an encoder that can convert the noisy data into pure data. The autoencoder will automatically learn to encode from the data without manual intervention; therefore, the autoencoder can be classified as an unsupervised learning algorithm. A GPR signal containing noise is expressed as

$$\tilde{x} = x + noise \tag{1}$$

where $\tilde{x}$ refers to the signal that is eroded by noise; x is the original signal; and, *noise* represents the noise. The purpose of noise attenuation is to remove the *noise* from $\tilde{x}$.

An autoencoder consists of two operators, namely an encoder and a decoder, where the encoder function can be expressed as

$$h = f_\theta(\tilde{x}) = \sigma_1(\mathbf{W}\tilde{x} + \mathbf{b}) \tag{2}$$

and the decoder function can be expressed as

$$y = f_{\theta'}(h) = \sigma_2(\mathbf{W}'h + \mathbf{b}') \tag{3}$$

where $\tilde{x}$ denotes the input GPR data with noise; h represents the latent features (latent vectors) of the input data; y signifies the recovered GPR data without noise; $\mathbf{W}$ and $\mathbf{W}'$ are the arrays representing the convolution kernel and deconvolution kernel, respectively; $\mathbf{b}$ and $\mathbf{b}'$ are also arrays representing the biases of the encoder and decoder, respectively; and, σ_1 and σ_1 are the activation functions of the encoder and decoder, respectively.

For the noise-containing GPR data that are represented by $\tilde{x}$, the purpose of the encoder is to process $\tilde{x}$, thereby generating a latent vector h with low-dimensional features. Subsequently, the encoder is forced to learn the latent vector of the input data, so that the decoder can accept the latent vector h to recover the clean GPR data by minimizing different loss functions. In this paper, the mean square error is used as the loss function:

$$L(\mathbf{x}, \mathbf{y}) = MSE = \frac{1}{m}\sum_{i=1}^{m}(\mathbf{x}_i - \mathbf{y}_i)^2 \tag{4}$$

where m represents the size of the image; that is, the image is $m = N_x \times N_y$, and x is the input data for the encoder, while y is the output data for decoder, where i is the subscript index of the image. This loss function enables the CDAEs to denoise effectively. First, the convolution and deconvolution kernels and biases are randomly initialized. Second, the reconstructed signal y_t is obtained. Third, the loss function is obtained. Finally, the Adam optimizer is used to update and obtain the optimum convolution kernel, deconvolution kernel, and biases of the CDAEs that minimize the loss function. Let $\theta = \{\mathbf{W}, \mathbf{b}\}$ be the set of DDAE parameters. According to the Adam optimizer [34], θ can be updated, as follows:

$$\theta_{t+1} = \theta_t - \frac{\eta}{\sqrt{\hat{v}(t) + \epsilon}}\hat{m}(t) \tag{5}$$

where η is the learning rate. The terms $\hat{v}(t)$ and $\hat{m}(t)$ are the bias-corrected first and second moment, which can be obtained using the formula $v_t/1 - \beta_2$ and $m_t/1 - \beta_1$. The terms m_t and v_t are the exponentially moving averages that are determined while using the formula $m_t = \beta_1 m_{t-1} + (1 - \beta_1)g_t$ and $v_t = \beta_2 v_{t-1} + (1 - \beta_2)g_t^2$. The term g_t is the gradient along time t, whereas β_1 and β_1 are the exponential decay rates for the first and second moments, respectively. The optimum $\beta_1, \beta_2, \epsilon$, and η parameters are $0.9, 0.999, 0.0$, and 0.001, respectively.

2.2. Convolutional Layer and Max-Pooling Operation

An autoencoder neural network can be composed of either densely connected layers to form a densely connected network or convolutional layers to form a convolutional neural network (convnet). The densely connected layers receive feature spatial locations as the input data and then learn global patterns from the input data. In contrast, the convolutional layers learn local translation-invariant patterns; that is, after learning a certain pattern at a certain position, a convnet can recognize the pattern anywhere instead of learning the pattern anew if it appears at a new location, which is how a densely connected network operates. In addition, the convolutional network can learn the spatial hierarchy of patterns: the first convolutional layer will learn smaller local patterns, such as edges, the second convolutional layer will learn larger patterns that contain the features of the first layer, and so on. This allows for a convolutional network to effectively learn increasingly complex and abstract visual concepts. We select the convnet in this paper according to the merits that were obtained from the above analysis.

Figure 1 depicts the convolutional working principle of GPR profile data, a type of single-channel (single-depth) dataset, where $depth_0$ represents the input depth of the input feature map. For GPR profile data, the input depth is 1. A convolution works by sliding these windows of size 3×3 or other sizes over the 3D input feature map, stopping at every possible location and extracting the 3D patch of surrounding features (shape (height, width, $depth_0$)). Each such 3D patch is then transformed (via a tensor product with the same learned weight matrix, called the convolution kernel) into a 1D vector of shape ($depth_1$). All of these vectors are then spatially reassembled into a 3D output map of shape (height, width, $depth_2$), where $depth_2 = depth_1$, which is, the depth of the convolutional layer filter. By filling the data throughout the final output feature map, we can obtain an output feature map that is the same size as the input feature map. Every spatial location in the output feature map corresponds to the same location in the input feature map (for example, the lower-right corner of the output contains information regarding the lower-right corner of the input).

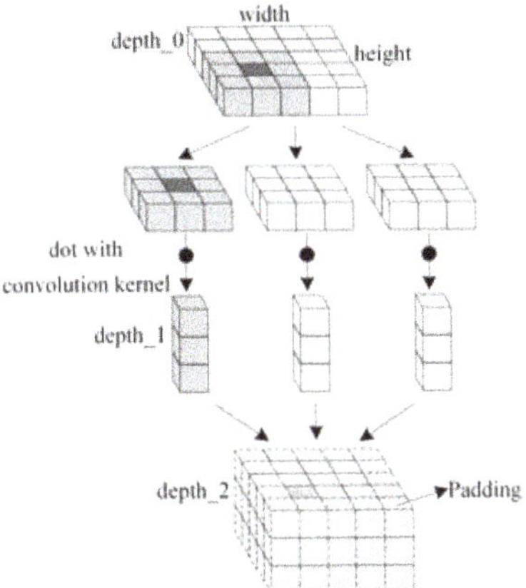

Figure 1. Representation of convolution.

Max-pooling includes extracting windows from the input feature map and outputting the maximum value in each channel. This process is conceptually similar to convolution, but the difference is that it does not transform local patches through a learned linear transformation (convolution kernel); the max-pooling operation is performed with 2×2 windows and a stride of 2 to downsample the feature maps by a factor of 2.

In order to more intuitively understand the features of convolution and max-pooling operations, we select GPR profile data with a size of 64×64 as the input and pass the input data through three convolutional layers with filter depths of 16, 32, and 64, whose convolution kernel size is 5×5, and each convolutional layer is followed by a max-pooling layer with 2×2 windows and a stride of 2. Subsequently, using the rectified linear unit (ReLU) function that is shown in (6) as the activation function of the convolutional layers, which can avoid the vanishing gradient phenomenon as compared to other activation functions, (such as *Sigmoid* or *tanh*), we can obtain the feature maps of each intermediate activation, as shown in Figure 2.

$$ReLU(x) = \max(0, x) \tag{6}$$

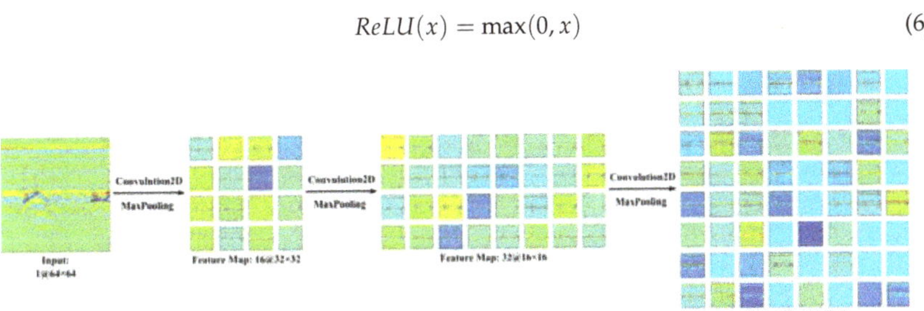

Figure 2. Feature maps of each layer.

Figure 2 presents the output feature map of each layer. The output of the first layer represents a collection of various edge information detectors; and, at this stage, the activation output will retain almost all of the information in the original data. However, the features extracted from the data at a given layer become increasingly abstract and less visually interpretable with the increasing depth of the layer. Deeper activations will carry decreasingly less information regarding specific visual inputs, and increasingly more information about the target itself. The deep neural network effectively acts as an information extraction pipeline, repeatedly transforming the input original GPR data, thereby filtering out irrelevant information and amplifying and refining useful information.

It should also be noted that the output of some filters is blank, as demonstrated in Figure 2. This means that the pattern encoded by the filter cannot be found in the input data, but not that the kernel is dead, which is a normal phenomenon and it will not affect the performance of the network. It is worth noting that the sparsity of an activation increases with the depth of the layer.

2.3. Implementation of CDAEs

Figure 3 shows the network structure of a conventional CDAE. For the encoder part, the complete GPR profile data comprise considerable feature information; hence, if the entire profile dataset is used for the training data, more profile data will be required, which will overwhelm the available computational resources. Therefore, the profile data can be divided into several small blocks by sliding the window to convert the entire profile dataset into fragmented profile data by randomly arranging the slider data. The size of the slider is 32 × 32 and the sliding step is 4. In this way, each window slider contains less feature information. By adopting multiple complete sections of window slider information, the waveform feature information can be more effectively extracted without requiring excessive amounts of data, and heterogeneous images can be combined to boost the sample size for improved noise attenuation performance. Subsequently, fragmented profile data are selected in batches with a size of 64 and they are passed through three convolutional layers in succession. A connected max-pooling layer is encountered after each convolutional layer. The data from the last max-pooling layer are expanded in one dimension and then passed into the fully connected layer as the input data. Finally, a 1D latent vector is output to complete the learning process of the encoder with regard to the data features. The decoder part can be understood as the reverse operation of the encoder: its purpose is to restore the latent vector and ultimately obtain the profile with the same dimensions as the encoder input, thereby completing the denoising of the GPR data.

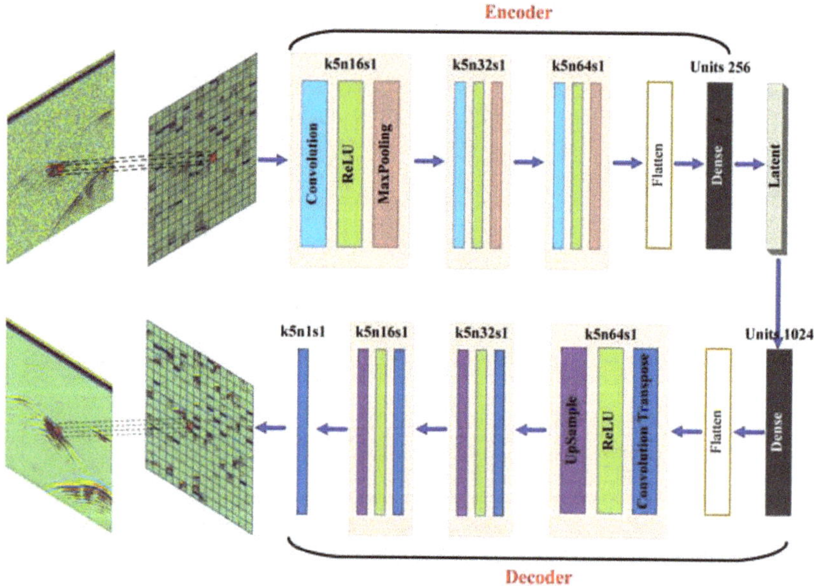

Figure 3. Structure of CDAEs.

The above content elaborates on the detailed principle and essence of conventional CDAEs. During the implementation of conventional CDAEs, we found that these autoencoders may encounter problems, such as overfitting and the representational bottlenecks and vanishing gradients that are commonly found in deep learning. Therefore, it is necessary to modify the structure of conventional CDAEs to further improve their performance.

Algorithm 1 is the CDAE process. First, we segment the GPR datasets into patches $\tilde{x}$ and segment the GPR data without noise into x in the same way.

Algorithm 1 The Training Process for CDAEs

Input: $\tilde{x}, x$
Output: *CDAEs with parameter* **W** *and* **b**
  1: Initialization $epoch = 0, CDAEs$ *with parameter* **W** *and* **b**
  2: **while** $epoch < epochs$ **do**
  3:     $latent$=Encoder.fit($\tilde{x}$)
  4:     $\tilde{x}$=Decoder.fit($latent$)
  5:     mse=MSE($x, \tilde{x}$)
  6:     Update **W** *and* **b** using backpropagation
  7: **end while**
  8: **return** *CDAEs with parameter* **W** *and* **b**

---

3. Network Structure Modification Strategy

3.1. Overfitting Problem and Dropout Regularization Layer

The basic problem of machine learning is to strike a balance between optimization and generalization. By adjusting the hyperparameters, such as the weight value and bias of the model, to obtain better performance on the training data, we will make this process an optimization and use the adjusted model to predict any unknown data. Determining the process of model performance on unknown data is called generalization. Of course, the goal is to achieve good generalization, but generalization is uncontrollable: the only way is to adjust the model that is based on the training data.

At the beginning of the training, optimization and generalization tasks are associated: the loss of verification data gradually decreases with the loss of training data. At this stage, this model is said to be underfit and, thus, further development is necessary. The network, at this time, does not model all of the relevant patterns in the training data. However, after a certain period of epochs on the training data, the generalization task ceases to improve, and the validation metrics plateau and then begin to degrade; at the same time, the generalization loss no longer decreases, but instead begins to increase. In this case, the model is characterized by overfitting; that is, the model begins to learn patterns specific to the training data. The usual solution is to obtain more training data in order to prevent the model from learning irrelevant patterns or from memorizing the training data. However, the training data are limited by both the number of samples and the computer hardware. Therefore, it is necessary to optimize the model from other perspectives to avoid overfitting.

For the model that was learned by the neural network, when given some training data and network architecture, there will be many weights and biases to explain the data, which is, the network is nonunique. The principle of the dropout regularization layer is to randomly drop out (i.e., set to zero) a number of output features of the layer during training. The purpose of this operation is to introduce noise into the output values of a layer to interfere with the network, thereby disrupting happenstance nonsignificant patterns and reducing the overfitting of the model by eliminating the memorization of the neural network on the training datasets [35,36].

3.2. Local Receptive Field and Atrous Convolution

The local receptive field, which is an important concept in a convnet, is defined as the size of the input layer area that determines an element of the output result of a certain layer in a convnet. In mathematical language, the local receptive field is a mapping of an element of output in a certain layer to an input layer in a convnet.

Figure 4a depicts the 1-atrous convolution of size 3 × 3, which is the same as the convolution operation. Likewise, Figure 4b illustrates the 2-atrous convolution of size 3 × 3; however, although the actual convolution kernel size is still 3 × 3, since the atrous number is 2, the image block has a size of 7 × 7, which indicates that the local receptive field is analogous to a 7 × 7 convolution kernel. In Figure 4, only the red dots have weights for the convolution operation, while the weights of the remaining points are all 0, so no operation is performed. From this figure, there are only nine weight values for the actual convolution operation, but the local receptive field of the convolution has increased to 7 × 7. Therefore, for data that need additional features, atrous convolution can be used to obtain a larger local receptive field and reduce the memory consumption of the computer.

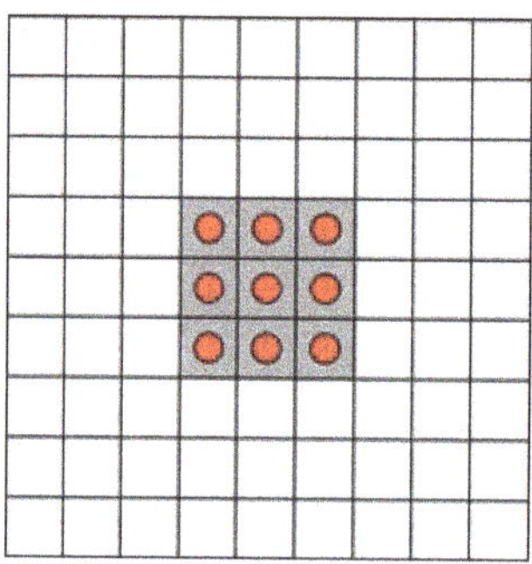

(**a**) 1-Atrous convolution of size 3 × 3

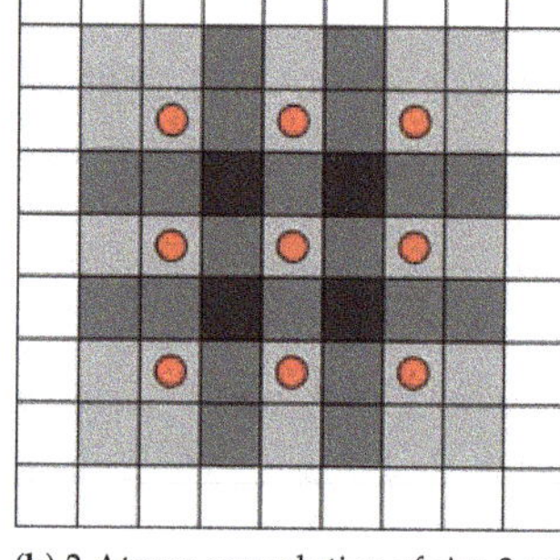

(**b**) 2-Atrous convolution of size 3 × 3

Figure 4. Atrous convolution.

3.3. Loss of Detailed Information and Residual Connections

First, in the conventional sequential model, each successive presentation layer is built on top of the previous layer, which means that the model can only access the information that is contained in the activation of the previous layer. If the features of one layer have too few dimensions, then the model will be constrained by how much information the activations of this layer can contain. Generally, each output will be a previous operation as an input; accordingly, if one operation results in the loss of information, all of the downstream operations will fail to recover the lost information. In other words, any loss of information is permanent. This concept is considered to be representational bottleneck in deep learning.

Second, gradient backpropagation is the main algorithm used to train deep neural networks. Its working principle is to backpropagate the feedback gradient signal from the output layer to the shallower layer. If this feedback signal must propagate through the deep stack, then the signal may be attenuated or even lost entirely, which makes the network training almost unchanged or impossible to train. This problem is called the vanishing gradient problem.

To solve these two problems, the residual-connection structure was proposed. The residual connection involves making the output of the shallower layer available as the input of the deeper layer, effectively creating a shortcut in the sequential network. However, instead of simply linking it to the subsequent activation, the previous output is added to the subsequent activation. Because any loss of information is permanent, the earlier information is reinjected downstream, and residual connections may partially avoid representational bottlenecks for deep learning models. In addition, the introduction of the path's remaining connections may transmit information in parallel to a purely linear main layer stack, thereby facilitating the deep layers of the stack by any propagation gradient [37,38].

For the denoising attenuation problem of GPR data, the loss of information only occurs during the process of transforming profile data into feature latent vectors. Therefore, the sliding block part can be omitted, and only the autoencoder structure needs to be modified. Figure 5 shows the modified autoencoder structure.

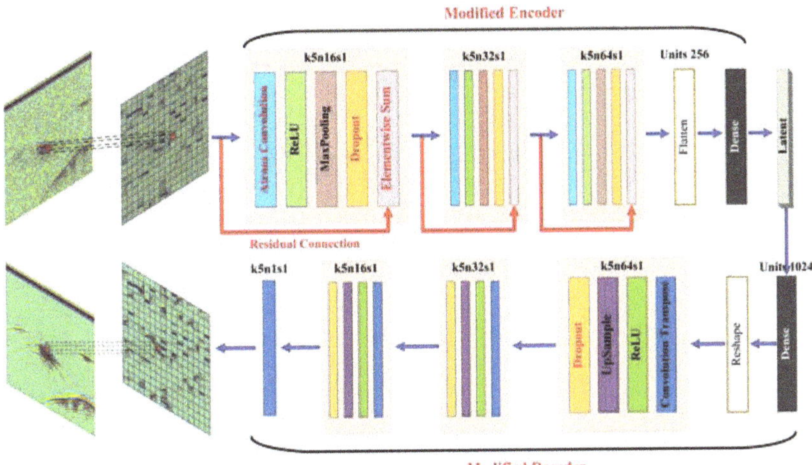

Figure 5. Structure of CDAEsNSO.

4. Selection of Model Parameters and Model Testing

4.1. Training Dataset and Validation Dataset

We randomly generated 200 sets of GPR profiles, including 100 stochastic irregular underground anomaly models and 100 stochastic tunnel lining anomaly models, in order to train a CDAE model. The number, shape, size, location, and physical parameters (such as permittivity and conductivity) of the anomalies in the two model sets are all completely randomly generated. By adding Gaussian noise of different levels to the GPR anomaly profiles and using the sliding window method, each GPR profile is divided into multiple image blocks to obtain more detailed waveform features. In addition, because the GPR profile data contain considerable amounts of redundant information, using each entire profile for the training dataset would introduce substantial redundancy. The sliding window method can avoid this phenomenon and obtain more training data with fewer profiles to meet the training requirements of a neural network by combining heterogeneous images to boost the sample size for improved noise attenuation performance. Figure 6 depicts rhe specific dataset establishment process. When using this method to construct a dataset, in order to avoid the possibility that the window signal loses the original signal information when the data are divided into window blocks, which are mainly hyperbolas spanning the entire "bigger" original image, we need to select the window size flexibly according to the frequency of the signal. The size of the sliding window block in this paper is 32×32 and the sliding step is 4. By preprocessing 200 sets of GPR profiles with noise and randomly arranging sliding window blocks, we obtained 649,800 samples of data as the input for the autoencoder. We employ the mini-batch stochastic gradient descent method, together with the Adam optimizer, to minimize the above loss.

In order to test the generalization ability of the model, the 649,800 samples were divided into three parts: 2000 datasets were randomly selected as the validation dataset to participate in the validation and evaluation during the iterative model training; 2000 datasets were randomly selected as evaluation data to verify and evaluate the final model result; and, the remaining datasets were used as training datasets to iteratively train the model. We use synthetic data to train the network model and apply the network to unknown data (synthetic data or measured data).

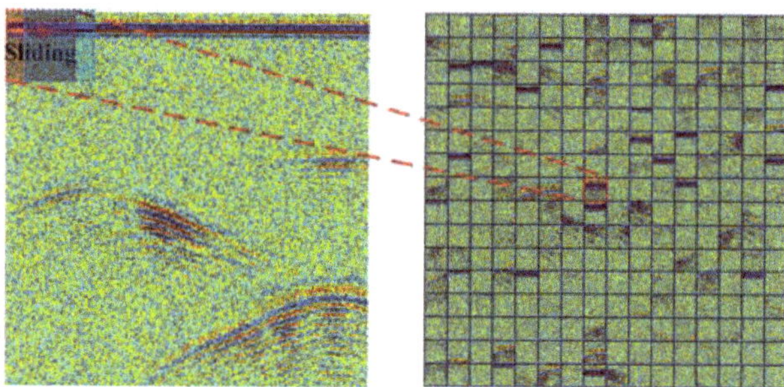

Figure 6. Sliding window process.

4.2. Structure of the Model

CDAEs consist of both encoders and decoders, and Table 2 describes their hierarchical structure. In the encoder part, we input GPR profile datasets with dimensions of $32 \times 32 \times 1$, and we obtain a 1D latent vector with a length of 256 after training. Subsequently, the GPR profile features are transformed into a set of low-dimensional latent vectors, where Conv2D represents the 2D convolutional layer in Table 2; the input receives a 4D tensor; and, the output shape of each layer represents the total number of samples, the image size and the image depth. In the decoder part, the input receives a 1D latent vector with a length of 256, and then GPR profile datasets with dimensions of $32 \times 32 \times 1$ are finally obtained after training. At this point, the reconstruction of the low-dimensional latent vector is complete, where Conv2DTr represents the 2D deconvolution layer; the input receives a 1D vector; and, the output shape of each layer represents the total number of samples, the image size and the image depth.

Table 2. Autoencoder structure.

Encoder Part		Decoder Part	
Layer (Type)	Output Shape	Layer (Type)	Output Shape
Input Layer (Input)	(Batch, 32, 32, 1)	Input Layer (Input)	(Amount, 256)
Conv2D	(Batch, 32, 32, 16)	Dense	(Batch, 1024)
Max-Pooling	(Batch, 16, 16, 16)	Reshape	(Batch, 4, 4, 64)
Conv2D	(Batch, 16, 16, 32)	Conv2DTr	(Batch, 4, 4, 64)
Max-Pooling	(Batch, 8, 8, 32)	Upsampling	(Batch, 8, 8, 64)
Conv2D	(Batch, 8, 8, 64)	Conv2DTr	(Batch, 8, 8, 32)
Max-Pooling	(Batch, 4, 4, 64)	Upsampling	(Batch, 16, 16, 32)
Flatten	(Batch, 1024)	Conv2DTr	(Batch, 16, 16, 16)
Dense (Output)	(Batch, 256)	Upsampling	(Batch, 32, 32, 16)
		Conv2DTr (Output)	(Batch, 32, 32, 1)

Connecting the encoder and the decoder end to end, the output of the encoder is used as the input of the decoder, as shown in Figure 3. The GPR profile data are first converted into a low-dimensional latent vector by the encoder, and then the low-dimensional latent vector is finally reconstructed into GPR profile data by the decoder. So far, the construction of the CDAE network structure has been completed.

4.3. Selection of Parameters

Table 2 demonstrates that the number of filters used by the convolutional layers are 16, 32 = m and 64; the length of the 1D latent vector is 256; the convolution kernel size is 5×5; and, the number of training epochs is set to 10. After repeatedly testing the model performance and the above parameters, the recommended parameters for the best optimization are given in Table 3. In this table, the "Loss" represents the training loss during the training process, and the "Validation Loss" represents the loss obtained by using the data that are not involved in the training process to verify the network performance.

Table 3. Parameters of the Autoencoder.

Filter	Size of Kernel	Length of Latent	Loss	Validation Loss
(16, 32, 64)	(3 × 3)	256	2.1365×10^{-3}	1.5857×10^{-3}
(16, 32, 64)	(5 × 5)	256	1.4921×10^{-4}	7.6509×10^{-4}
(16, 32, 64)	(3 × 3)	512	5.5231×10^{-3}	4.6564×10^{-3}
(16, 32, 64)	(5 × 5)	512	3.8327×10^{-3}	1.8842×10^{-3}
(32, 64, 128)	(3 × 3)	256	1.2146×10^{-2}	9.5504×10^{-3}
(32, 64, 128)	(5 × 5)	256	9.6551×10^{-3}	6.7461×10^{-3}
(32, 64, 128)	(3 × 3)	512	2.8746×10^{-2}	2.0010×10^{-2}
(32, 64, 128)	(5 × 5)	512	1.6854×10^{-2}	1.2517×10^{-2}

Analyzing the parameters presented in Table 3 demonstrates that for a GPR profile with noise, due to its low feature dimension, when selecting a larger filter depth, there will be significant performance degradation, and a larger local receptive field will be required; when the latent vector is excessively large, redundant data will be encountered. Therefore, the feature dimension of the GPR profile determines the optimal parameters of CDAEs.

5. Strategy of Modifying the Network Structure

5.1. Adding the Dropout Regularization Layer

In order to reduce the overfitting problem caused by the neural network memorizing the training data, a dropout regularization layer is added after each convolutional layer and deconvolution layer of CDAEs, and some output features of this layer are randomly dropped out to disrupt happenstance nonsignificant patterns and reduce the overfitting of the model by removing the training data memorized by the neural network, as shown in Figure 5. Figure 7 presents a comparison of the training loss and verification loss after 10 epochs of CDAEs. The blue line represents the loss of CDAEs; and, the red line represents the loss of the dropout CDAEs (D-CDAEs). In this paper, the dropout ratio is set to 0.2, which means that the output values of this layer are excluded at a ratio of 0.2.

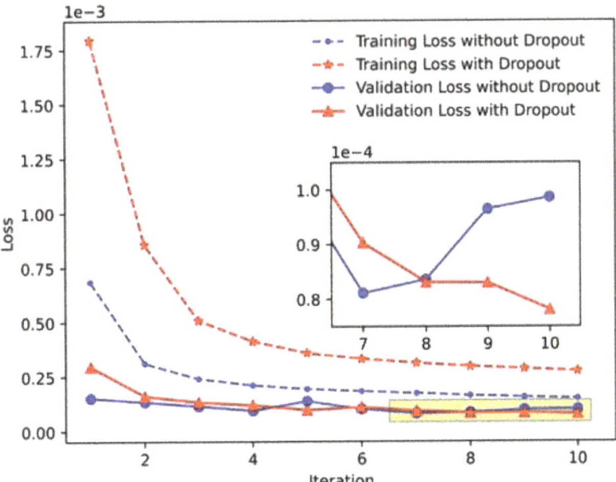

Figure 7. Loss function values of CDAEs and D-CDAEs.

Figure 7 indicates that the training loss of CDAEs is smaller than that of D-CDAEs, which is due to the characteristics of the dropout regularization layer. Each time the data pass through the dropout regularization layer, the data will be dropped out according to the established ratio, which is equivalent to introducing noise and it has a certain impact on the training loss. From the overall trend, the training losses of CDAEs and D-CDAEs present decreasing trends. In addition, D-CDAEs have a smaller and more stable validation loss when compared with CDAEs. To more accurately distinguish the details of the two curves, we partially enlarge the curves in Figure 7. The validation loss of CDAEs reached its minimum value at the seventh training epoch and then began to increase, suggesting that the model started to overfit at this time. The verification loss of D-CDAEs decreased gradually during the iterative training without overfitting. Therefore, the addition of the dropout regularization layer can increase the training loss to a certain extent while avoiding the overfitting phenomenon.

5.2. Replacing Convolution with Atrous Convolution

A tunnel lining model containing an "H-beam" was established to illustrate the influence of using atrous convolution instead of convolution on the attenuation of noise [39], as shown in Figure 8. The dimensions of the model are 5.0 m × 2.5 m. From top to bottom, the model consists of steel, a waterproof board, and an H-beam. The window length of the profile is 70 ns, and the antenna excites a Ricker wavelet with a frequency of 400 MHz. The finite difference time-domain (FDTD) algorithm is used for forward modeling [40].

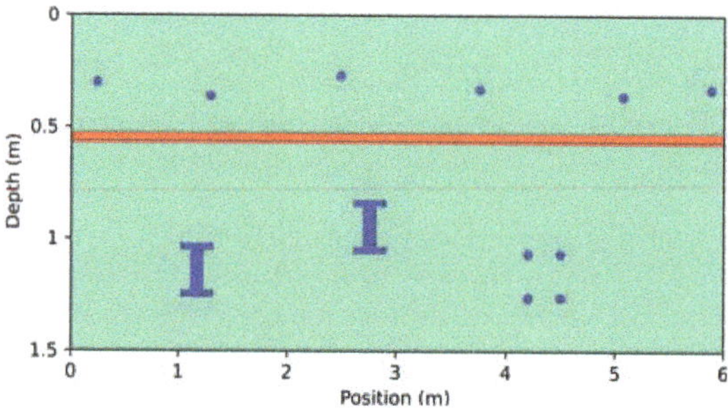

Figure 8. Tunnel lining model.

Figure 9a depicts the GPR forward profile (raw data without noise). Multiple strong reflection hyperbolas from the steel are detected at the top. The interface of the waterproof board is also clearly discernible, but the reflected wave energy is relatively weak. A strong reflection from the H-beam can be seen at 30–40 ns. In addition, a significant number of multiples (i.e., waves that are reflected multiple times) are observed due to the interactions of the anomalies with electromagnetic waves, which complicates the profile information greatly. Random Gaussian noise was added to the GPR profile to form a noisy profile, as shown in Figure 9b (data contaminated by noise). Because the existence of noise seriously affects the quality of the profile data, the multiples and the weak reflections from the waterproof board are completely submerged by the noise, and the strong reflections from the anomalies are also damaged. To compare the noise level that is introduced by these methods more intuitively, we introduce the SNR shown in (7) as the evaluation criterion:

$$SNR = 10\log_{10}\left[\frac{\sum_{x=1}^{N_x}\sum_{y=1}^{N_y}(f(x,y))^2}{\sum_{x=1}^{N_x}\sum_{y=1}^{N_y}(f(x,y)-\hat{f}(x,y))^2}\right] \tag{7}$$

Among them, N_x and N_y represent the size of the data (that is, the data is N_x timesN_y); $f(x,y)$ represents the original data; and $\hat{f}(x,y)$ represents the data with noise. From a statistical perspective, in regard to the difference between the original data and the evaluation data, the smaller the noise level, the greater the effective SNR. Therefore, a larger SNR value indicates better noise attenuation performance.

Each GPR profile has large amounts of redundant data and, hence, a larger local receptive field is necessary to obtain more feature information. Therefore, we proposed using atrous convolution to replace convolution. CDAEs and AD-CDAEs were used to denoise the profile in Figure 9b, and the noise attenuation results are shown in Figure 10a,b, respectively; the residuals after noise attenuation are shown in Figure 11a,b, respectively.

Analyzing both Figures 10b, 11a,b and 12c reveals that the two models can effectively attenuate the noise on the GPR profile; in particular, the reflected waves from the waterproof board and the multiples are all reconstructed. However, CDAEs also destroy the information of the original profile while removing the noise, and the damage to the waveforms is somewhat obvious. In contrast, as shown in the residual profile, AD-CDAEs retain more waveform information than CDAEs and maintain higher fidelity with respect to the effective feature information of the radar wavefield.

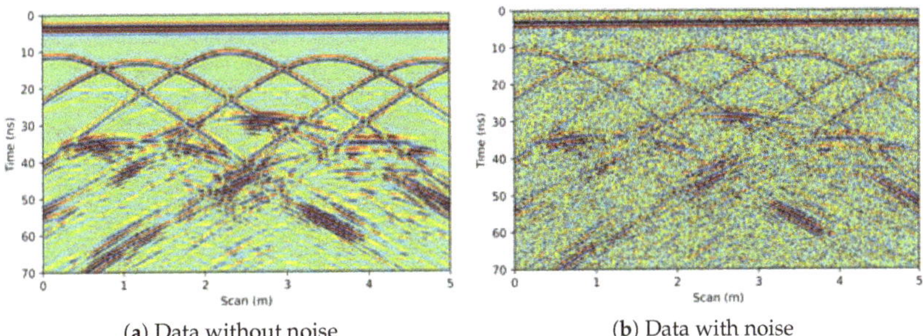

(a) Data without noise   (b) Data with noise

Figure 9. Synthetic profile data of the tunnel lining.

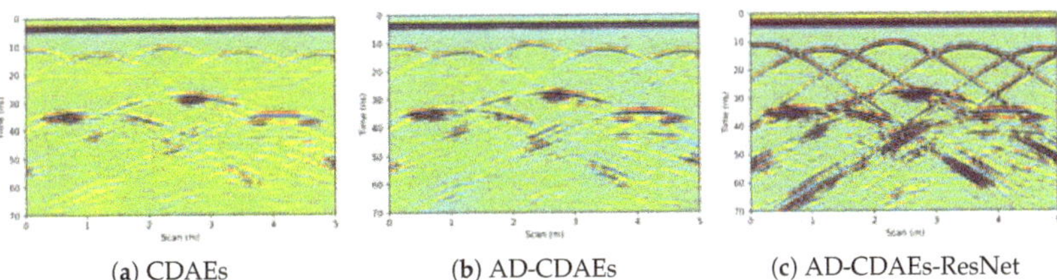

(a) CDAEs   (b) AD-CDAEs   (c) AD-CDAEs-ResNet

Figure 10. Noise attenuation result for the GPR profile of the tunnel lining.

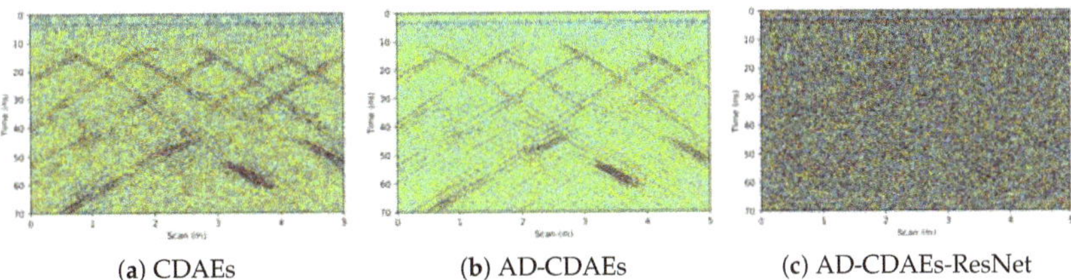

(a) CDAEs   (b) AD-CDAEs   (c) AD-CDAEs-ResNet

Figure 11. Noise attenuation residual for the GPR profile of the tunnel lining.

5.3. Modifying the Network Structure by Residual Connections

Because deep learning suffers from the problems of representational bottlenecks and vanishing gradients, CDAEs are prone to destroying useful feature information while removing Gaussian noise; this tendency causes a large amount of data to be lost and, thus, CDAEs do not meet the noise attenuation requirements of GPR profiles. We used residual connections to further modify the structure of the network and obtained the modified atrous-dropout-CDAEs-residual network (AD-CDAEs-ResNet) that is presented in Figure 5, which is the final version of CDAEsNSO. The second convolutional layer is no longer limited to receiving the output from only the first convolutional layer, but it can also receive the original profile data, which carry the most authentic information. Introducing a path to transport purely linear information helps to propagate gradients through arbitrarily deep stacks of layers, effectively preventing the loss of information and the problem of vanishing gradients. To verify the noise attenuation effect of the modified network structure on the GPR profile, we again took the H-beam model (Figure 8) as an example. Figures 12c and 13c show the noise attenuation result and the residual of AD-CDAE-ResNet, respectively.

These results show that the reflected waves from the waterproof board and the multiples are well preserved, and that the information in the original profile can be retained with high fidelity, indicating that AD-CDAEs-ResNet can effectively remove random noise and resolve the problems of representational bottlenecks and vanishing gradients. To more intuitively compare the noise attenuation effects of several autoencoders, the SNR is calculated before and after the attenuation of noise, and Table 4 shows the results.

Table 4. SNR comparison.

Strategy	Noise Profile	CDAEs	AD-CDAEs	AD-CDAEs-ResNet
SNR	11.3708	17.0944	19.7689	34.4409

Comparing the SNRs presented in Table 4 exposes that CDAEs have the worst noise attenuation effect among the three methods due to the substantial removal of effective information. AD-CDAEs can partly improve the noise attenuation effect by increasing the size of the local receptive field. Ultimately, the AD-CDAEs-ResNet achieves the best noise attenuation effect, because it solves the problems of representational bottlenecks and vanishing gradients.

5.4. Comparison with Other Typical Noise Attenuation Methods

We selected several commonly used noise attenuation algorithms to highlight that CDAEs can better distinguish between noise data and effective signals and have a better effect on noise attenuation. Because AD-CDAEs-ResNet is not based on a fixed-base transformation method, it can be autonomous. The self-adjustment is carried out according to the characteristics of various signals, so the signal has less damage, whereas the method of using a fixed-base transformation means more damage to the signal. Based on this principle, we chose the commonly used method based on fixed-base transformation. In addition, we also selected a signal-based SR strategy to process the data. Similarly, the sparse representation strategy cannot be self-adjusted according to the characteristics of various signals. For this reason, we selected the wavelet transform method and K-singular value decomposition (K-SVD) method according to these methods in Table 1 to process the profile data in Figure 8. Figure 12 shows the noise attenuation result, and Figure 13 shows the noise residual. Table 5 shows the detailed SNR data.

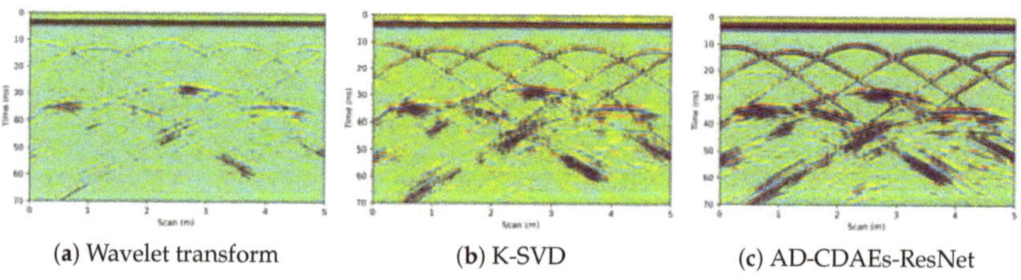

(a) Wavelet transform   (b) K-SVD   (c) AD-CDAEs-ResNet

Figure 12. Noise attenuation result for the GPR profile of the tunnel lining compared with other methods.

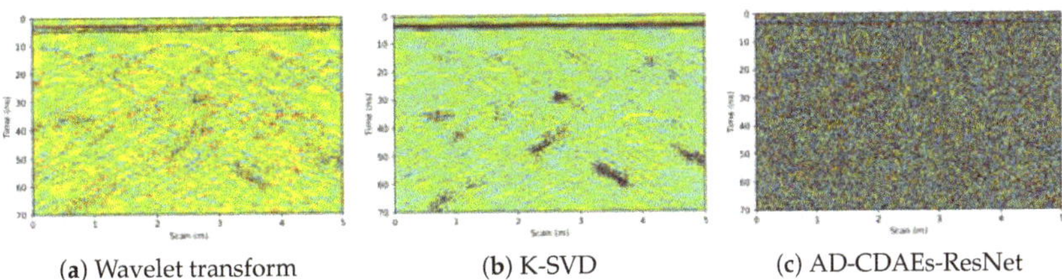

(**a**) Wavelet transform   (**b**) K-SVD   (**c**) AD-CDAEs-ResNet

Figure 13. Noise attenuation residual for the GPR profile of the tunnel lining compared with other methods.

Table 5. SNR comparison with other methods.

Strategy	Noise Profile	Wavelet Transform	K-SVD	AD-CDAEs-ResNet
SNR	11.3708	19.1945	24.5208	34.4409

The results shown in Figures 12 and 13 indicate that, by comparing the noise attenuation results and noise residuals of AD-CDAEs-ResNet and other methods, the wavelet transform method that is based on a fixed basis destroys the effective signal, although it has a certain degree of noise suppression. Because the K-SVD method is based on a dictionary learning strategy, a better overcomplete dictionary is obtained through dictionary learning, so a better noise attenuation effect is obtained, and the damage to the signal is also correspondingly reduced. However, from the point of view of the noise residual, this method is still more serious to the effective signal, especially the position with strong signal energy (direct wave and strong reflection area), and it even causes a distortion of the direct wave signal. Finally, the AD-CDAEs-ResNet algorithm that is proposed in this paper obtains the optimal effect, and it can effectively distinguish the noise signal from the effective signal, and it perfectly suppresses the noise without causing a distortion of other signals. In addition, from the comparison of the SNR values presented in Table 5, the same conclusion can be obtained.

We list the SNRs of these algorithms in the form of a histogram to summarize the abovementioned noise attenuation algorithms and compare the improvement of SNRs by various algorithms, as shown in Figure 14. The histogram conveniently provides a comparison and description.

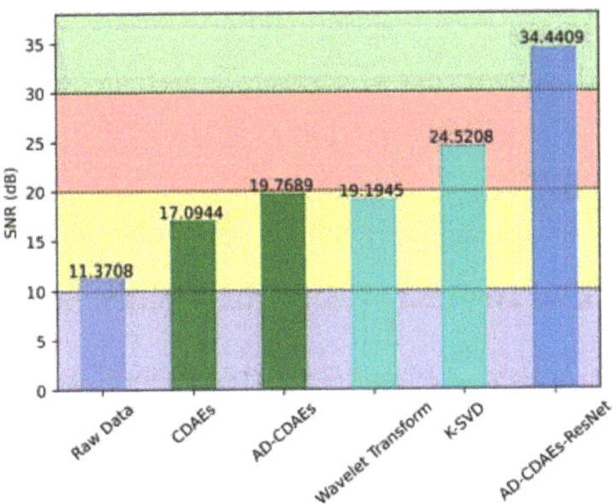

Figure 14. SNR histogram of various methods.

6. Results

6.1. Synthetic Data

To illustrate the adaptability of AD-CDAEs-ResNet to complex models, Gaussian random noise and Gaussian spike impulse noise, we designed the tunnel lining model that is shown in Figure 15 with the dimensions of 5.0 m × 2.5 m. From top to bottom, there are two linings with a relative permittivity of 9, an uncompacted part filled with water, and cavities. The window length of the profile is 70 ns, and the antenna excites a Ricker wavelet with a frequency of 400 MHz. The FDTD algorithm is used for forward modeling.

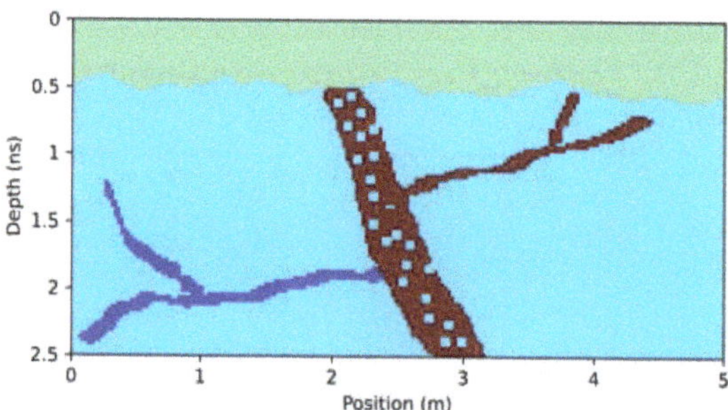

Figure 15. The model of the complex tunnel lining.

Figure 16a presents the GPR forward profile. We can identify the reflected waves, diffracted waves, and multiples generated by the interfaces between the surrounding rock and the two linings, the water-bearing fractures, and the water-free fractures. The positions, sizes, and shapes of the geological anomalies can be roughly estimated by analyzing the shapes of the radar reflections. Gaussian random noise and high-interference Gaussian spike impulse noise were added to the profile to obtain the noisy GPR profile that is shown in Figure 16b. Because of the influence of this Gaussian noise, the waves reflected from the second lining are submerged by noise, and there are 5 instances in which vertical Gaussian impulse spikes disturb the data section; these phenomena seriously affect the interpretation of the data.

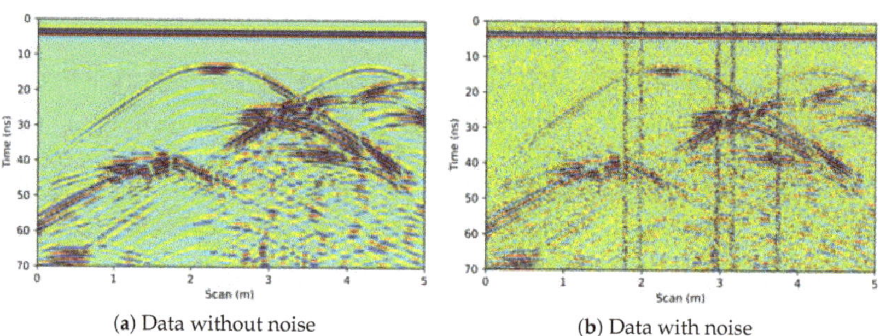

(**a**) Data without noise   (**b**) Data with noise

Figure 16. Synthetic profile data of a complex tunnel lining.

6.2. Field Data

We selected a pipeline field dataset to validate the effectiveness of AD-CDAEs-ResNet to illustrate the practicality of the proposed algorithm. The data were collected using the SIR-4000 GPR with an antenna that excites a wavelet with a frequency of 400 MHz as the excitation signal. The data were collected from Huangxiu Agricultural Culture Park, Yueyang city, Hunan Province. We used the spot measurement mode with a spot distance of 0.05 m for the measurement. A total of 111 channels of data were collected, and the recording time was 60 ns. Figure 17 shows the radar survey lines and field measurements.

Figure 17. Photo showing the field collection of data.

7. Discussion

Figure 18a shows the noise attenuation result that was obtained by applying AD-CDAEs-ResNet to the mixed noise profile containing both Gaussian spike impulse noise and Gaussian noise. This figure indicates that AD-CDAEs-ResNet can effectively remove mixed noise while reliably preserving information of the reflected waves and multiples. Figure 18b presents the residual of AD-CDAEs-ResNet, indicating that the proposed algorithm does less damage to the information in the original profile and maintains better fidelity. The SNRs before and after the attenuation of noise are 9.7392 dB and 27.1753 dB, respectively, verifying the effectiveness of the algorithm.

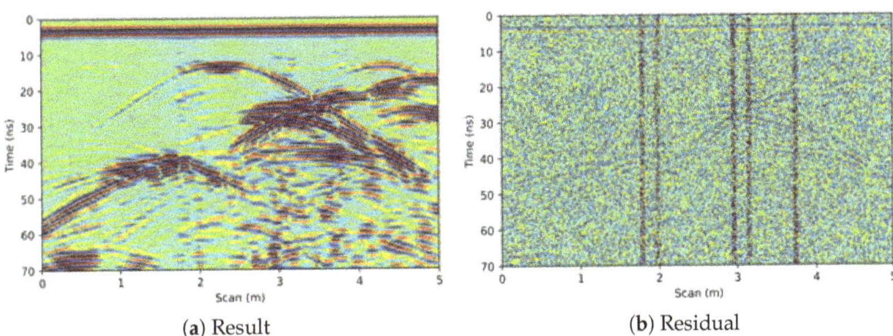

(a) Result   (b) Residual

Figure 18. Noise attenuation results and residual for the GPR profile of the complex tunnel lining.

Figure 19a shows a section of the GPR data that were acquired in the field. Two groups of Gaussian spike impulse noise are circled, and the reflected waves from the pipelines are not particularly clear. Figure 19b presents the noise attenuation result of AD-CDAEs-ResNet. Hyperbolic diffracted waves from the upper and lower interfaces of pipelines are readily detectable, and the layer interfaces at approximately 56 *ns* can also be identified. Figure 19c illustrates the residual of AD-CDAEs-ResNet, containing almost all noise and very little information pertaining to reflected waves. From these residuals, it is evident that the three groups of Gaussian spike impulse noise are obviously removed, which confirms that the proposed noise attenuation algorithm does less damage to the information in the original profile and achieves an improved noise attenuation effect.

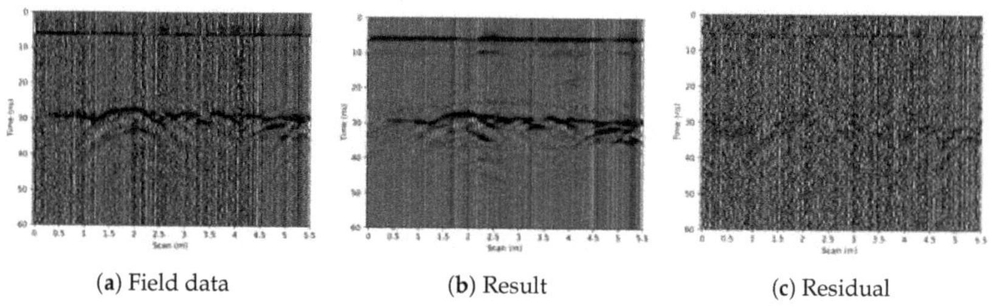

(a) Field data   (b) Result   (c) Residual

Figure 19. Noise attenuation result and residual of the GPR profile from the field data.

8. Conclusions

When compared with data domain transformation-based methods and SR-based techniques, convolutional denoising autoencoders (CDAEs) that are based on deep learning can adjust themselves according to the features of signals. CDAEs represent a kind of unsupervised learning neural network that can adapt to the denoising algorithm. However, when CDAEs are directly applied to attenuate the noise of GPR data, they encounter various problems, including overfitting, the size of the local receptive field, and the representation bottlenecks and vanishing gradients that are typical of deep learning. Therefore, the authors proposed some network structure optimization strategies, such as the addition of a dropout regularization layer, an atrous convolution layer and a residual-connection structure, and obtained a new GPR noise attenuation algorithm, namely CDAEsNSO.

CDAEsNSO, which was proposed based on CDAEs, can effectively remove random noise and Gaussian spike impulse noise from GPR data. Moreover, the proposed algorithm does little damage to useful waveforms, such as information of reflected waves, diffracted waves, and multiples, in the original profile and maintains high data fidelity, effectively improving the noise attenuation effect. At the same time, this method also has certain limitations. For example, a large amount of data is required as learning samples during network training, and the network is a computationally intensive operation during the training process, which requires a higher level of computer equipment. In terms of network training, it is also more time consuming than other algorithms. However, once the network training is completed, the model can be used directly to achieve end-to-end processing operations. In the final processing stage, when compared with other algorithms, it does not need to be recalculated.

Therefore, the efficiency of data processing is higher.

The GPR profiles contain considerable amounts of redundant information. Nevertheless, the detailed features in a whole profile are sparse. Therefore, we proposed the sliding window method to process GPR profile data to obtain training datasets by combining heterogeneous images to boost the sample size for improved noise attenuation performance. This method not only avoids the redundancy of information while obtaining more detailed waveform characteristics, but also utilizes fewer GPR data to obtain more training datasets that meet the training requirements of CDAEs.

Author Contributions: Conceptualization, D.F., X.W. (Xiangyu Wang) and X.W. (Xun Wang); methodology, X.W. (Xun Wang); validation, X.W. (Xiangyu Wang) and X.W. (Xun Wang); formal analysis, X.W. (Xiangyu Wang); resources, S.D. and H.Z.; writing—original draft preparation, X.W. (Xiangyu Wang); writing—review and editing, X.W. (Xiangyu Wang), X.W. (Xun Wang) and H.Z.; visualization, X.W. (Xiangyu Wang); supervision, D.F. and X.W. (Xun Wang); project administration, D.F.; funding acquisition, D.F. All authors have read and agreed to the published version of the manuscript.

Funding: This research was funded by the National Natural Science Foundation of China under grant numbers 42074161 and 41774132. The APC was funded by the Free Exploration Project of Central South University under Grant 2020zzts185.

Institutional Review Board Statement: Not applicable.

Informed Consent Statement: Not applicable.

Data Availability Statement: The code uses Python 3.8 + TensorFlow 2.0 version and is based on the MIT license. The Python code used to reproduce the results and all the physical models and GPR profile data in this article can be publicly accessed through https://github.com/nephilim2016/AutoEncoder-for-GPR-Denoise.

Acknowledgments: The authors would like to thank Yi Lei, the School of Geosciences and Info-Physics, Central South University, and the Key Laboratory of Metallogenic Prediction of Nonferrous Metals for giving us the opportunity to collect the field data. They would also like to thank the support of TensorFlow and Keras for this research. In addition, they would also like to thank American Journal Experts (AJE) and Cory R. for editing this article.

Conflicts of Interest: The authors declare no conflict of interest.

Abbreviations

The following abbreviations are used in this manuscript:

GPR	ground penetrating radar
SNR	signal-noise ratio
CDAEs	convolutional denoising autoencoders
CDAEsNSO	convolutional denoising autoencoders with network structure optimization
AD-CDAEs	atrous-dropout convolutional denoising autoencoders
ResCDAEs	residual-connection convolutional denoising autoencoders
ReLU	rectified linear unit
K-SVD	K-singular value decomposition

References

1. Dong, Z.; Ye, S.; Gao, Y.; Fang, G.; Zhang, X.; Xue, Z.; Zhang, T. Rapid detection methods for asphalt pavement thicknesses and defects by a vehicle-mounted ground penetrating radar (GPR) system. *Sensors* **2016**, *16*, 2067. [CrossRef]
2. Khamzin, A.K.; Varnavina, A.V.; Torgashov, E.V. Utilization of air-launched ground penetrating radar (GPR) for pavement condition assessment. *Constr. Build. Mater.* **2017**, *141*, 130–139. [CrossRef]
3. Diamanti, N.; Annan, A.P.; Redman, J.D. Concrete bridge deck deterioration assessment using ground penetrating radar (GPR). *J. Environ. Eng. Geophys.* **2017**, *22*, 121–132. [CrossRef]
4. Solla, M.; Lagüela, S.; Fernández, N. Assessing rebar corrosion through the combination of nondestructive GPR and IRT methodologies. *Remote Sens.* **2019**, *11*, 1705. [CrossRef]
5. Rasol, M.A.; Pérez-Gracia, V.; Fernandes, F.M.; Pais, J.C.; Santos-Assunçao, S.; Santos, C.; Sossa, V. GPR laboratory tests and numerical models to characterize cracks in cement concrete specimens, exemplifying damage in rigid pavement. *Measurement* **2020**, *158*, 107662. [CrossRef]
6. Liu, X.; Dong, X.; Xue, Q. Ground penetrating radar (GPR) detects fine roots of agricultural crops in the field. *Plant Soil.* **2018**, *423*, 517–531. [CrossRef]
7. Lai, W.; Chang, R.K.; Sham, J.F. Blind test of nondestructive underground void detection by ground penetrating radar (GPR). *J. Appl. Geophys.* **2018**, *149*, 10–17. [CrossRef]
8. Terrasse, G.; Nicolas, J.M.; Trouvé, E. Application of the Curvelet Transform for clutter and noise removal in GPR data. *IEEE J. Sel. Top. Appl. Earth Observ. Remote Sens.* **2017**, *10*, 4280–4294. [CrossRef]
9. He, T.; Shang, H. Direct-wave denoising of low-frequency ground-penetrating radar in open pits based on empirical curvelet transform. *Near Surf. Geophys.* **2020**, *18*, 295–305. [CrossRef]
10. Gan, L.; Zhou, L.; Liu, S.M. A de-noising method for GPR signal based on EEMD. *Appl. Mech. Mater. Trans. Tech. Publ. Ltd.* **2014**, *687*, 3909–3913. [CrossRef]
11. Li, J.; Liu, C.; Zeng, Z.F.; Chen, L.N. GPR Signal Denoising and Target Extraction with the CEEMD Method. *IEEE Geosci. Remote Sens. Lett.* **2015**, *12*, 1615–1619.
12. Ostoori, R.; Goudarzi, A.; Oskooi, B. GPR random noise reduction using BPD and EMD. *J. Geophys. Eng.* **2018**, *15*, 347–353. [CrossRef]
13. Zhang, X.; Feng, X.; Zhang, X. Dip Filter and Random Noise Suppression for GPR B-Scan Data Based on a Hybrid Method in f-x Domain. *Remote Sens.* **2019**, *11*, 2180. [CrossRef]
14. Kobayashi, M.; Nakano, K. A denoising method for detecting reflected waves from buried objects by ground-penetrating radar. *Electr. Commun. Jpn.* **2013**, *96*, 1–13. [CrossRef]
15. Huang, W.L.; Wu, R.S.; Wang, R. Damped dreamlet representation for exploration seismic data interpolation and denoising. *IEEE Trans. Geosci. Remote Sens.* **2018**, *56*, 3159–3172. [CrossRef]

16. Qiao, X.; Fang, F.; Zhang, J. Ground Penetrating Radar Weak Signals Denoising via Semi-soft Threshold Empirical Wavelet Transform. *Ingénierie Syst. Inf.* **2019**, *24*, 207–213. [CrossRef]
17. Chen, L.; Zeng, Z.; Li, J. Research on weak signal extraction and noise removal for GPR data based on principal component analysis. *Glob. Geol.* **2015**, *18*, 196–202.
18. Huang, W.; Wang, R.; Chen, X. Damped sparse representation for seismic random noise attenuation. In *SEG Technical Program Expanded Abstracts 2017*; Society of Exploration Geophysicists: Tulsa, OK, USA, 2017; pp. 5079–5084.
19. Deng, L.; Yuan, S.; Wang, S. Sparse Bayesian learning-based seismic denoise by using physical wavelet as basis functions. *IEEE Geosci. Remote Sens. Lett.* **2017**, *14*, 1993–1997. [CrossRef]
20. Bi, W.; Zhao, Y.; An, C. Clutter elimination and random-noise denoising of GPR signals using an SVD method based on the Hankel matrix in the local frequency domain. *Sensors* **2018**, *18*, 3422. [CrossRef]
21. Liu, L.; Ma, J. Structured graph dictionary learning and application on the seismic denoising. *IEEE Trans. Geosci. Remote Sens.* **2018**, *57*, 1883–1893. [CrossRef]
22. Bai, X.; Peng, X. Radar image series denoising of space targets based on Gaussian process regression. *IEEE Trans. Geosci. Remote Sens.* **2019**, *57*, 4659–4669. [CrossRef]
23. Zhang, J.; Zhang, H.; Li, Y. A Morphological component analysis based on mixed dictionary for signal denoising of ground penetrating radar. *Int. J. Simul. Process. Model.* **2019**, *14*, 431–441. [CrossRef]
24. Li, H.; Wang, R.; Cao, S. Mathematical morphological filtering for linear noise attenuation of seismic data. *Geophysics* **2016**, *81*, V159–V167. [CrossRef]
25. Huang, W.; Wang, R.; Zhang, D. A method for low-frequency noise suppression based on mathematical morphology in microseismic monitoring. *Geophysics* **2017**, *82*, V369–V384. [CrossRef]
26. Li, J.; Le Bastard, C.; Wang, Y.; Gang, W.; Ma, B.; Sun, M. Enhanced GPR Signal for Layered Media Time-Delay Estimation in Low-SNR Scenario. *IEEE Geosci. Remote Sens. Lett.* **2016**, *13*, 299–303. [CrossRef]
27. Gondara, L. Medical image denoising using convolutional denoising autoencoders. In Proceedings of the 2016 IEEE 16th International Conference on Data Mining Workshops (ICDMW), Barcelona, Spain, 12–15 December 2016; pp. 241–246.
28. Zhou, H.; Feng, X.; Dong, Z.; Liu, C.; Liang, W. Application of Denoising CNN for Noise Suppression and Weak Signal Extraction of Lunar Penetrating Radar Data. *Remote Sens.* **2021**, *13*, 779. [CrossRef]
29. Grais, E.M.; Plumbley, M.D. Single channel audio source separation using convolutional denoising autoencoders. In Proceedings of the 2017 IEEE global conference on signal and information processing (GlobalSIP), Montreal, QC, Canada, 14–16 November 2017; pp. 1265–1269.
30. Meng, Z.; Zhan, X.; Li, J. An enhancement denoising autoencoder for rolling bearing fault diagnosis. *Measurement* **2018**, *130*, 448–454. [CrossRef]
31. Ashfahani, A.; Pratama, M.; Lughofer, E. DEVDAN: Deep evolving denoising autoencoder. *Neurocomputing* **2020**, *390*, 297–314. [CrossRef]
32. Zhu, W.; Mousavi, S.M.; Beroza, G.C. Seismic signal denoising and decomposition using deep neural networks. *IEEE Trans. Geosci. Remote Sens.* **2019**, *57*, 9476–9488. [CrossRef]
33. Saad, O.M.; Chen, Y. Deep denoising autoencoder for seismic random noise attenuation. *Geophysics* **2020**, *85*, V367–V376. [CrossRef]
34. Kingma, D.P.; Ba, J. Adam: A method for stochastic optimization. *arXiv* **2014**, arXiv:1412.6980.
35. Dahl, G.E.; Sainath, T.N.; Hinton, G.E. Improving deep neural networks for LVCSR using rectified linear units and dropout. In Proceedings of the 2013 IEEE International Conference on Acoustics, Speech and Signal Processing, Vancouver, BC, Canada, 26–30 May 2013; pp. 8609–9613.
36. Srivastava, N.; Hinton, G.; Krizhevsky, A. Dropout: A simple way to prevent neural networks from overfitting. *J. Mach. Learn. Res.* **2014**, *15*, 1929–1958.
37. He, K.; Zhang, X.; Ren, S. Deep residual learning for image recognition. In Proceedings of the IEEE Conference on Computer Vision and Pattern Recognition, Las Vegas, NV, USA, 27–30 June 2016; pp. 770–778.
38. Wu, S.; Zhong, S.; Liu, Y. Deep residual learning for image steganalysis. *Multimed. Tools Appl.* **2018**, *77*, 10437–10453. [CrossRef]
39. Lyu, Y.; Wang, H.; Gong, J. GPR Detection of Tunnel Lining Cavities and Reverse-time Migration Imaging. *Appl. Geophys.* **2020**, *17*, 1–7. [CrossRef]
40. Zhang, B.; Dai, Q.; Yin, X. A new approach of rotated staggered grid FD method with unsplit convolutional PML for GPR. *IEEE J. Sel. Top. Appl. Earth Obs. Remote Sens.* **2015**, *9*, 52–59. [CrossRef]

 *remote sensing*

Article

GPR Clutter Reflection Noise-Filtering through Singular Value Decomposition in the Bidimensional Spectral Domain

Rui Jorge Oliveira [1,2,3,*], **Bento Caldeira** [1,2,3], **Teresa Teixidó** [4] and **José Fernando Borges** [1,2,3]

[1] Instituto de Ciências da Terra, ICT, Universidade de Évora, Apartado 94, 7002-554 Évora, Portugal; bafcc@uevora.pt (B.C.); jborges@uevora.pt (J.F.B.)
[2] Departamento de Física, Escola de Ciências e Tecnologia, Universidade de Évora, Apartado 94, 7002-554 Évora, Portugal
[3] Earth Remote Sensing Laboratory, EaRSLab, Universidade de Évora, Apartado 94, 7002-554 Évora, Portugal
[4] Instituto Andaluz de Geofísica, Campus Universitario de Cartuja, Universidad de Granada, 18071 Granada, Spain; tteixido@ugr.es
* Correspondence: ruio@uevora.pt

Abstract: Usually, in ground-penetrating radar (GPR) datasets, the user defines the limits between the useful signal and the noise through standard filtering to isolate the effective signal as much as possible. However, there are true reflections that mask the coherent reflectors that can be considered noise. In archaeological sites these clutter reflections are caused by scattering with origin in subsurface elements (e.g., isolated masonry, ceramic objects, and archaeological collapses). Its elimination is difficult because the wavelet parameters similar to coherent reflections and there is a risk of creating artefacts. In this study, a procedure to filter the clutter reflection noise (CRN) from GPR datasets is presented. The CRN filter is a singular value decomposition-based method (SVD), applied in the 2D spectral domain. This CRN filtering was tested in a dataset obtained from a controlled laboratory environment, to establish a mathematical control of this algorithm. Additionally, it has been applied in a 3D-GPR dataset acquired in the Roman villa of Horta da Torre (Fronteira, Portugal), which is an uncontrolled environment. The results show an increase in the quality of archaeological GPR planimetry that was verified via archaeological excavation.

Keywords: applied geophysics; digital signal processing; enhancement of 3D-GPR datasets; clutter noise removal; spectral filtering

Citation: Oliveira, R.J.; Caldeira, B.; Teixidó, T.; Borges, J.F. GPR Clutter Reflection Noise-Filtering through Singular Value Decomposition in the Bidimensional Spectral Domain. *Remote Sens.* **2021**, *13*, 2005. https://doi.org/10.3390/rs13102005

Academic Editors: Yuri Álvarez López and María García Fernández

Received: 4 May 2021
Accepted: 15 May 2021
Published: 20 May 2021

Publisher's Note: MDPI stays neutral with regard to jurisdictional claims in published maps and institutional affiliations.

1. Introduction

1.1. Ground-Penetrating Radar Surveys in Archaeologic Environment

The main feature in applied geophysics for subsurface exploration is the use of remote sensing techniques. This feature becomes more important in archaeological environments as a previous stage to excavations and can play an important role in site delimitation, being able to make heritage protection endeavors more effective. The data processing is usually performed by fully automatic routines that produce acceptable results when the physical conditions of the terrain are favorable but fail under adverse conditions creating doubt regarding the validity of the geophysical techniques used in archaeological exploration arises.

Ground-penetrating radar (GPR) is one of the most common electromagnetic methods used in Archaeology. It has become popular since 1970 [1] due to its advantages of quick data acquisition and high resolution. The range and resolution are functions of the antenna frequency used, which typically varies between 200 MHz and 1.6 GHz in archaeological environment [2]. This method allows us to determine the spatial distribution of structures buried in the ground, such as walls, ditches, floors, cavities, and even water levels. For a high detection, two conditions should be given: a high dielectric contrast between the buried structures and a lower background noise. Currently, the standard processing

177

flow [3] operate well when the datasets have good-medium quality, but specific routines are required for the datasets with low signal/noise ratio.

1.2. The Noise Problem in the GPR Datasets

This study focuses on those cases where the archaeological context allows us to infer that there is a strong possibility that buried structures are present, but the acquired datasets are not capable of detecting it. In these adverse conditions of low contrast, the elimination, as far as possible, of the clutter background noise on radargrams (usually designated B-scans) without significantly affecting the signal level is a crucial factor to increasing the amplitude of the useful signal. Therefore, a filter approach operating in the spectral domain using automatic data selection techniques is proposed.

Considering that the GPR signal contains two types of information: the useful signal that arrives at the receiver antenna after being reflected by buried structures; and the noise that arrives at the same time but is produced by undefined mechanisms and paths. The coherent noise as a coupled air wave, ground surface fluctuations, diffractions, environmental wave interferences, can be removed with standard filtering processes [3]. However, the clutter-scattering reflections produced by non-target objects, e.g., soil heterogeneities, stones and archaeological collapses [4] are difficult to remove without creating artefacts or producing suppressions in the useful signal.

In archaeological sites this noise is very frequent and expressive. When it is in excess in GPR signals, it is not possible to distinguish between it and the part corresponding to the buried structures (the useful signal), because its amplitudes and frequencies are similar to this useful signal components. For example, the presence of drops or corners can produce as many reflections as those originating from structures that may exist, preventing a good assessment of the subsurface content. The problem related to the identification and selection of the useful signal in GPR datasets is the main target of this research. In next sections we will present an effective algorithm that removes the clutter noise from the 3D-GPR datasets and increases the quality of the GPR images.

1.3. GPR Filtering Approaches

Regarding clutter noise removal techniques in GPR datasets, there are several approaches to perform it, which can be grouped into three main categories [5].

The first category corresponds to the classical filtering methods used to eliminate most of the coherent noise. E.g., average trace or moving average subtraction [6], time-gating with zero padding [7], entropy-based [8], Kalman filtering [9], F–K filtering [10], multiresolution wavelet analysis [11,12], wavelet filtering [13] and directional total variation minimization filter [14].

The second category comprises filters based on parametric or statistical-based methods that require reference datasets, making them dependent on the assumptions used to estimate the clutter parameters and their statistical features [15–17].

The third category is classified as subspace technique [15] that explores the statistical properties of the data decomposing it into different subspaces, such as the useful information and the noise. Examples of this type of approach are eigen image filtering [18], independent component analysis [19], principal component analysis (PCA) [20], singular value decomposition (SVD) [20] and robust principal component analysis [21,22].

1.4. Proposed CRN Filter

The proposed method, framed in SVD-based filtering approach [5], intends to establish a new methodology to eliminate clutter reflection noise present in the B-scans as effectively as possible without modifying the useful GPR signal and as automatic.

The method considers the 2D spectral domain to perform the selection using SVD. This approach is intended to complement the common processing operations of GPR datasets. As a complementary operation to others, it is assumed that the data must be filtered out of random noise, migrated, and deconvolved.

To develop and configure this method, a laboratory experiment was prepared. The procedure consists of the identification of clutter reflection noise under controlled conditions using data obtained in a laboratory model. Then, the extraction of clutter signal was performed through the study of the signal in the 2D spectral domain and using the SVD factorization technique. The methodology was also applied to field data (uncontrolled conditions) in 3D-GPR datasets.

2. Implementation of the CRN Filter in GPR Datasets

2.1. General Overview

This research began with some laboratory experiments to study the influence of geophysical parameters on the direct problem, for the calculation of synthetic B-scans [23]. In the laboratory, a model (Figure 1a) was built to recreate a field scenario on a small scale [24]. In this, a B-scan was acquired for use in processing operations (Figure 1b), using a 1.6 GHz antenna. In parallel, a synthetic model was created in a numerical environment (Figure 1c) for later calculation of the synthetic B-scan (Figure 1d). These steps were implemented in MATLAB using the matGPR package (release 2) [25].

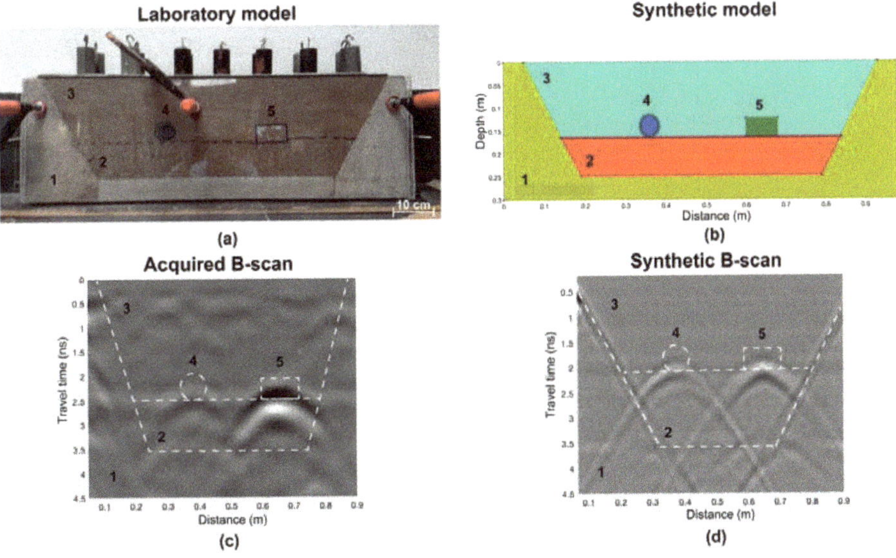

Figure 1. (**a**) Model created in the laboratory using soil from an archaeological site in which objects were buried are: 1—cement slopes and base; 2—soil, layer 1; 3—soil, layer 2; 4—hollow plastic cylinder; 5—solid aluminum prism. (**b**) Synthetic model. (**c**) Acquired B-scan. (**d**) Synthetic B-scan. The B-scans are similar in terms of the object reflections. The differences lie in the background noise present in the acquired data.

Comparing both datasets, in the acquired B-scan there is an amount of background noise while in synthetic one this is absent. Evidently, the difference is because the synthetic model a homogeneous medium has been considered (a limitation of the model parametrization). From which we can deduce that the soil embedded of the laboratory model produces clutter reflections when it is scanned by a 1.6 MHz GPR antenna. The attempt to filter the acquired data to bring them closer to the synthetic ones was not possible with conventional operations to process GPR data (Figure 2). This standard flow mainly consists of the correction of the surface position, deconvolution and infinite impulse response (IRR) filters [3]. This limitation motivated the study of new and more effective data filtering methodology.

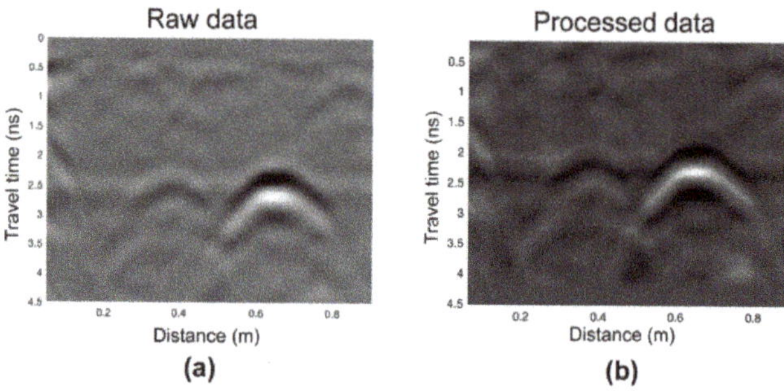

Figure 2. (**a**) Acquired B-scan (raw data) obtained in the laboratory model. (**b**) Processed B-scan using conventional flow (correction of the surface position, deconvolution and IRR filters). The standard processing does not remove the clutter reflections corresponding to the disperse elements of the lab soil.

The main goal rises to distinguish between useful and clutter reflections. Related literature suggests several approaches applied to images with noise using spatial filtering. These methods can be applied pixel by pixel in the raster image [26,27] that allow only a local control of the input and does not require any information about the general structure of the image, or in the spectral domain of the data using the Fourier transform [27,28] which allows to consider the structure of the image.

The CRN filter operates in the spectral domain. In the next subsections we describe the main steps taken to design it to apply to GPR datasets.

2.2. Bidimensional Spectral Analysis of a GPR Dataset

From a B-scan, which consists a set of GPR traces as a function of the position, x (x-axis), and the travel time, t (y-axis), after applying the 2D Fourier transform, the position is converted into wavenumber, k, and the time is converted into frequency, f. Figure 3a,b shows the 2D Fourier transform of the acquired B-scan (Figure 2a). In both the magnitude and phase spectra it is possible to verify that the useful signal is in the internal area delimited by the red lines. To better identify the area of interest, a 3D representation of the magnitude spectrum was generated (Figure 3c). This allows the classification of the spectrum in larger amplitudes corresponding to the coherent reflections, and the smaller ones to the clutter reflections. The outlier observed in the center of the magnitude spectrum corresponds to the wavenumber values near zero.

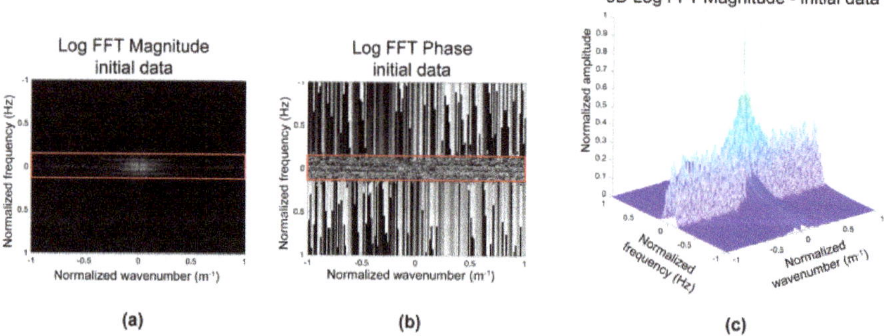

Figure 3. (**a**) Magnitude and (**b**) phase spectra of the acquired B-scan. (**c**) 3D magnitude spectrum of the same B-scan. The area inside the red lines indicates the location of useful signal in the spectra. In the 3D view, an outlier value is observed in the center of the spectrum, which corresponds to the k values near zero.

2.3. Parametrization and Application of the Filter Matrix

To assess how the removal of a given spectra sector impacts on the original signal, an elliptical shape (Figure 4a) was created covering the entire light-colored band observed in the magnitude spectrum. To remove the anomalous values around $k = 0$, a small circular area had to be created around the center of the spectrum. Thus, the total area that parameterizes the filter corresponds to the interior of the elliptical shape, except the central part defined by the circular shape. The filter to remove clutter signals will be applied as a bandpass filter, which maintains the unit values and eliminates the null values in the area (Figure 4b).

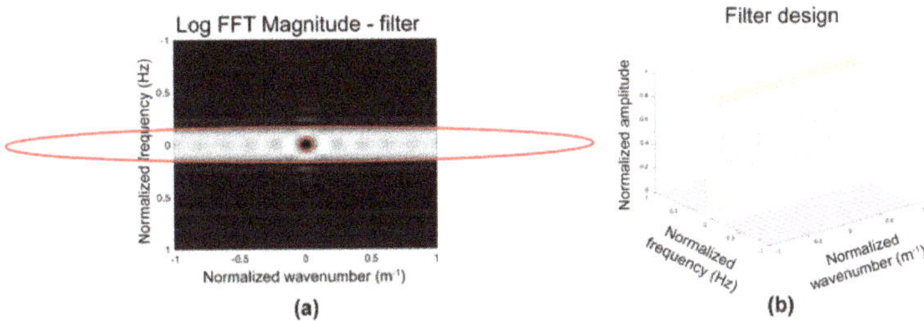

Figure 4. (**a**) Magnitude spectrum with the elliptical red lines corresponding to the filter design. (**b**) Filter design of this dataset. This bandpass filter has unit value in the area to be maintained and null value in the area to be eliminated.

This filter (Figure 4b) can be considered to be a matrix where the parameters are the frequency range limits and the 2D finite impulse response (FIR) of the filter. To determine the matrix coefficients a dedicated MATLAB code has been made using the Image Processing Toolbox routines [29], where the outputs are the filter-matrix coefficients. The next step is the application of the 2D Fourier transform to the matrix that contains the filter coefficients and its multiplication by the transformed GPR data. Therefore, the inverse of Fourier transform is applied to restore the initial space-time domain. The results can be observed in Figure 5.

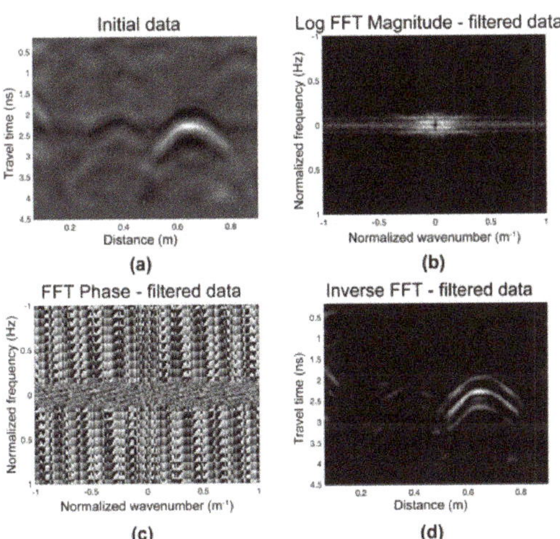

Figure 5. Results from the spectral filtering approach. (**a**) Input data in the space-time domain. (**b**) Magnitude spectrum of the filtered data. (**c**) Phase spectrum of the filtered data. (**d**) Filtered data in the space-time domain. The applied filter was effective to erase the CRN. However, this is highly dependent on the user parametrization.

The proposed approach is very effective, but in this case the matrix parameterization process was made by trial-and-error. This was easily implemented under controlled conditions due to the knowledge of the location of the buried objects and the corresponding reflections in the B-scan. However, in a scenario more complex or with larger amounts of clutter reflections, this task may be compromised due to the high dependence of the parameterization on the user. For this reason, the possibility of applying mathematical techniques for the automatic classification of the GPR data to reduce the user dependence on this filtering approach was investigated.

2.4. Using Singular Value Decomposition (SVD) in the CRN Filtering

There is a great diversity of multivariate statistical techniques that allow the classification of a dataset into their principal components, where each one receives a weight. Principal component analysis (PCA) is a technique that allows us to estimate the number of principal components present for a given dataset by calculating its dominant vectors [30–32]. In the PCA, the covariance matrix of the data contains the values related to the spectral decomposition. It is a square matrix (X), which in the case of being symmetric, then they the singular values are the absolute values of the eigenvalues of X. Taking this into account, after applying the 2D Fourier transform to the GPR data, it become symmetrical (Figure 3a), so that the PCA can be applied to classify these types of datasets.

The numerical implementation of PCA to matrix-based image data is carried out using SVD algorithms, which can be applied to matrices of real numbers and complex numbers. SVD is a very effective and stable factorization technique to diagonalizing matrices [30–32], because the system can be decomposed into a set of linearly independent components, each with an associated weight. This technique is generally used in data compression [33–36] and is very useful for filtering out noise [35,37].

According to [32], when SVD is applied to a dataset defined by a X matrix ($M \times N$; $M > N$), the reconstruction can be performed through the linear combination of the calculated parameters after the selection of the principal components to be used in reconstruction (1).

$$X = USV^T \tag{1}$$

U is an $M \times N$ orthogonal matrix, where the columns represent the left singular vectors that are eigenvectors of XX^T. V is an $N \times N$ orthogonal matrix, where the columns represent the right singular vectors that are eigenvectors of $X^T X$. S is an $M \times N$ matrix, where the diagonal values represent the singular values of X, which are different from zero; these values are sorted ordered in descending order and sum to 1.

The reconstruction of a system decomposed by SVD is carried out through the linear combination of the principal components, and some noise can be excluded without compromising the structural integrity of the image itself [32,38]. For example, in the SVD of an image, the singular values specify the luminance of the layers image and the respective pairs of singular vectors specify the geometry of each layer [30,31]. The principal components of an image with largest amplitudes usually are associated with larger singular values [32].

For a GPR dataset, the numerical environment corresponds to an amplitude matrix defined by real numbers. When the 2D Fourier transform is applied, the matrix becomes defined by complex numbers, and its spectra have symmetry. These two characteristics are compatible with the theoretical assumptions behind the SVD technique [32]. Therefore, we have the expectation that it will be effective in this type of data.

A test was carried out in MATLAB using the *svd* function. The input data are the 2D Fourier transform acquired B-scan. When applying SVD, the singular values and vectors of the matrix are calculated and stored in the variables U, S, and V. From S (the singular values), the weights of each principal component are calculated.

The graphical projection of the weights of each principal component will allow us to determine which ones will have greater expression in the dataset and to decide which components should be considered in the reconstruction of the data, to the detriment of the exclusion of the other components. In Figure 6, it is possible to verify that the first component has a much larger weight (approximately 65%) than the other components (<10% each). To determine which components should be considered in the reconstruction of the initial matrix, while maintaining only the useful signal of the B-scan and excluding the clutter noise, it is necessary to perform some tests by trial-and-error to ascertain the effect of removing components from the initial dataset.

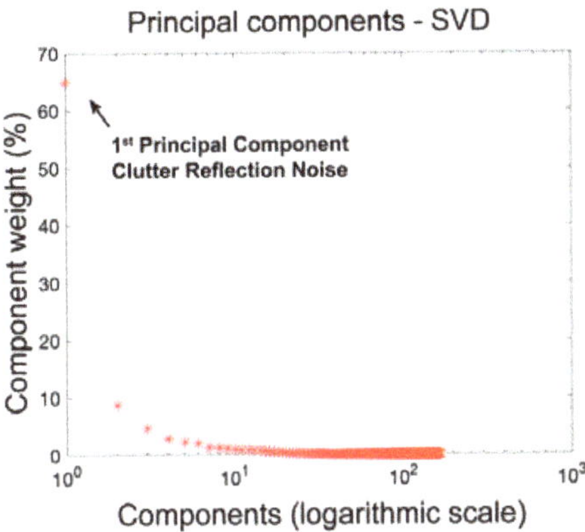

Figure 6. Principal components of a B-scan obtained by SVD after the application of the 2D Fourier transform. The first principal component is dominant (65%). The remaining components have a weight less than 10% each.

The first test consists of the reconstruction of the initial data matrix considering only the first principal component to ascertain whether this component, since it has a much larger weight than the other components, corresponds to the useful signal. To verify the effect of excluding a set of principal components, the data must be restored in the space-time domain. The magnitude spectrum (3D view) will also be represented.

Keeping only the first principal component (Figure 7a,b), the filtered data shows an amplitude distribution that does not resemble the initial B-scan (Figure 2a). The magnitude spectrum reveals that most of the information has been eliminated; only remains the part corresponding to the center frequency. This result suggests that the hypothesis of the correspondence of the first principal component to the useful signal of the B-scan can be ruled out.

The second test has been the reconstruction of the initial matrix keeping only the first two principal components. The filtered data (Figure 7c,d) shows that there is also no similarity with the initial data, and the range of amplitudes also shows that most of the information it contains has been eliminated.

Following the tests, keeping the first three principal components, the filtered data already begin to show similarities with the initial data (Figure 7e,f), and the magnitude spectrum also begins to show similarities with the initial magnitude spectrum.

The next test consists of restoring the initial matrix while keeping all the principal components, excluding only the first one. The results show that the filtered data do not have the CRN that was observed in the initial data (Figure 8a), and the magnitude spectrum (Figure 8b) maintains the morphology of the initial spectrum (Figure 3c).

The result shown in Figure 8 reveals that all the principal components except the first one, with about 35% weight (less than the first) are sufficiently to represent the data. Thus, the removal of the first principal component does not compromise the structural integrity of the dataset. It can be verified that the erased component corresponds to the clutter reflections noise contained in B-scan and the remain principal components corresponds to useful reflections signal.

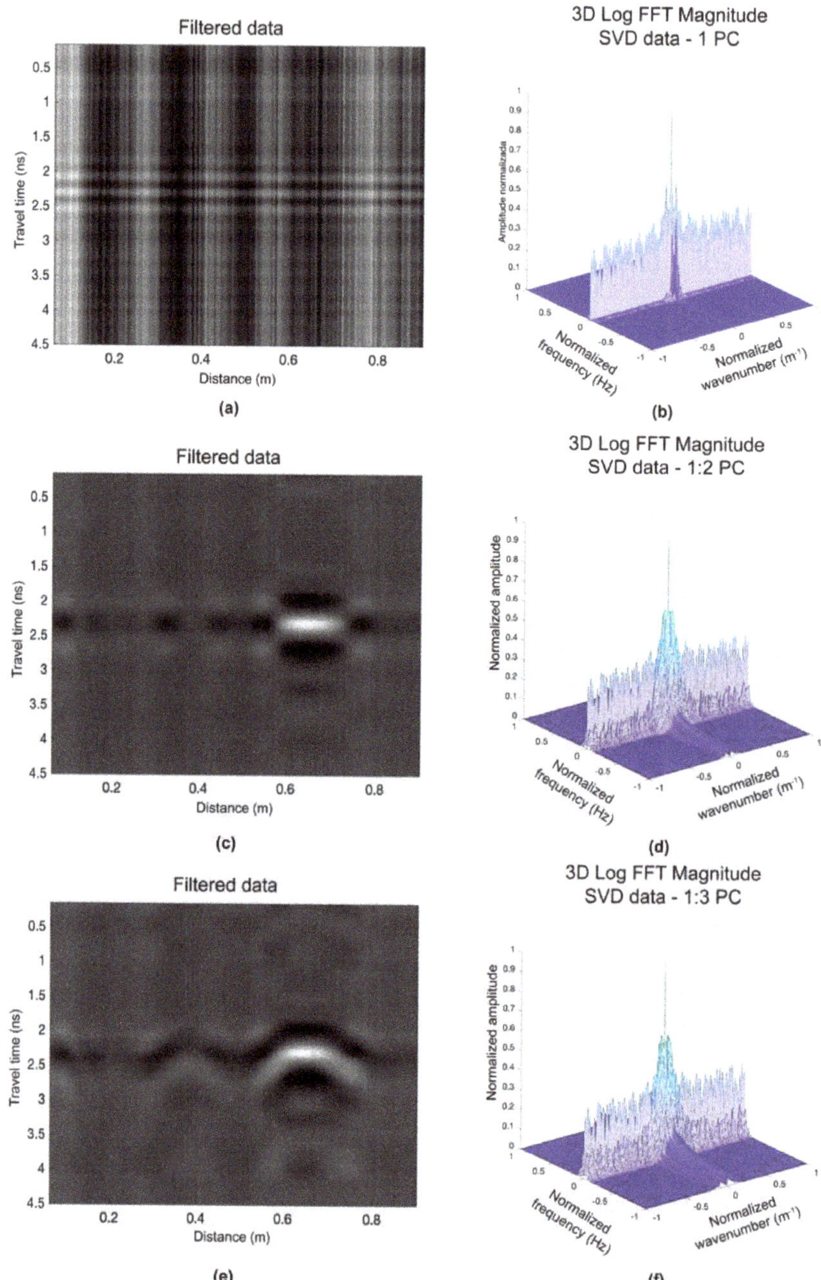

Figure 7. Results of the performed tests to verify the effect of the data restoration and its magnitude spectra considering only some principal components. (**a,b**) Filtered data keeping only the first principal component. (**c,d**) Filtered data keeping the first two principal components. (**e,f**) Filtered data keeping the first three principal components.

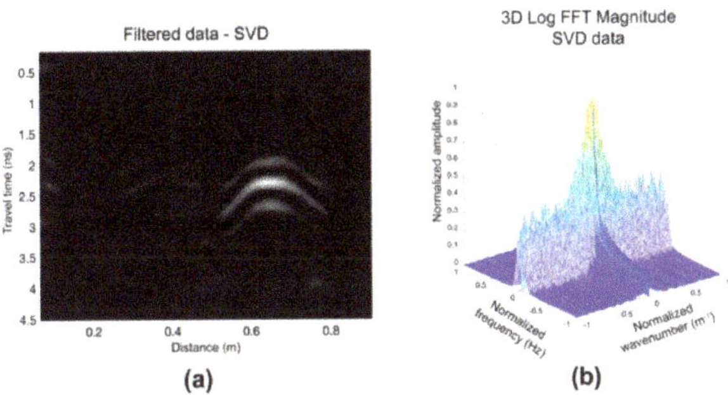

Figure 8. Results of the test to verify the effect of the initial data restoration considering all the principal components except the first one. (**a**) Filtered data. (**b**) 3D magnitudes of the filtered data. By removing the first principal component, which corresponds to the CRN, the restored signal corresponds to the useful information contained in the B-scan.

The next step will be to eliminate the amplitudes around $k = 0$. These amplitudes, which are generated by low frequencies (i.e., located at the center of the spectra), are not included in the noise since they are part of the data themselves. Their elimination must be carried out by trial-and-error until it can be seen that their removal has been successful. The numerical implementation consists of the creation of a unitary matrix, in which the center value is null (a binary filter), in the space-time domain. This implementation occurs in the application of the 2D Fourier transform, so that the filter is applied in the spectral domain through the multiplication of the filter by the data filtered by SVD.

To visualize the results of the whole CRN filter approach, it is necessary to restore the data into space-time domain. Figure 9 allows us to verify that this CRN filter has been successful. Only the reflections corresponding to the objects buried in the laboratory model were maintained.

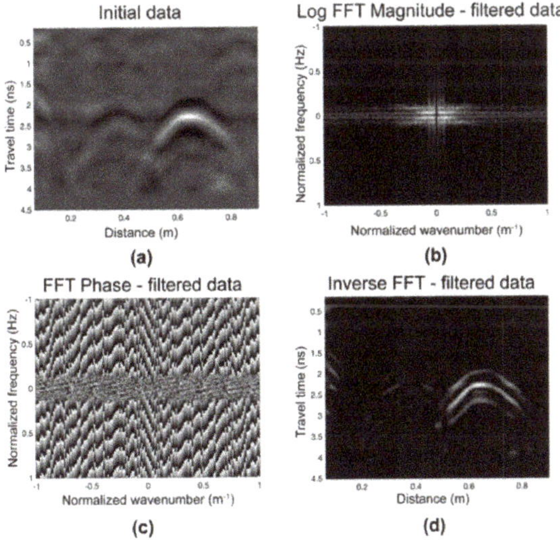

Figure 9. Results from the application of the filter in the spectral domain. (**a**) Input data in the space-time domain. (**b**) Magnitude spectrum of the filtered data. (**c**) Phase spectrum of the filtered data. (**d**) Resulting data CRN filtered in the space-time domain.

3. Application of the CRN Filter to a 3D-GPR Dataset (Uncontrolled Environment)

A definitive test of this approach will be carried out with a 3D-GPR dataset. This test intends to verify whether the methodology improves the GPR planimetry of the buried structures. The considered 3D-GPR dataset was acquired in the Roman villa of Horta da Torre (Fronteira, Portugal), obtained with a 400 MHz antenna. Before the implementation, the profiles were partially processed with a standard flow [3] where the applied operations mainly consisted of background and vertical filtering, deconvolution, and gain adjustments. Then, the data were prepared to be processed in MATLAB, to apply the CRN proposed filtering iteratively. After that, the processed data were exported again to the commercial GPR-software to generate the 3D-GPR images (time depth-slices).

To evaluate the effectiveness of CRN filtering approach, two datasets were compared at the same depth-slices. The first one is the 3D-GPR dataset processed with the better standard flow (labelled as *Standard* in Figure 10), and the other is the same processed 3D-GPR dataset where the CRN filtering has been added (labelled as *CRNF* in Figure 10).

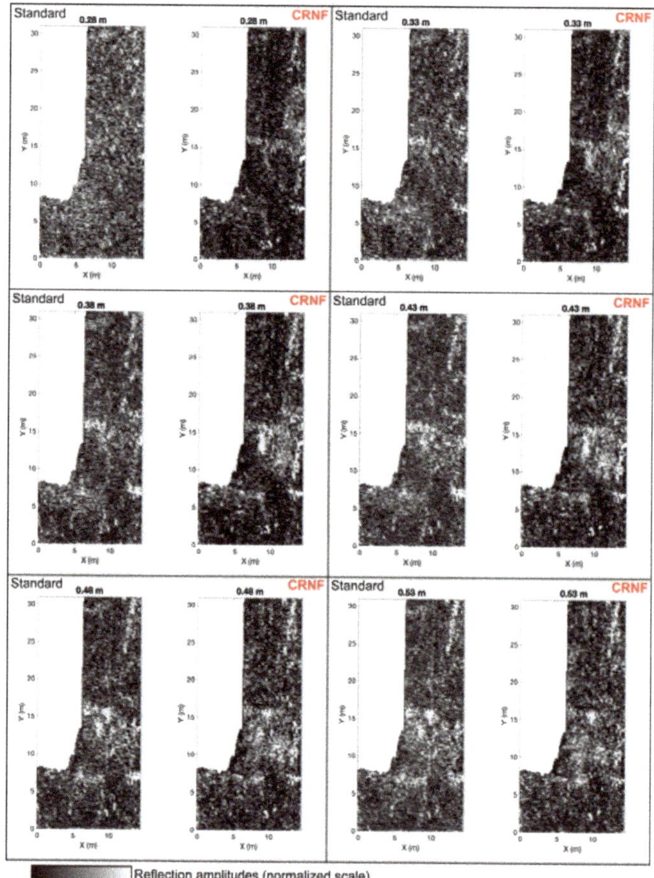

Figure 10. Comparison between the same GPR depth-slices without (Standard) and with spectral filtering (CRNF). The CRN filtered results allow us to observe that there was an increase in the contrast between the background and the coherent reflections of buried structures.

The analysis of the depth-slices of both datasets allows us to verify that the CRN filtering highlights reflection alignments that correspond to buried wall-type structures, and the visible alignments in both datasets were also verified as a better definition in the dataset. These observations allow us to conclude that the proposed filtering approach increases the detection of the buried structures.

To enhance the effectiveness of the CRN filter, cover surfaces [39] were generated from both datasets, considering all reflections between the depths 0.28 and 0.63 m (Figure 11a). In this 2D and 3D cover views, an improvement of the planimetry for the CRNF dataset can be verified again. In Figure 11b, the CRNF dataset was overlaid to an aerial image of the site, which allowed us to point out the excavated structures (yellow lines) and the alignments that can be seen in CRNF dataset (blue lines, unexcavated). The planimetry overlay obtained by the CRNF dataset with the archaeological excavation shows that in addition to increasing the geometric sharpness, the CRN filtering has not produced any artefacts or suppression of important reflections.

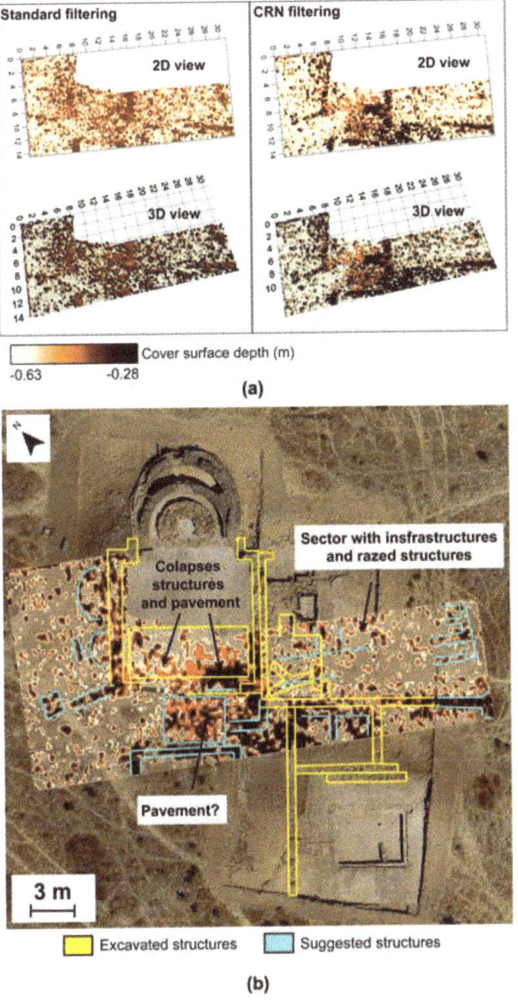

Figure 11. (**a**) Cover surfaces from Standard and CRNF datasets (2D and 3D views). (**b**) Cover surface from CRNF dataset superimposed on an aerial image obtained by a drone (provided by Geodrone). A schematic representation of the visible structures (yellow lines) and the alignments obtained by the geophysical prospection (blue lines, unexcavated) is included. The CRNF results show that the structures visible in the Standard dataset become more defined and better individualized.

4. Discussion of the Results

The results obtained from the laboratory and field datasets show that the proposed approach is effective for filtering CRN in GPR datasets. The selection of the noise next to the signal, using the SVD technique applied in the 2D spectral domain, allowed the implementation in a customized and automated way.

In the laboratory test, a GPR profile was obtained with a 1.6 GHz antenna. The considered model was created in controlled conditions. In this study, there is a prior knowledge about the buried structures as well as their locations.

The first test carried out consisted of a manual filter parametrization by trial-and-error. Despite the good results, this scheme is highly dependent on the user decision, which prevents a possible automation of this procedure. Therefore, the next step was to study an automatic way to filter the clutter noise, to produce results similar to the obtained in the manual approach. SVD applied in the 2D spectral domain allowed to perform it effectively. However, the isolated use of SVD did not solve the noise problem completely. It is necessary to filter the amplitudes around $k = 0$, located in the center of the spectra, to reduce the banded graphical effect due to the high wavelength reflections. In this step, the user must use a trial-and-error approach that consists of the parametrization of the size of the filter and the cut off value until its successful removal. This is the weakest part of the approach, because it does not allow for a complete automation.

The application of the proposed filtering methodology to the laboratory dataset allowed us to isolate the reflections corresponding to the buried structures and exclude the remaining reflections corresponding to the CRN. The filtered B-scan shows that this processing considerably increased the contrast between the reflections that corresponds to buried structures and the remaining reflections; only the latter remained with values tending towards zero. The increase of the contrast allows us to enhance the perception about the existence of buried structures in the subsurface.

The approach was also applied to a 3D-GPR field data, obtained with a 400 MHz antenna in an archaeological site. This was later excavated exposing walls and floors, which was crucial to confirm the correspondence with the reflections alignments visible in the GPR results and to verify that the applied filtering approach is effective in removing the background noise while maintaining only the information corresponding to those structures.

The produced results were compared with the standard processing results, after the production of depth-slices at same depths of the 3D reflectivity model and the cover surfaces of each dataset. The graphical representation of both allows us to observe that the proposed approach can provide more useful information, with an automatic parametrization, than the obtained by classical processing. In the data obtained by the CRNF approach, the reflections corresponding to buried structures are visible at all the considered depths, showing high contrast between the background and the alignments of the reflections that correspond to buried structures that seem to have better definition. In the data obtained by standard processing, it is possible to interpret the alignments of the reflections as buried structures. However, the contrast and graphical resolution are not as clear as in the case of the proposed methodology. This comparison may seem unfair; the standard processing depends on the user experience; thus, another filter configuration can produce better results than the obtained with the standard parametrization used. Thus, we highlight that the great advantage of the proposed approach is the automation that SVD allows the introduction of the CRNF, i.e., a user with less experience in data processing can obtain a more accurate result.

Despite the effectiveness of the approach, there are some limitations that have been noted. It is necessary to remove the amplitude values around $k = 0$ manually, as described to the laboratory case. The good results are not so evident when we analyze the individual processed B-scans. The improve is verified after considering the depth-slices extracted from 3D reflectivity model (Figures 10 and 11).

As mentioned before, this approach is complementary to the conventional processing operations used for GPR datasets. It cannot be used without first applying the common procedures. Several tests were provided to study at what stage of processing this approach works best. If it is applied before, although the amplitude values of the useful signal have not been modified, as the noise values no longer exist, then the GPR signal is anatomically different, which corrupts the application of the remaining processing operations.

The main advantage of this technique compared to others that already exist is the way in which useful information is identified in a dataset with high amount of CRN. The use of a filtering procedure combining the SVD technique with the advantages of the 2D Fourier transform has a better adaptation than the classic filtering methods that fail if the requirements for its implementation are not verified [6–14]. Nor do they depend on prior knowledge of the structure or standard information as with statistical-based filtering methods [15–17]. The proposed method depends on the data itself [15] and its application is in a two-dimensional way, unlike the most common approaches that usually consider the 1D Fourier transform, which results in a not complete effectiveness in the data selection. Using the 2D Fourier transform will improve de data selection. This approach is based on the one applied to improve noisy images by transform it into the 2D spectral domain and the selection of the useful information using parametrization based on the symmetry characteristics of the transformed dataset. Selecting the data of interest using the SVD technique turns this operation more automated from the user. The user only needs to choose the principal component that corresponds to the noise to be eliminated. This knowledge is achieved after the spectral study of the GPR data and the effect of principal components removal in the data. This procedure does not modify the structural integrity of the data.

The continuation of this study should focus on the complete automation of this GPR data filtering approach, namely the removal of the amplitudes around $k = 0$.

5. Conclusions

This study proposes a new methodology to perform clutter reflection noise (CRN) filtering present in ground-penetrating radar (GPR) datasets, which the standard processing flows usually cannot mitigate effectively. This approach has the advantage of the use of the 2D spectral domain, through the 2D Fourier transform, to turns easier the filter parametrization, due to its symmetry characteristics and the constant computational effort whatever the amount of data.

The high dependency of the user to parametrize the spectral filtering decreased using singular value decomposition factorization technique. This study allowed to conclude that the CRN corresponds to the first principal component of the GPR dataset. It is dominant, about 60–65% of the transformed data. Their exclusion, i.e., restoring the initial matrix considering all the principal components except the first one, eliminates most of the CRN of the B-scans considered.

The approach was effectively tested in a laboratory GPR dataset, with controlled conditions. The tests have peaked with the application to 3D-GPR dataset obtained in the Roman villa of Horta da Torre (Fronteira—Portugal, an uncontrolled environment) which produced better results with the CRN filtering methodology than using the standard processing flow. This allowed to increase the geometric sharpness of the GPR planimetry and has not produced any numerical artefacts.

Author Contributions: Conceptualization, R.J.O., B.C. and T.T.; methodology, R.J.O., B.C. and T.T.; software, R.J.O.; validation, R.J.O., B.C., T.T. and J.F.B.; formal analysis, R.J.O., B.C., T.T. and J.F.B.; investigation, R.J.O., B.C., T.T. and J.F.B.; resources, R.J.O., B.C., T.T. and J.F.B.; data curation, R.J.O., B.C., T.T. and J.F.B.; writing—original draft preparation, R.J.O.; writing—review and editing, R.J.O., B.C., T.T. and J.F.B.; visualization, R.J.O., B.C., T.T. and J.F.B.; supervision, R.J.O., B.C., T.T. and J.F.B. All authors have read and agreed to the published version of the manuscript.

Funding: This work has been partially supported by the research project "Innovación abierta e inteligente en la EUROACE" with the reference 0049_INNOACE_4_E, by the European Union through the European Regional Development Fund included in COMPETE 2020, and through the Portuguese Foundation for Science and Technology (FCT) projects UIDB/04683/2020-ICT (Institute of Earth Sciences) and SFRH/BSAB/143063/2018.

Institutional Review Board Statement: Not applicable.

Informed Consent Statement: Not applicable.

Data Availability Statement: Data sharing not applicable.

Acknowledgments: The aerial orthophotography from Roman villa of Horta da Torre was provided by Geodrone.

Conflicts of Interest: The authors declare no conflict of interest.

References

1. Vickers, R.S.; Dolphin, L.T. A Communication on archaeological radar experiment at Chaco Canyon, New Mexico. *MASCA Newsl.* **1975**, *11*, 6–8.
2. Conyers, L.B.; Goodman, D. *Ground-Penetrating Radar: An Introduction for Archaeologists*; AltaMira Press: Walnut Creek, CA, USA, 1997.
3. Jol, H.M. (Ed.) *Ground Penetrating Radar Theory and Applications*; Elsevier Science: Amsterdam, The Netherlands, 2009.
4. Bi, W.; Zhao, Y.; An, C.; Hu, S. Clutter Elimination and Random-Noise Denoising of GPR Signals Using an SVD Method Based on the Hankel Matrix in the Local Frequency Domain. *Sensors* **2018**, *18*, 3422. [CrossRef] [PubMed]
5. Chen, G.; Fu, L.; Chen, K.; Boateng, C.; Ge, S. Adaptive Ground Clutter Reduction in Ground-Penetrating Radar Data Based on Principal Component Analysis. *IEEE Trans. Geosci. Remote Sens.* **2019**, *57*, 3271–3282. [CrossRef]
6. Nobes, D. Geophysical surveys of burial sites: A case study of the *Oaro urupa*. *Geophysics* **1999**, *64*, 357–367. [CrossRef]
7. Hamran, S.; Gjessing, D.; Hjelmstad, J.; Asrholt, E. Ground penetrating synthetic pulse radar: Dynamic range and modes of operation. *J. Appl. Geophys.* **1995**, *33*, 7–14. [CrossRef]
8. Solimene, R.; D'Alterio, A. Entropy-based clutter rejection for intrawall diagnostics. *Int. J. Geophys.* **2012**, *2012*, 41804. [CrossRef]
9. Kempen, K.; Sahli, H. Signal processing techniques for clutter parameters estimation and clutter removal in GPR data for landmine detection. In Proceedings of the 11th IEEE Signal Processing Workshop on Statistical Signal Processing, Singapore, Singapore, 6–8 August 2001; pp. 158–161.
10. Haysashi, N.; Sato, M. F-K filter designs to supress direct waves for Bistatic ground penetrating radar. *IEEE Trans. Geosci. Remote Sens.* **2010**, *48*, 1433–1444. [CrossRef]
11. Mohideen, S.; Perumal, S.; Sathik, M. Image de-noising using discrete wavelet transform. *Int. J. Comput. Sci. Netw. Secur.* **2008**, *8*, 213–216.
12. Jeng, Y.; Lin, C.; Li, Y.; Chen, C.; Huang, H. Application of multiresolution analysis in removing ground-penetrating radar noise. In Proceedings of the CSPG CSEG CWLS Convention, Calgary, AB, Canada, 4–8 May 2009; pp. 416–419.
13. Shi, X.; Yang, Q. Supperssion the direct wave noise in GPR data via the 2-D physical wavelet frame. In Proceedings of the International Conference on Transportation, Mechanical, and Electrical Engineering (TMEE), Changchun, China, 16–18 December 2011; pp. 1161–1164.
14. Wang, W.; Zhao, B.; Liu, X.; Fang, G. Total-variation improved split Bregman method for ground penetrating radar image restoration. *J. Appl. Geophys.* **2013**, *99*, 146–153. [CrossRef]
15. El-Shenawee, M.; Pappaport, C. Monte Carlo simularions for clutter statistics in minefields: AP-mine-like-target buried near a dielectric object beneath 2-D random rough ground surfaces. *IEEE Trans. Geosci. Remote Sens.* **2002**, *40*, 1416–1426. [CrossRef]
16. Soldovieri, F.; Lopera, O.; Lambot, S. Combination of advanced inversion techniques for an accurate target localization via GPR for demining applications. *IEEE Trans. Geosci. Remote Sens.* **2011**, *49*, 451–461. [CrossRef]
17. Lo Monte, F.; Soldovieri, D.; Erricolo, M.; Wicks, M. Radio frequency tomography for below ground imaging and surveillance of targets under cover. In *Effective Surveillance for Homeland Security: Balancing Technology and Social Issues*; Taylor & Francis/CRC Press: Boca Raton, FL, USA, 2013.
18. Freire, S.; Ulrych, T. Application of singular value decomposition to vertical seismic profiling. *Geophysics* **1988**, *53*, 778–783. [CrossRef]
19. Vuksanovic, B.; Bostanudin, N. Clutter removal techniques for GPR images in structure inspection tasks. *Proc. SPIE* **2012**, *8334*, 833413.
20. Liu, C.; Song, C.; Lu, Q. Random noise de-noising and direct wave eliminating based on SVD method for ground penetrating radar signals. *J. Appl. Geophys.* **2017**, *144*, 125–133. [CrossRef]
21. Kalika, D.; Knox, M.; Collins, L.; Torrione, P.; Morton, K. Leveraging robust principal component analysis to detect buried explosive threats in handheld ground-penetrating radar data. *Proc. SPIE* **2015**, *9454*, 94541D.

22. Song, X.; Xiang, D.; Zhou, K.; Su, Y. Improving RPCA-based clutter suppression in GPR detection of antipersonnel mines. *IEEE Geosci. Remote Sens. Lett.* **2017**, *14*, 1338–1442. [CrossRef]
23. Oliveira, R.J.; Caldeira, B.; Teixidó, T.; Borges, J.F. Influência da forma de onda emitida na comparação de radargramas. *Comun. Geol.* **2016**, *103*, 149–154.
24. Bitri, A.; Grandjean, G. Frequency-wavenumber modelling and migration of 2D GPR data in moderately heterogeneous dispersive media. *Geophys. Prospect.* **1998**, *46*, 287–301. [CrossRef]
25. Tzanis, A. matGPR Release 2: A freeware MATLAB®package for the analysis & interpretation of common and single offset GPR data. *FastTimes* **2010**, *15*, 17–43.
26. Hirani, A.N.; Totsuka, T. Combining Frequency and Spatial Domain Information for Fast Interactive Image Noise Removal. In Proceedings of the 23rd Annual Conference on Computer Graphics and Interactive Techniques-SIGGRAPH 96, New Orleans, LA, USA, 4–9 August 1996; pp. 269–276.
27. Fialka, O.; Čadík, M. FFT and convolution performance in image filtering on GPU. *Proc. Int. Conf. Inf. Vis.* **2006**, *5*, 609–614.
28. Gonzalez, R.C.; Woods, R.E. *Digital Image Processing*; Prentice Hall: Upper Saddle River, NJ, USA, 2002.
29. Lim, J.S. *Two-Dimensional Signal and Image Processing*; Prentice Hall: Englewood Cliffs, NJ, USA, 1990.
30. Andrews, H.C.; Patterson, C.L. Singular Value Decompositions and Digital Image Processing. *IEEE Trans. Acoust.* **1976**, *24*, 26–53. [CrossRef]
31. Kamm, J.; Nagy, J.G. Kronecker product and SVD approximations in image restoration. *Linear Algebra Appl.* **1998**, *284*, 177–192. [CrossRef]
32. Sadek, R.A. SVD Based Image Processing Applications: State of The Art. *Contrib. Res. Chall.* **2012**, *3*, 26–34.
33. Ganic, E.; Eskicioglu, A.M. Robust DWT-SVD domain image watermarking: Embedding data in all frequencies. *Proc. Multimed. Secur. Work.* **2004**, *4*, 166–174.
34. Konstantinides, K.; Natarajan, B.; Yovanof, G.S. Noise estimation and filtering using block-based singular value decomposition. *IEEE Trans. Image Process.* **1997**, *6*, 479–483. [CrossRef] [PubMed]
35. Sadek, R.A. Blind synthesis attack on SVD based watermarking techniques. In Proceedings of the 2008 International Conference on Computational Intelligence for Modelling Control & Automation, Vienna, Austria, 10–12 December 2008.
36. Xu, X.; Dexter, S.D.; Eskicioglu, A.M. A hybrid scheme for encryption and watermarking. *Secur. Steganography Watermarking Multimed. Contents VI* **2004**, *5306*, 725–736.
37. Yang, J.; Lu, C. Combined Twhniques of Singular Value Decomposition and Vector Quantization for Image Coding. *IEEE Trans. Image Process.* **1995**, *4*, 1141–1146. [CrossRef] [PubMed]
38. Moonen, M.; Van Dooren, P.; Vandewalle, J. A singular value decomposition updating algorithm for subspace tracking. *SIAM J. Matrix Anal. Appl.* **1992**, *13*, 1015–1038. [CrossRef]
39. Peña, J.A.; Teixidó, T. Cover Surfaces as a new technique for 3D GPR image enhancement. Archaeological applications. In *RNM104—Nformes*; Universidad de Granada–Instituto Andaluz de Geofísica: Granada, Spain, 2013; pp. 1–10.

remote sensing

Article

Underground Pipeline Identification into a Non-Destructive Case Study Based on Ground-Penetrating Radar Imaging

Nicoleta Iftimie *, Adriana Savin , Rozina Steigmann and Gabriel Silviu Dobrescu

National Institute of Research and Development for Technical Physics, 47 D. Mangeron Blvd,
700050 Iasi, Romania; asavin@phys-iasi.ro (A.S.); steigmann@phys-iasi.ro (R.S.); gdobrescu@phys-iasi.ro (G.S.D.)
* Correspondence: niftimie@phys-iasi.ro; Tel.: +40-232-430-680

Abstract: Ground-penetrating radar (GPR) has become one of the key technologies in subsurface sensing and, in general, in nondestructive testing (NDT), since it is able to detect both metallic and nonmetallic targets. GPR has proven its ability to work in electromagnetic frequency range for subsoil investigations, and it is a risk-reduction strategy for surveying underground various targets and their identification and detection. This paper presents the results of a case study which exceeds the laboratory level being realized in the field in a real case where the scanning conditions are much more difficult using GPR signals for detecting and assessing underground drainage metallic pipes which cross an area with large buildings parallel to the riverbed. The two urban drainage pipes are detected based on GPR imaging. This provides an approximation of their location and depth which are convenient to find from the reconstructed profiles of both simulated and practical GPR signals. The processing of data recorded with GPR tools requires appropriate software for this type of measurement to detect between different reflections at multiple interfaces located at different depths below the surface. In addition to the radargrams recorded and processed with the software corresponding to a GPR device, the paper contains significant results obtained using techniques and algorithms of the processing and post-processing of the signals (background removal and migration) that gave us the opportunity to estimate the location, depth, and profile of pipes, placed into a concrete duct bank, under a structure with different layers, including pavement, with good accuracy.

Keywords: ground-penetrating radar; nondestructive testing; pipelines detection; modeling; signal processing

check for updates

Citation: Iftimie, N.; Savin, A.; Steigmann, R.; Dobrescu, G.S. Underground Pipeline Identification into a Non-Destructive Case Study Based on Ground-Penetrating Radar Imaging. *Remote Sens.* **2021**, *13*, 3494. https://doi.org/10.3390/rs13173494

Academic Editors: Yuri Álvarez López and María García Fernández

Received: 29 July 2021
Accepted: 30 August 2021
Published: 3 September 2021

Publisher's Note: MDPI stays neutral with regard to jurisdictional claims in published maps and institutional affiliations.

1. Introduction

One of the most effective and powerful nondestructive testing (NDT) employed in road surveys nowadays is the ground-penetrating radar (GPR), due to its high flexibility of usage and reliability of results. A reliable risk-reduction strategy to pipe examination is the key for ensuring the sustainable development and improvement of the life time of urban water supply and drainage system. The drainage pipes are critical endowment for a smart city as a precursor for reaching a sustainable development and having a limited life time. The pipes age with functioning time being buried deep underground and can lead to significant safety hazards as water dissipation and soil contamination. These possible disadvantages affect day-to-day use and the long-lasting life of urban pipes. Among NDT inspection techniques is laser scanning, which is geospatial method that can detect only spatial distribution visible in the acquisition, while ultrasound elastic waves, a geophysical method, is capable to detect only quantitative data about failures and cannot achieve more characteristics of pipes [1].

GPR is a geophysical technique based on very short electromagnetic pulses (1–20 ns) propagation within radiofrequency band, typically between 10 MHz to few GHz, to map profile and underground features [2,3]. GPR has numerous characteristics, providing a high resolution, strong anti-interference ability, and high efficiency. It is also a nondestructive technique; consequently, GPR has been extensively used in many fields, such as geological

193

exploration, water conservancy engineering, and urban construction [4,5]. The properties depending on frequency (dielectric permittivity (ε), electrical conductivity (σ) and magnetic permeability (μ)) play a significant role in the electromagnetic energy dissipation in a medium containing complex discontinuities [3–5].

Until now, GPR surveys are widely employed as a noninvasive detection tool to detect unknown targets in deep underground and is useful in the localization of electromagnetic (EM) discontinuities in the subsurface with high resolution [6,7].

Interesting applications fields of GPR are measurements for underground targets location (ex. cables, landmine and UXO, drainage pipes) as in archaeology [8], civil engineering [9,10], military applications [11–13], etc. The transducers realize the coupling of energy within the near field by evanescent and propagating waves [14].

The GPR equipment record the time between the sending of the impulse and its receiving after the scattering the emitted waves which undergo several propagation processes. The main registered is reflected by interfaces, while some is also scattered and returns to the surface, and the signals are presented in B-scan radargram [11,15,16].

The main interest is to detect as many characteristics as is possible to image a buried object [14], and to extract its clear image from ground discontinuities [17,18]. With GPR technology underground, drainage pipelines can be detected and it has been used also in civil engineering to evaluate major structural damage, such as holes and cavities in roads, plates, and bridge decks [19,20]. Estimating the location of damaged pipes is significant for service performance and sustainable management.

A new infrastructure has been developed in Iasi, Romania, on the terraces of the riverbed that crosses the city and includes an upper terrace (underground various targets and drainage pipes detection in this proposed), a lower terrace (detection of pipes for sewer leaks [21]), and a medium terrace (investigations related to spill basins and civil protection dam [22]. The land area along to the riverbed is particularly important, both for the design of civil constructions and for infrastructure. The problem becomes even more important given the action of seismic movements, as these lands have different behavior from the usual situation. Furthermore, for the case of concrete/asphalted tracks, the fatigue life of the urban pipes located under the road/pedestrian area represents a problem of high importance, because they are close to urban heating system and utility water ducts.

This paper presents a case study employing GPR signals for detecting and assessing underground drainage metallic pipes which cross an area with large buildings parallel to the riverbed. The research exceeds the case study at the laboratory level and was realized in the field in a real case where the scanning conditions are much more difficult due to the fact that the area is not perfectly straight and the weather conditions are not always favorable. Reflections of electromagnetic waves occur and are created in places or layers of the ground where a variation of electrical or magnetic properties occurs. In areas with different water content and with buried pipes and tunnels, there is a variation in the speed of propagation of radar waves, and strong reflections occur. With the help of techniques and algorithms for processing and post-processing the appropriate signals (migration and background removal), the noises are eliminated and the shape and depth of the investigated objects are rendered to a significant extent. The work was performed in the frame of a large project (starting from 2015 to present) with the aim of transforming the area around river bank into an ecological agreement park. The project targeted all areas that were at risk by restoring the infrastructure with different types of buried pipes for water transport or sewage leakage, overflow basins, and the civil protection dam [21,22].

This paper presents the results recorded with a GPR tool to detect drainage pipes with unknown approximate position buried under a bike track on the river bank, and a succession of digital signal processing and post-processing methods applied both to A-scan and B-scan that provide an easy way to read and interpret the results. The proposed methods and the algorithms are considered for the recognition and detection of a concrete duct bank containing two drainage pipes for hot water transportation (turn-return). The situation has been simulated using the finite-difference time-domain (FDTD) [2] software in order to interpret the signals recorded by receiving GPR antenna, bowtie type, working at 400 MHz [23–25].

2. Materials and Methods

2.1. Generic GPR with Bowtie Antenna and Propagation Waves

The electromagnetic methods as GPR are non-invasive where individual measurements are quasi real-time, and due to the relatively low frequencies used, have the advantage of penetrating electromagnetic waves at great depths in the ground and obtaining scattering information from buried bodies at great depths. As it is known, the resolution is defined as the capacity of the measurement system to discriminate individual elements embedded in a different medium [11]. The penetration depth decreases with the increase in the conductivity of the medium and, for higher frequency, high resolution and lower penetration depths are obtained. Figure 1 presents a standard GPR principle consist of a transmitting (Tx) and a receiving (Rx) antennas placed in a shielded case which is displaced over the surface to be scanned [26,27]. The GPR emits EM waves that penetrate in the ground in the form of an ellipse (*inset bottom* Figure 1).

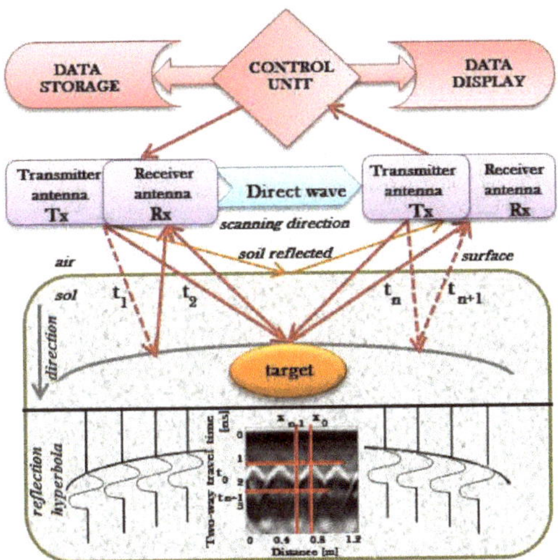

Figure 1. Process of a generic GPR system (*inset bottom*-schematic diagram of GPR reflection hyperbola generation and real signal GPR B-scan).

After being recorded and processed by the control unit embedded in a GPR system, the reflected waves were mixed into a reflection signal A-scan, measuring the interval between emission and reception of the signal delivered by the reception transducer (Figure 1). It was predicted that the time of flight t of the GPR signal will be double—forward and backward to the buried target at z depth, and the remaining constant during the survey [2,28,29] will increase as the distance between transmitter and receiver increases (x) (*inset bottom* Figure 1). Because the area scanned with GPR had mostly multi-layer targets, the speed had to be calibrated according to layer n [2,30]

The hyperbola in the radar signature when the radar is moving along X-axis (propagating medium is considered homogeneous) [31] is given by

$$R = \sqrt{z_0^2 + (X - x_0)^2} \tag{1}$$

where (x_o, z_o) is a perfect point scattered in the 2D plane, X is the synthetic aperture vector, and R is the path length vector (from antenna to scattered). As mentioned in [21], simulation techniques that comprise single frequency models, time domain models, ray tracing, integral techniques, and discrete element methods may be useful in foreseen the results to be obtained from in-field measurements. The FDTD technique is one of the simulation methods that is most suitable with GPR surveys [32,33].

The propagating waves (homogeneous waves—the wavenumber is real [34]) in the near field of the transducer determined the coupling of the energy into the ground. The nonmagnetic soil ($\mu_s = \mu_0$) in which the pipes are buried has electrical properties, relative permeability (ε_{rs}), and conductivity (σ_s), and the field induced by antenna has the features of the rectangular coil [7]. For a region free of sources, the Helmholtz equation is useful according to [35].

The field generated by the emission coil feed by a current with frequency [36] can be expressed using dyadic Green's function [37] and integral method

$$\overline{E}_0(\overline{r}) = j\omega\mu_2 \int_{Vsource} \overset{\leftrightarrow}{G}_{12}(\overline{r},\overline{r}\prime)\overline{J}(\overline{r}\prime)d\overline{r}\prime, \tag{2}$$

where μ_2 is magnetic permeability of the medium 2 (soil as Figure 1 depicts), $\overset{\leftrightarrow}{G}_{12}$ is component of dyadic Green's function matrix. The electric conductivities considered are σ_f for the target and respective σ_2 for the stratified soil [38] and the total electric field become

$$\overline{E}_2(\overline{r}) + j\omega\mu_2\sigma_2 \int_{Vbody} \overset{\leftrightarrow}{G}_{22}(\overline{r},\overline{r}\prime)\overline{E}_2(\overline{r}\prime)\left[\frac{\sigma_f(\overline{r}\prime)}{\sigma_2} - 1\right]d\overline{r}\prime = \overline{E}_0(\overline{r}), \tag{3}$$

and perturbation field in air in the presence of conductive target is according with [39]

$$\overline{E}_1(\overline{r}) = j\omega\mu_2\sigma_2 \int_{Vbody} \overset{\leftrightarrow}{G}_{21}(\overline{r},\overline{r}\prime)\overline{E}_2(\overline{r}\prime)\left[\frac{\sigma_f(\overline{r}\prime)}{\sigma_2} - 1\right]d\overline{r}\prime \tag{4}$$

where $\overset{\leftrightarrow}{G}_{12}$, $\overset{\leftrightarrow}{G}_{22}$ are components of dyadic Green's function matrix [38].

2.2. Geophysical Surveys in Pavement Assessment and Drainage Water Pipes Detection

The GPR equipment used is Utility Scan Standard System (Geophysical Survey Systems, Inc. GSSI, Nashua, NH, USA) (Figure 2a) [40], which had a 400-MHz antenna, allowing a penetration depth until 4.5 m depending on the moisture of soil. The front wheel of utility scanner had an encoder, allowing a displacement precision determination of ±1 mm, and the scan intervals assured by the GSSI System software was 100 scans/m. The sampling rate was 0.04 ns, with the quantization of the signal being made on 16 bits [40]. The equipment was set up to record A-scan at each 10 cm, and the time window for which the signals were obtained is 32 ns. The control unit contained a function that allowed the testing of the terrain dielectric by recording a data set and then performing its migration. By knowing the soil permittivity, the penetration depth could also be implicitly known. For example, the user manual showed that a profile of at least 3 m in length is collected over well-known objects and that they have to go through those objects at a right angle [40]. By using the up and down arrows, we could adapt the dielectric value for that hyperbolas profiles crack-up to points.

The dielectric constant of the soil in the survey area was established at $\varepsilon_r = 12$, in the basis of previous investigations [21,22] knowing the exactly type of soil. Due to the weather conditions (rainy), the value of the dielectric constant was taken from a table from the user manual of GSSI equipment and tested with TEST_DIEL function [41].

A region of [2000 × 600] cm from the Bahlui river bank was surveyed (Figure 2b) in the immediate vicinity of the riverbed. The urban drainage pipes for hot water was taken into study (Figure 3). The pipes had an unknown approximate position, being buried under a bike track. We assumed that they were buried parallel with the riverbed. Due to its orientation and practical survey procedures, the scanning was performed in

3 parallel traces with a 2000-cm length, separated between them with 100 cm, as seen in Figure 3a. The scanning of directions orthogonal on the direction of riverbank was facilitated. Figure 3c presents a photo of the testing zone when the new pipes were brought to be replaced.

(a)

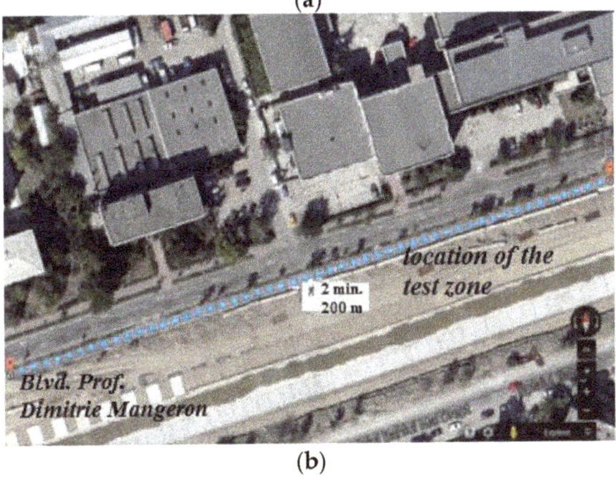

(b)

Figure 2. Experimental set-up: (**a**) GPR survey system with a 400-MHz antenna; (**b**) test area of the underground urban drainage pipes.

The urban drainage pipes had an approximate 70-cm diameter and, for hot water transportation, had 15-cm wall thickness of insulation (nonwoven glass fiber). In order to simplify the data presentation, a zone of [600 × 500] cm was selected and the scanning was effectuated in 6 transversal traces with 600-cm length, separated between them with 100 cm (the traces were effectuated in both directions, see Figure 3b).

2.3. Signal Processing: A-Scan and B-Scan

The mean value to A-scan data set were crucially assured to be close to zero, so that the amplitude probability distribution from A-scan data set was symmetric to the mean value [2].

$$A\prime_n = A_n - \frac{1}{N} \sum_{n=1}^{N} A_n,$$ (5)

where A_n are values of raw data set, A'_n are values of processed data set, n is the data set number, and N represents total number of data sets.

The filtering operation is given by

$$A\prime_n = A_n + \frac{A_n - A\prime_{n-1}}{K},$$ (6)

where A'_n is averaged value, A_n is the current value.

The K factor will be chosen to take values from n to N or a fixed value, and will contribute to the average value. Averaging has no effect on discontinuities.

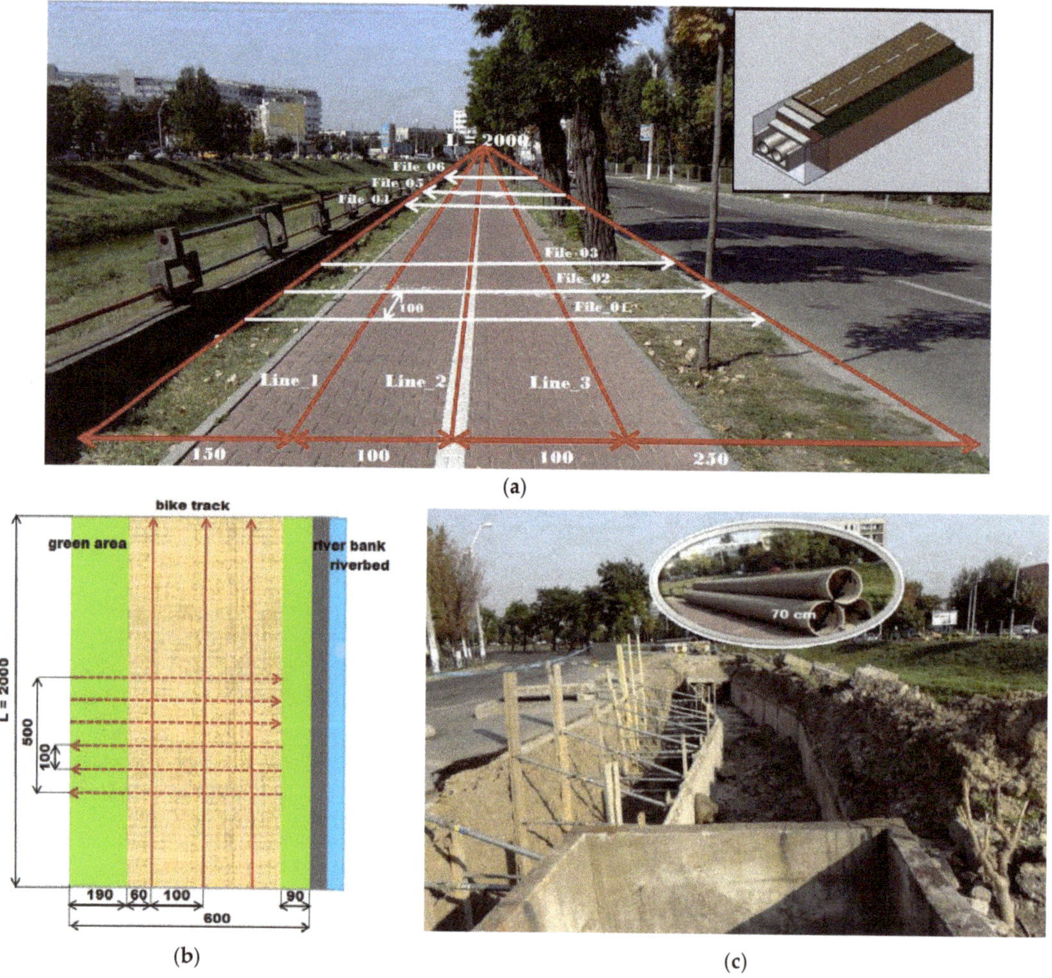

Figure 3. Surveyed zone: (**a**) the investigated region on both directions on drainage pipes (inset—configuration of the surveyed zone); (**b**) scanning scheme on both directions; (**c**) the region where the two urban drainage pipes should be repaired or replaced (inset-pipes to be replaced); (**c**) picture was taken several days after GPR surveying); (all dimensions are in cm).

Thus, a window of L pixels was defined, and the mean of the pixels in it from all the pixels in this window was subtracted. The window advanced and the procedure was repeated until the entire image is solved, as

$$g(x,y) = f(x,y) - \frac{1}{L} \sum_{i=-L/2}^{i=L/2} f(x+i,y), \tag{7}$$

where g is filtered image, f is raw data, and L is the window size.

Using nonlinear optimization of the decomposition technique, [41] improves GPR imaging by simultaneously determining spatial variations in size and delaying soil reflection.

Considering $A_i(x)$, the spatially soil reflection amplitude, and $B_i(x)$, the time delay of the soil reflection apex over segment i^{th}, then $A_i(x)$ and $B_i(x)$ could be approximated as a sum

$$A_i(x) = \sum_{n=0}^{4} a_{i_n} T_n(x); \quad B_i(x) = \sum_{n=0}^{4} b_{i_n} T_n(x) \, , \tag{8}$$

where $T_n(x)$ are the Chebyshev polynomials established by the recursive relation.

$$T_{n+1}(x) = 2T_n(x) - T_{n-1}(x), \quad n > 1 \, , \tag{9}$$

with $T_0(x) = 1$ and $T_1(x) = x$.

3. Results

3.1. Application of GPR Data Raw and FDTD Simulations in the Detection and Replace of Water Pipes

As we showed in previous researches [11,21,22], a GPR device with the corresponding control unit can record a continuous image of the subsurface, which indicates the presence, depth, and the layout of soil features required in classification, characterization, and surveying of soil as well as detection and identification of various buried targets.

GPR waves are modified by the subsurface layer and the recorded radar data sets a contrast in electrical and magnetic properties; those changes can then be detected, represented, and characterized. A GPR data set recording delivers high-resolution information that is able to use at the interpretation and the extrapolation of information obtained with algorithms and pre-processing techniques. Figure 4 presents the scan on longitudinal directions of a zone where two urban drainage pipes with known diameter placed in a concrete duct are buried. Only one reflection peak can be seen, given by direct-coupling by the A-scan raw, which takes place when the antenna is lightly displaced from the soil [42–44]. In this case, the direct waveform transmitting and receiving antenna in connected with the surface to produce a mixed waveform. A signal (A-scan) with a deep reflection to record above concrete described in Figure 4a. It can also be seen that, in the case of real data set, the signal is very noisy, including clutters [21]. B-scans were obtained from 55 raw A-scan types, using the specific signal processing, similar to those of ultrasound examinations, as presented in Figures 4 and 5.

Figure 4 shows the results which were divided by subheadings to assure a succinct and accurate evaluation of the data set recording, their interpretation, as well as the preliminary conclusions that could be drawn. At the distance of 44 m from the starting point (Figure 4b), a signal with the form of a distorted hyperbola and a peak pointing upwards is observed at the depth of 20.2 ns. This is indicating the fact that the drainage pipes with the axes parallel with the scanning direction change their orientation.

Figure 5a,b shows the presence of the two drainage pipes for hot water transportation, with an 85-cm diameter (70-cm diameter of the pipe and 15-cm protective layer), both buried in concrete duct ($\varepsilon_s = 8$) [41], the top of pipes being at a 20.2-ns depth. It first uses the GPRMax software to produce GPR synthetic datasets through FTDT simulations [31]. The simulated data was processed with a code in Matlab 2020b (MathWorks, Inc., Natick, MA, USA). In Matlab, we used functions for removing noise by adaptive filtering, for example, "wiener2", which filters the grayscale image using a pixel-wise adaptive low-pass

Wiener filter, and a pixelwise adaptive Wiener method based on statistics estimated from a local neighborhood of each pixel. The waves penetrating the ground are propagated along the scanning line and produce EM pulses at manually chosen intervals and detect buried targets. The next reflected EM pulse can be incorporated within a radargram B-scan to produce an underground 2D image. Figure 6 presents the simulation of the experimental set-up and surveying conditions using GPRMax 2D. The layout of the survey is previewed with GPRMaxGV—Gnuplot viewer—a free plot script auto formatter developed by Goran Bekic [45]. The pipelines are difficult to identify in noisy profiles, which means that their GPR patterns are sensitive to noises. Introducing the raw radargrams as the test set into the simulation model, a positive section, corresponds favorable with the object locations. A reliable detection depth of GPR is determined by the central frequency of the electromagnetic wave and the attribute of the subsurface formation, while detection resolution is limited by the wave frequency and signal bandwidth [7].

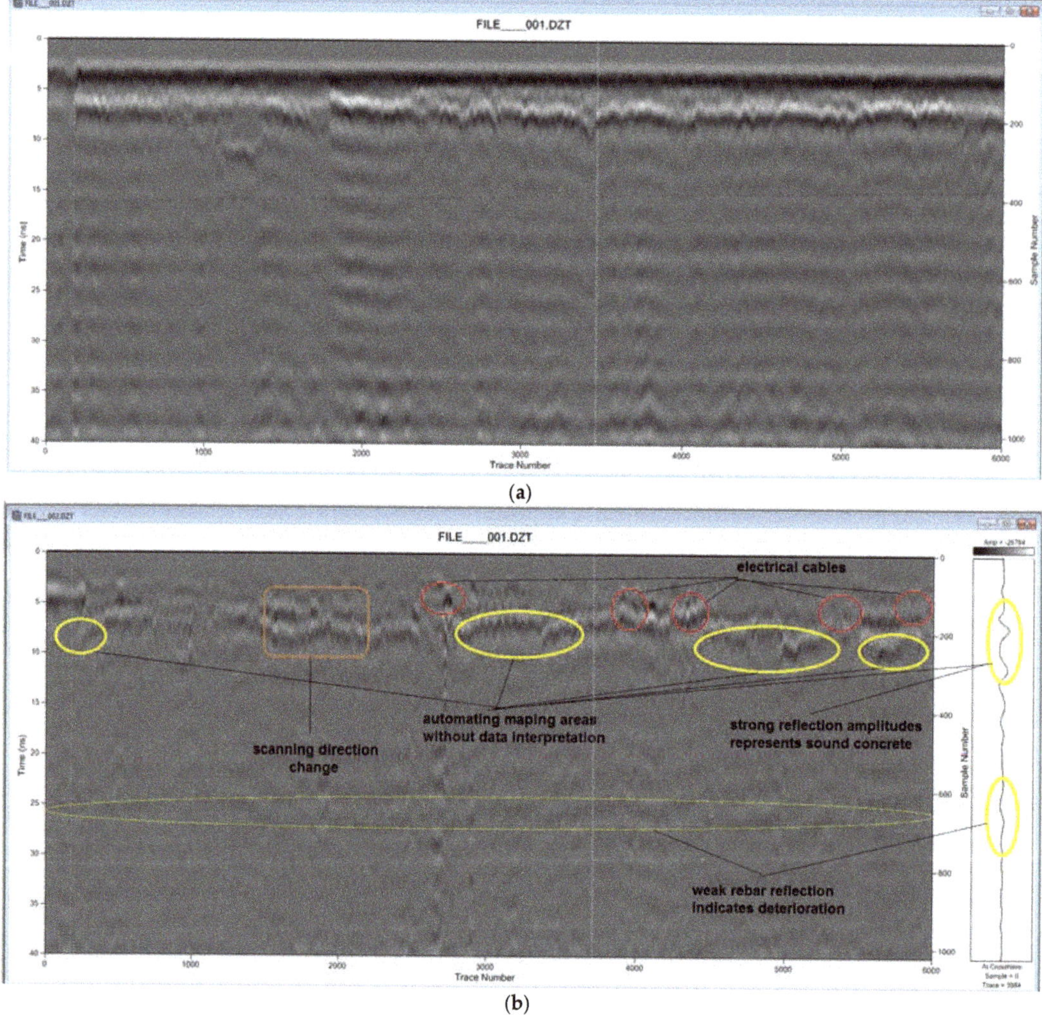

Figure 4. (a) GPR raw data containing a profile with the air wave, (b) GPR radargrams of surveyed zone—parallel traces and partition for data interpretation.

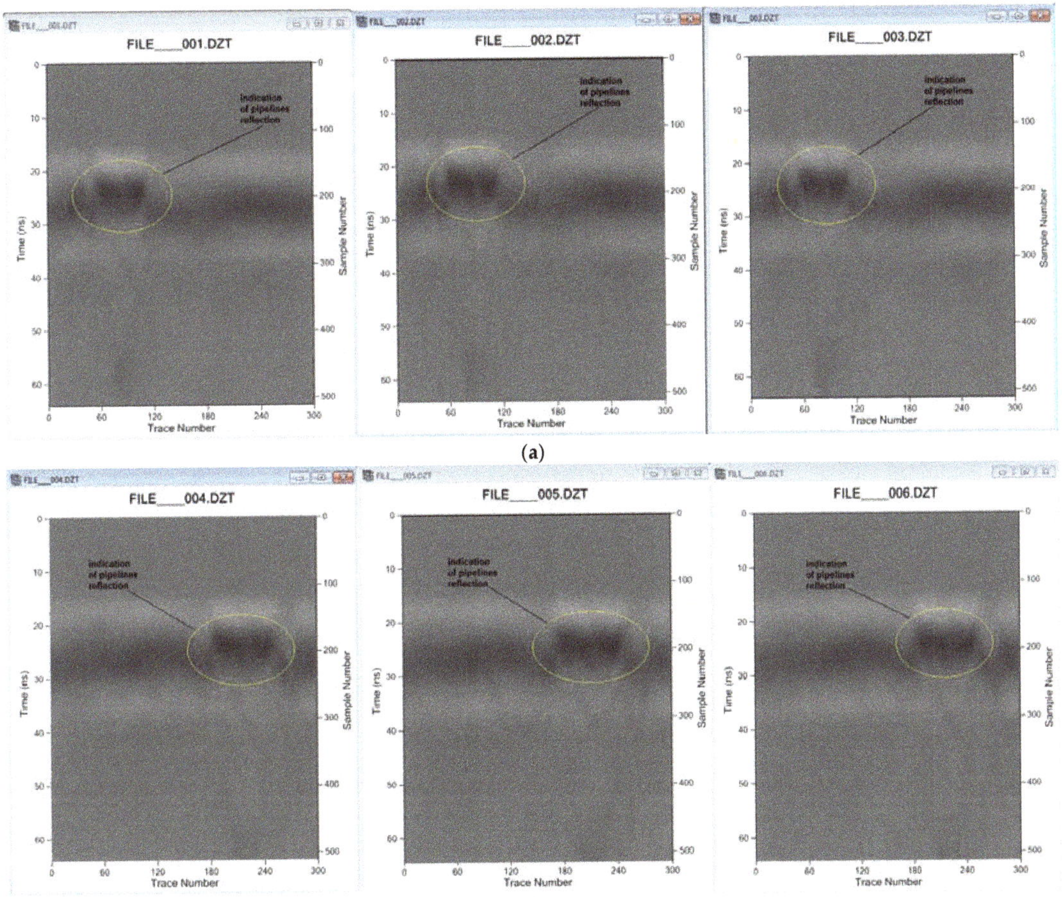

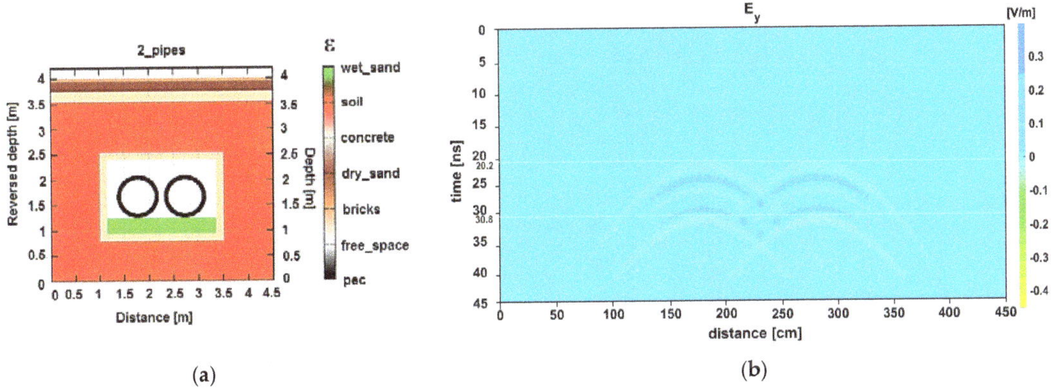

Figure 5. Raw data imaging at scan of zone transversal on drainage pipes (see Figure 3a): (**a**) File_001_002_003 forward scan, (**b**) File_004_005_006 backward scan.

Figure 6. Simulation using GPRMax 2D (**a**) the simulated geometry and visualization using GPRMaxGV Gnuplot viewer; (**b**) Signal processing of simulated data.

3.2. A-Scan and B-Scan Results

An original A-scan is presented in Figure 7a, while Figure 7b presents the processed A-scan where the mean value was zero and the noise was decreased according to Equation (6). The value of K was selected as 1.02 according with [21] on the basis of probability of detection principle and of the characteristics of reception, where K is a measure of interpretation to eliminate the conflict between two sequential acquisition of data and to keep at least 2% error, according to [6]. Considering a set of five samples containing a B-scan, a series of techniques of signal processing can be taken into account. Usually, the clutters hamper the imaging of GPR data. In order to make the evaluation of GPR radargrams as accurate and correct as possible, extraction of no-longer-desired signals as retransmission of wave from Tx to Tr or reflections in the air–soil interface must be effectuated. This is named background removal; good results can be obtained using a subtract mean trace procedure [46].

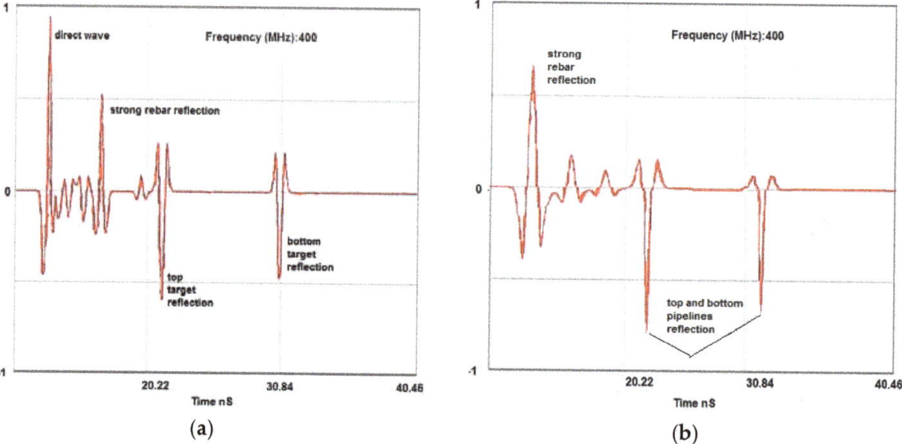

Figure 7. A-scan after filtering of raw experimental data delivered by GPR system: (**a**) original GPR record (raw data with direct wave); (**b**) after filtered signal (show both rebar and the pipelines reflections).

After the weighting coefficients $(\overline{a_i}, \overline{b_i})$ was calculated, the amplitude $Ai(x)$ and the delay $Bi(x)$ of segment i^{th} were determined with Equation (5). Parameter calculation was carried out with a nonlinear least square error minimization function, in the Matlab 2020b Optimization Toolbox. If the optimization does not converge for a data set, we must use a recursive approach. Imaging procedures can be used to concentrate the energy existing in a point target's hyperbolic arc which returns to a unique point [41]. This technique, named migration, allows the establishment of the depths of underground objects. Figure 8 present B-scan results processing of one file over the scanned area. For a B-scan original image from the inspected area, the algorithm presented above [21] is applied. Figure 8a shows the results after background removal with L = 20 pixels (length of sliding Hanning window). Figure 8b presents the result after application of migration technique using the Kirchhoff method with 0.15 m/ns. Top and bottom reflections of concrete duct bank, in which the two drainage pipes are located, could be remarked at approximately 20.2 ns, 30.8 ns from the data set. The parallel line with scanning direction at 30.8 ns describes the reflections on the remnant water from the pipes and can be seen on the processed image from the experimental data. The rest of parallel lines were due to numerous reflections from the soil scanned zone and on the interface concrete duct bank of the urban drainage pipes. In the same area, we can notice the signals whose shape may indicate the presence of buried electrical cables (for example, at 98.6 m, 128.5 m, 153.3 m, 169 m, etc.). On the basis of raster scan results, the GPR-slice function from matGPR software (open-source software) was used to obtain a 3-D volume of GPR data and to enable its visualization in the form of opaque or translucent slices [47,48] In GPR operation, the high resolution in depth is obtained by utilizing a transmitted signal of wideband. The high resolution was obtained by coherently processing scattered electromagnetic waves which were measured while the device scanned along a line of terrain. The resolution after the postprocessing depended on the focusing capacity of the migration algorithm.

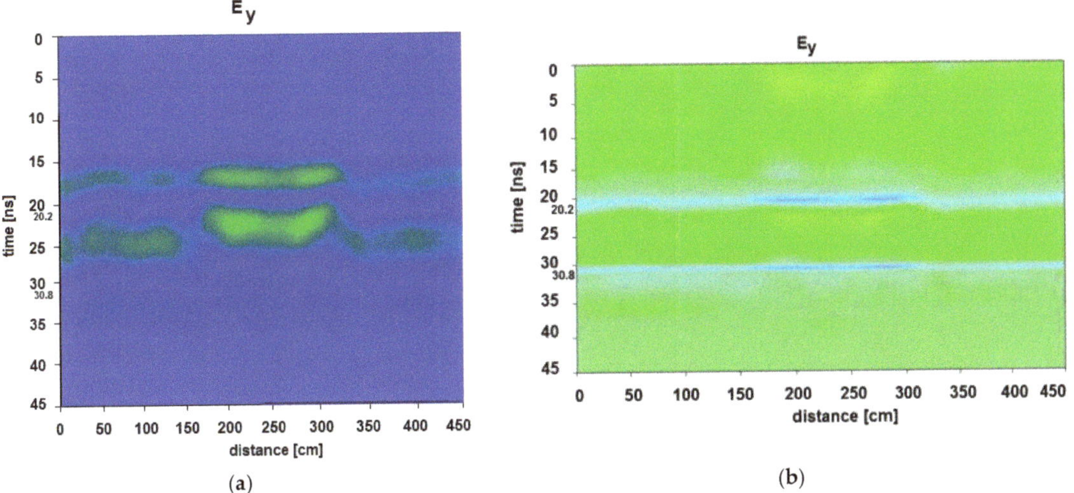

(a)   (b)

Figure 8. B-scan raw data processed after using both techniques: (**a**) ground removal; (**b**) migration.

It can be observed that, with each processing step, an improvement of resolution was obtained, helping to emphasize the delimitation of bottom and upper level of the pipes.

B-scan radargrams were concatenated as a result of the parallel line positioned at 20.2 ns, allowing the representation of a 3D dataset under the form of planes (isometric surfaces) [49–52] with equal signal, generating and displaying an orthographical designing of the subsurface equal to 55% of the maximum signal included in the 3D data set volume (Figure 9).

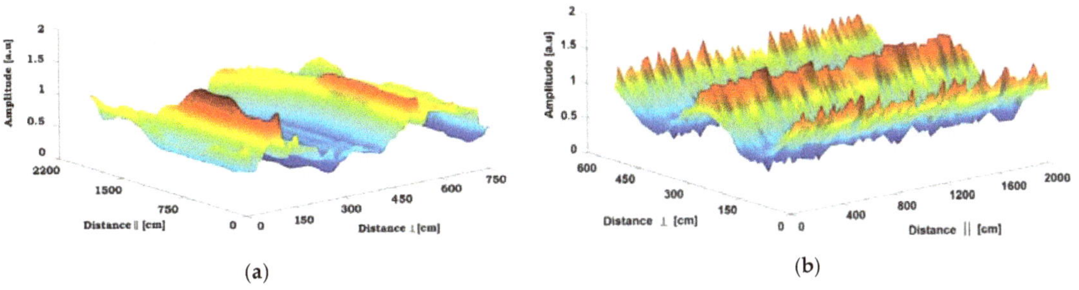

(a)   (b)

Figure 9. Drainage water pipes reconstruction: (**a**) before and (**b**) after processing.

4. Discussions

Most researches [12,13,15] on object detection only determined whether the objects were detectable and where the objects were. Other researches have determined the sizes of rebars and small-scale voids [3,5,8,42] inside concrete, but investigations inside the complicated underground sections become difficult. Giannikis et al. [33] attempted to estimate the buried object size (buried ammunitions, including landmine), but they only utilized numerically produced data. Elsewhere, Grimberg et al. [11] determined the location and size subsurface buried ammunitions, including antitank mines. Jin et al. [20] developed a machine learning framework based on wavelet scattering networks to analyze GPR data for subsurface pipeline identification with smaller diameters. In our previous works, we also detected pipes for sewer leaks [21] and investigated spill basins and civil protection dams [22]. This paper presents the results of a case study employing GPR signals for detecting and assessing underground drainage metallic pipes which cross an area with large buildings parallel to the riverbed. The research exceeds the case study at the laboratory level, which is accomplished in the field in a real case where the scanning conditions are far more difficult. After surveying the zones, the GPR radargram recording signals of drainage pipes with an unknown approximate position were buried under a bike track on the river bank. It used a succession of digital signal processing and post-processing methods applied both to A-scans and B-scans which provide an easy way to read and interpret the results. The simulations were made with FDTD software to aid interpretation of the signals recorded by receiving a GPR antenna. The profile of the drainage pipes was obtained and their diameters could then be estimated. Numerical modelling and signal post-processing algorithms of GPR can lead to a better understanding of the operating principle of the radar detection tools [47,53]. GprMax allows the simulation of real cases of GPR in order to gain an idea of what is expected during surveys and to improve complex signal processing and interpretation skills until receiving the real data [9,10]. The propagation waves theory is applied to the GPR detection of drainage pipes, and the signal processing technique is used for A-scans and B-scans of recorded dataset [21,22]. In our study, simulations were made with FDTD software and survey data were recorded by GPR system, working at 400 MHz where the location, depth, and profile of pipes could be determined.

5. Conclusions

Using GPR raw data and techniques and algorithms of processing and post-processing of the signals (background removal and migration), the obtained results provided the opportunity to estimate the location, depth, and profile of pipes, placed into a concrete replaceable duct bank. High-frequency electromagnetic waves can recognize underground objects in depth due to sharp energy attenuation. Accepting the noise sensitivity of pipelines and the failure in recognizing the deep pipeline, the techniques and algorithms used in these study case presents promising applicability in both simulated and practical GPR signals. The ground removal and migration methods were introduced and evaluated for comparison study of GPR B-scan image processing and gave us the opportunity to estimate location, depth, and profile of pipes. Processing of the acquired data was carried out with Matlab software, which is in the endowment of our laboratory and not in a standard application of a commercial software suite available and used by geophysicists, because of the technical features of our survey. Based on Matlab software, the B-scan data were converted into images, thus highlighting the depth and position of the buried pipes and the shape that can be evaluated with good accuracy. Future research will focus on the knowledge of the physical, mechanical, and chemical properties of the lands in the investigates area as well as the analysis of environmental pollution risks by testing areas, including runways, platforms, perimeter road, and road handling, which represent decisive factors both in the design of civil engineering and for infrastructure works.

Author Contributions: Conceptualization, N.I., A.S. and R.S.; methodology, N.I., A.S. and R.S.; software, R.S. and G.S.D.; validation, N.I. and A.S.; investigation, N.I. and G.S.D.; writing—original draft preparation, N.I. and A.S.; writing—review and editing, N.I., A.S., R.S. and G.S.D.; supervision, N.I. and A.S.; All authors have equal contributions. All authors have read and agreed to the published version of the manuscript.

Funding: This paper is partially supported and the APC was funded by Romanian Ministry of Research, Innovation and Digitization under Nucleus Program PN 19 26 01 02.

Institutional Review Board Statement: Not applicable.

Informed Consent Statement: Not applicable.

Data Availability Statement: Not applicable.

Conflicts of Interest: The authors declare no conflict of interest.

References

1. Ghavamian, A.; Mustapha, F.; Baharudin, B.T.; Yidris, N. Detection, localisation and assessment of defects in pipes using guided wave techniques: A review. *Sensors* **2018**, *18*, 4470. [CrossRef] [PubMed]
2. Daniels, D.J. *Ground Penetrating Radar*, 2nd ed.; IEEE Press: Stevenage Herts, UK, 2004.
3. Pedret Rodés, J.; Martínez Reguero, A.; Pérez-Gracia, V. GPR Spectra for Monitoring Asphalt Pavements. *Remote Sens.* **2020**, *12*, 1749. [CrossRef]
4. Diamanti, N.; Annan, A.P.; Redman, J.D. Concrete bridge deck deterioration assessment using ground penetrating radar (GPR). *J. Environ. Eng. Geophys.* **2017**, *22*, 121–132. [CrossRef]
5. Dong, Z.; Ye, S.; Gao, Y.; Fang, G.; Zhang, X.; Xue, Z.; Zhang, T. Rapid detection methods for asphalt pavement thicknesses and defects by a vehicle-mounted ground penetrating radar (GPR) system. *Sensors* **2016**, *16*, 2067. [CrossRef]
6. Daniels, D.J. A review of GPR for landmine detection. *Sens. Imaging* **2006**, *7*, 90–123. [CrossRef]
7. Jol, H.M. *Ground Penetrating Radar Theory and Applications*; Elsevier: Amsterdam, The Netherlands, 2009; pp. 3–40.
8. Solla, M.; Lorenzo, H.; Rial, F.I.; Novo, A. GPR evaluation of the Roman masonry arch bridge of Lugo (Spain). *NDT E Int.* **2011**, *44*, 8–12. [CrossRef]
9. Massarelli, C.; Campanale, C.; Uricchio, V.F. Ground Penetrating Radar as a Functional Tool to Outline the Presence of Buried Waste: A Case Study in South Italy. *Sustainability* **2021**, *13*, 3805. [CrossRef]
10. Demirci, S.; Yigit, E.; Eskidemir, I.H.; Ozdemir, C. Ground penetrating radar imaging of water leaks from buried pipes based on back-projection method. *NDT E Int.* **2012**, *47*, 35–42. [CrossRef]
11. Grimberg, R.; Savin, A.; Iftimie, N.; Leitoiu, S.; Andreescu, A. GPR for UXO recognition. *Stud. Appl. Electromagn. Mech.* **2011**, *35*, 381–388.
12. Xavier, N.N.; Mercedes, S.; Paula, G.P.; Lorenzo, H. GPR Signal Characterization for Automated Landmine and UXO Detection Based on Machine Learning Techniques. *Remote Sens.* **2014**, *6*, 9729–9748.
13. Gonzalez-Huici, M.A.; Giovanneschi, F. A combined strategy for landmine detection and identification using synthetic GPR responses. *J. Appl. Geophys.* **2013**, *99*, 154–165. [CrossRef]
14. Smith, G.S.; Petersson, L.E.R. On the Use of Evanescent Electromagnetic Waves in Detection and Identification of Objects Buried in Lossy Soil. *IEEE Trans. Antenna Propag.* **2000**, *48*, 1295–1300. [CrossRef]
15. Ozdemir, C.; Demirci, S.; Yigit, E. Practical algorithms to focus B-scan GPR images: Theory and application to real data. *Prog. Electromagn. Res. B* **2008**, *6*, 109–122. [CrossRef]
16. Lestari, A.A.; Yarovoy, A.G.; Ligthart, L.P. Analysis and Design of Improved Antennas for GPR. *Subsurf. Sens. Technol. Appl.* **2002**, *3*, 295–326. [CrossRef]
17. Soil Taxonomy—A Basic System of Soil Classification for Making and Interpreting Soil Surveys. Available online: http://www.nrcs.usda.gov/Internet/FSE_DOCUMENTS/nrcs142p2_051232.pdf (accessed on 15 July 2021).
18. Hipp, J.E. Soil Electromagnetic parameters as functions of frequency. Soil density and soil moisture. *Proc. IEEE* **1974**, *62*, 98–103. [CrossRef]
19. Alonso, R.; Pozo, J.; Buisán, S.T.; Álvarez, J.A. Analysis of the Snow Water Equivalent at the AEMet-Formigal Field Laboratory (Spanish Pyrenees) During the 2019/2020 Winter Season Using a Stepped-Frequency Continuous Wave Radar (SFCW). *Remote Sens.* **2021**, *13*, 616. [CrossRef]
20. Jin, Y.; Duan, Y. Wavelet scattering network-based machine learning for ground penetrating radar imaging: Application in pipeline identification. *Remote Sens.* **2020**, *12*, 3655. [CrossRef]
21. Iftimie, N.; Dobrescu, G.S.; Savin, A. Imaging Subsurface Water Pipe using GPR and Evanescent Waves: Experimental and Simulations Data. *Appl. Mech. Mater.* **2015**, *771*, 359–364. [CrossRef]
22. Iftimie, N.; Savin, A.; Danila, N.A.; Dobrescu, G.S. Radar pulses to image the subsurface using Ground Penetrating Radar (GPR). *IOP Conf. Ser. Mater. Sci. Eng.* **2019**, *564*, 012130. [CrossRef]
23. Oliveira, R.J.; Caldeira, B.; Teixidó, T.; Borges, J.F. GPR Clutter Reflection Noise-Filtering through Singular Value Decomposition in the Bidimensional Spectral Domain. *Remote Sens.* **2021**, *13*, 2005. [CrossRef]
24. Rodríguez-Santalla, I.; Gomez-Ortiz, D.; Martín-Crespo, T.; Sánchez-García, M.J.; Montoya-Montes, I.; Martín-Velázquez, S.; Barrio, F.; Serra, J.; Ramírez-Cuesta, J.M.; Gracia, F.J. Study and Evolution of the Dune Field of La Banya Spit in Ebro Delta (Spain) Using LiDAR Data and GPR. *Remote Sens.* **2021**, *13*, 802. [CrossRef]
25. Bianchini Ciampoli, L.; Gagliardi, V.; Ferrante, C.; Calvi, A.; D'Amico, F.; Tosti, F. Displacement monitoring in airport runways by persistent scatterers SAR interferometry. *Remote Sens.* **2020**, *12*, 3564. [CrossRef]

26. Martínez, J.; Rey, J.; Gutiérrez, L.M.; Novo, A.; Ortiz, A.J.; Alejo, M.; Galdón, J.M. Electrical resistivity imaging (ERI) and ground-penetrating radar (GPR) survey at the Giribaile site (upper Guadalquivir valley; southern Spain). *J. Appl. Geophys.* **2015**, *123*, 218–226. [CrossRef]

27. Leucci, G.; De Giorgi, L.; Gizzi, F.T.; Persico, R. Integrated geo-scientific surveys in the historical centre of Mesagne (Brindisi, Southern Italy). *Nat. Hazards* **2017**, *86*, 363–383. [CrossRef]

28. Benedetto, A.; Tosti, F.; Ciampoli, L.B.; D'Amico, F. An overview of ground-penetrating radar signal processing techniques for road inspections. *Signal Process.* **2017**, *132*, 201–209. [CrossRef]

29. Leucci, G. *Advances in Geophysical Methods Applied to Forensic Investigations: New Developments in Acquisition and Data Analysis Methodologies*; Springer: Berlin/Heidelberg, Germany, 2020; p. 200. ISBN 978-3-030-46241-3.

30. Wang, M.; Wen, J.; Li, W. Qualitative Research: The impact of Roat Orientation on Coarse Roots Detection Using Ground-Penetrating Radar (GPR). *BioResources* **2020**, *15*, 2237–2257. [CrossRef]

31. Özdemir, C.; Demirci, F.; Yigit, E.; Yilmaz, B. A Review on Migration Methods in B-Scan Ground Penetrating Radar Im-aging. *Math. Probl. Eng.* **2014**, *2014*, 16. [CrossRef]

32. Giannopoulos, A. GprMax2D/3D. User's Guide 2003. Available online: http://www.gprmax.com/code/UserGuideV2.pdf (accessed on 15 January 2021).

33. Giannakis, I.; Giannopoulos, A.; Warren, C.A. Realistic FDTD Numerical Modeling Framework of Ground Penetrating Radar for Landmine Detection. *IEEE J. Sel. Top. Appl. Earth Obs. Remote Sens.* **2016**, *9*, 37–51. [CrossRef]

34. Lowe, M.J. Matrix techniques for modeling ultrasonic waves in multilayered media. *IEEE Trans. Ultrason. Ferroe-Lectr. Freq. Control* **1995**, *42*, 525–542. [CrossRef]

35. Bladel, J.V. *Electromagnetic Fields*; McGraw-Hill: New York, NY, USA, 1964.

36. Grimberg, R.; Udpa, L.; Savin, A.; Steigmann, R.; Palihovici, V.; Udpa, S.S. 2D Eddy current sensor array. *NDT E Int.* **2006**, *39*, 264–271. [CrossRef]

37. Chew, W.C. *Waves and Fields in Inhomogeneous Media*; Van Nostrand Reinhold: New York, NY, USA, 1990.

38. Grimberg, R.; Udpa, L.; Savin, A.; Steigmann, R.; Bruma, A.; Udpa, S.S. Electromagnetic evaluation of soil condition. *Stud. Appl. Electromagn. Mech.* **2009**, *32*, 363–370.

39. Grimberg, R.; Udpa, L.; Udpa, S.S. Electromagnetic transducer for the determination of soil condition. *Int. J. Appl. Electromagn. Mech.* **2008**, *28*, 201–210. [CrossRef]

40. Geophysical Survey Systems, Inc. (GSSI) Site. Available online: http://www.geophysical.com/ (accessed on 15 January 2021).

41. Gunatilaka, A.H.; Baertlein, B.A.; Harvey, J.F.; Broach, J.T.; Dugan, R.E. Subspace decomposition technique to improve GPR imaging of antipersonnel mines. In *Detection and Remediation Technologies for Mines and Minelike Targets V*; International Society for Optics and Photonics: Orlando, FL, USA, 2000.

42. Pedret Rodes, J.; Pérez-Gracia, V.; Martínez-Reguero, A. Evaluation of the GPR frequency spectra in asphalt pavement as-sessment. *Constr. Build. Mater.* **2015**, *96*, 181–188. [CrossRef]

43. Solla, M.; Pérez-Gracia, V.; Fontul, S. A Review of GPR Application on Transport Infrastructures: Troubleshooting and Best Practices. *Remote Sens.* **2021**, *13*, 672. [CrossRef]

44. GPR-Repository. Available online: https://sites.google.com/site/gprbekic/gprutils/ (accessed on 15 January 2021).

45. De Chiara, F.; Fontul, S.; Fortunato, E. GPR Laboratory Tests for Railways Materials Dielectric Properties Assessment. *Remote Sens.* **2014**, *6*, 9712–9728. [CrossRef]

46. Andreas, T. MATGPR—Manual and Technical Reference. 2016. Available online: http://users.uoa.gr/~{}atzanis/matgpr/matgpr.html (accessed on 15 January 2021).

47. Giannopoulos, A. Modelling ground penetrating radar by GprMax. *Constr. Build. Mater.* **2005**, *19*, 755–762. [CrossRef]

48. Giannakis, I.; Giannopoulos, A. Time-synchronized convolutional perfectly matched layer for improved absorbing per-formance in FDTD. *IEEE Antennas Wirel. Propag. Lett.* **2014**, *14*, 690–693. [CrossRef]

49. Giannakis, I.; Giannopoulos, A.; Warren, C. Realistic FDTD GPR antenna models optimized using a novel linear/nonlinear Full-Waveform Inversion. *IEEE Trans. Geosci. Remote Sens.* **2019**, *57*, 1768–1778. [CrossRef]

50. Warren, C.; Giannopoulos, A. Creating finite-difference time-domain models of commercial ground-penetrating radar antennas using Taguchi's optimization method. *Geophysics* **2011**, *76*, G37–G47. [CrossRef]

51. Cheng, W.; Fang, H.; Xu, G.; Chen, M. Using SCADA to Detect and Locate Bursts in a Long-Distance Water Pipeline. *Water* **2018**, *10*, 1727. [CrossRef]

52. Cassidy, N.J.; Millington, T.M. The application of finite-difference time-domain modelling for the assessment of GPR in magneti-cally lossy materials. *J. Appl. Geophys.* **2009**, *67*, 296–308. [CrossRef]

53. Warren, C.; Giannopoulos, A.; Giannakis, I. GprMax: Open source software to simulate electromagnetic wave propagation for Ground Penetrating Radar. *Comput. Phys. Commun.* **2016**, *209*, 163–170. [CrossRef]

MDPI

St. Alban-Anlage 66

4052 Basel

Switzerland

Tel. +41 61 683 77 34

Fax +41 61 302 89 18

www.mdpi.com

Remote Sensing Editorial Office

E-mail: remotesensing@mdpi.com

www.mdpi.com/journal/remotesensing